PSEUDO-DIFFERENTIAL EQUATIONS AND STOCHASTICS OVER NON-ARCHIMEDEAN FIELDS

PURE AND APPLIED MATHEMATICS

A Program of Monographs, Textbooks, and Lecture Notes

MONOGRAPHS AND TEXTBOOKS IN PURE AND APPLIED MATHEMATICS

1. *K. Yano,* Integral Formulas in Riemannian Geometry (1970)
2. *S. Kobayashi,* Hyperbolic Manifolds and Holomorphic Mappings (1970)
3. *V. S. Vladimirov,* Equations of Mathematical Physics (A. Jeffrey, ed.; A. Littlewood, trans.) (1970)
4. *B. N. Pshenichnyi,* Necessary Conditions for an Extremum (L. Neustadt, translation ed.; K. Makowski, trans.) (1971)
5. *L. Narici et al.,* Functional Analysis and Valuation Theory (1971)
6. *S. S. Passman,* Infinite Group Rings (1971)
7. *L. Dornhoff,* Group Representation Theory. Part A: Ordinary Representation Theory. Part B: Modular Representation Theory (1971, 1972)
8. *W. Boothby and G. L. Weiss, eds.,* Symmetric Spaces (1972)
9. *Y. Matsushima,* Differentiable Manifolds (E. T. Kobayashi, trans.) (1972)
10. *L. E. Ward, Jr.,* Topology (1972)
11. *A. Babakhanian,* Cohomological Methods in Group Theory (1972)
12. *R. Gilmer,* Multiplicative Ideal Theory (1972)
13. *J. Yeh,* Stochastic Processes and the Wiener Integral (1973)
14. *J. Barros-Neto,* Introduction to the Theory of Distributions (1973)
15. *R. Larsen,* Functional Analysis (1973)
16. *K. Yano and S. Ishihara,* Tangent and Cotangent Bundles (1973)
17. *C. Procesi,* Rings with Polynomial Identities (1973)
18. *R. Hermann,* Geometry, Physics, and Systems (1973)
19. *N. R. Wallach,* Harmonic Analysis on Homogeneous Spaces (1973)
20. *J. Dieudonné,* Introduction to the Theory of Formal Groups (1973)
21. *I. Vaisman,* Cohomology and Differential Forms (1973)
22. *B.-Y. Chen,* Geometry of Submanifolds (1973)
23. *M. Marcus,* Finite Dimensional Multilinear Algebra (in two parts) (1973, 1975)
24. *R. Larsen,* Banach Algebras (1973)
25. *R. O. Kujala and A. L. Vitter, eds.,* Value Distribution Theory: Part A; Part B: Deficit and Bezout Estimates by Wilhelm Stoll (1973)
26. *K. B. Stolarsky,* Algebraic Numbers and Diophantine Approximation (1974)
27. *A. R. Magid,* The Separable Galois Theory of Commutative Rings (1974)
28. *B. R. McDonald,* Finite Rings with Identity (1974)
29. *J. Satake,* Linear Algebra (S. Koh et al., trans.) (1975)
30. *J. S. Golan,* Localization of Noncommutative Rings (1975)
31. *G. Klambauer,* Mathematical Analysis (1975)
32. *M. K. Agoston,* Algebraic Topology (1976)
33. *K. R. Goodearl,* Ring Theory (1976)
34. *L. E. Mansfield,* Linear Algebra with Geometric Applications (1976)
35. *N. J. Pullman,* Matrix Theory and Its Applications (1976)
36. *B. R. McDonald,* Geometric Algebra Over Local Rings (1976)
37. *C. W. Groetsch,* Generalized Inverses of Linear Operators (1977)
38. *J. E. Kuczkowski and J. L. Gersting,* Abstract Algebra (1977)
39. *C. O. Christenson and W. L. Voxman,* Aspects of Topology (1977)
40. *M. Nagata,* Field Theory (1977)
41. *R. L. Long,* Algebraic Number Theory (1977)
42. *W. F. Pfeffer,* Integrals and Measures (1977)
43. *R. L. Wheeden and A. Zygmund,* Measure and Integral (1977)
44. *J. H. Curtiss,* Introduction to Functions of a Complex Variable (1978)
45. *K. Hrbacek and T. Jech,* Introduction to Set Theory (1978)
46. *W. S. Massey,* Homology and Cohomology Theory (1978)
47. *M. Marcus,* Introduction to Modern Algebra (1978)
48. *E. C. Young,* Vector and Tensor Analysis (1978)
49. *S. B. Nadler, Jr.,* Hyperspaces of Sets (1978)
50. *S. K. Segal,* Topics in Group Kings (1978)
51. *A. C. M. van Rooij,* Non-Archimedean Functional Analysis (1978)
52. *L. Corwin and R. Szczarba,* Calculus in Vector Spaces (1979)
53. *C. Sadosky,* Interpolation of Operators and Singular Integrals (1979)
54. *J. Cronin,* Differential Equations (1980)
55. *C. W. Groetsch,* Elements of Applicable Functional Analysis (1980)

56. *I. Vaisman,* Foundations of Three-Dimensional Euclidean Geometry (1980)
57. *H. I. Freedan,* Deterministic Mathematical Models in Population Ecology (1980)
58. *S. B. Chae,* Lebesgue Integration (1980)
59. *C. S. Rees et al.,* Theory and Applications of Fourier Analysis (1981)
60. *L. Nachbin,* Introduction to Functional Analysis (R. M. Aron, trans.) (1981)
61. *G. Orzech and M. Orzech,* Plane Algebraic Curves (1981)
62. *R. Johnsonbaugh and W. E. Pfaffenberger,* Foundations of Mathematical Analysis (1981)
63. *W. L. Voxman and R. H. Goetschel,* Advanced Calculus (1981)
64. *L. J. Corwin and R. H. Szczarba,* Multivariable Calculus (1982)
65. *V. I. Istrăţescu,* Introduction to Linear Operator Theory (1981)
66. *R. D. Järvinen,* Finite and Infinite Dimensional Linear Spaces (1981)
67. *J. K. Beem and P. E. Ehrlich,* Global Lorentzian Geometry (1981)
68. *D. L. Armacost,* The Structure of Locally Compact Abelian Groups (1981)
69. *J. W. Brewer and M. K. Smith, eds.,* Emmy Noether: A Tribute (1981)
70. *K. H. Kim,* Boolean Matrix Theory and Applications (1982)
71. *T. W. Wieting,* The Mathematical Theory of Chromatic Plane Ornaments (1982)
72. *D. B. Gauld,* Differential Topology (1982)
73. *R. L. Faber,* Foundations of Euclidean and Non-Euclidean Geometry (1983)
74. *M. Carmeli,* Statistical Theory and Random Matrices (1983)
75. *J. H. Carruth et al.,* The Theory of Topological Semigroups (1983)
76. *R. L. Faber,* Differential Geometry and Relativity Theory (1983)
77. *S. Barnett,* Polynomials and Linear Control Systems (1983)
78. *G. Karpilovsky,* Commutative Group Algebras (1983)
79. *F. Van Oystaeyen and A. Verschoren,* Relative Invariants of Rings (1983)
80. *I. Vaisman,* A First Course in Differential Geometry (1984)
81. *G. W. Swan,* Applications of Optimal Control Theory in Biomedicine (1984)
82. *T. Petrie and J. D. Randall,* Transformation Groups on Manifolds (1984)
83. *K. Goebel and S. Reich,* Uniform Convexity, Hyperbolic Geometry, and Nonexpansive Mappings (1984)
84. *T. Albu and C. Năstăsescu,* Relative Finiteness in Module Theory (1984)
85. *K. Hrbacek and T. Jech,* Introduction to Set Theory: Second Edition (1984)
86. *F. Van Oystaeyen and A. Verschoren,* Relative Invariants of Rings (1984)
87. *B. R. McDonald,* Linear Algebra Over Commutative Rings (1984)
88. *M. Namba,* Geometry of Projective Algebraic Curves (1984)
89. *G. F. Webb,* Theory of Nonlinear Age-Dependent Population Dynamics (1985)
90. *M. R. Bremner et al.,* Tables of Dominant Weight Multiplicities for Representations of Simple Lie Algebras (1985)
91. *A. E. Fekete,* Real Linear Algebra (1985)
92. *S. B. Chae,* Holomorphy and Calculus in Normed Spaces (1985)
93. *A. J. Jerri,* Introduction to Integral Equations with Applications (1985)
94. *G. Karpilovsky,* Projective Representations of Finite Groups (1985)
95. *L. Narici and E. Beckenstein,* Topological Vector Spaces (1985)
96. *J. Weeks,* The Shape of Space (1985)
97. *P. R. Gribik and K. O. Kortanek,* Extremal Methods of Operations Research (1985)
98. *J.-A. Chao and W. A. Woyczynski, eds.,* Probability Theory and Harmonic Analysis (1986)
99. *G. D. Crown et al.,* Abstract Algebra (1986)
100. *J. H. Carruth et al.,* The Theory of Topological Semigroups, Volume 2 (1986)
101. *R. S. Doran and V. A. Belfi,* Characterizations of C*-Algebras (1986)
102. *M. W. Jeter,* Mathematical Programming (1986)
103. *M. Altman,* A Unified Theory of Nonlinear Operator and Evolution Equations with Applications (1986)
104. *A. Verschoren,* Relative Invariants of Sheaves (1987)
105. *R. A. Usmani,* Applied Linear Algebra (1987)
106. *P. Blass and J. Lang,* Zariski Surfaces and Differential Equations in Characteristic $p > 0$ (1987)
107. *J. A. Reneke et al.,* Structured Hereditary Systems (1987)
108. *H. Busemann and B. B. Phadke,* Spaces with Distinguished Geodesics (1987)
109. *R. Harte,* Invertibility and Singularity for Bounded Linear Operators (1988)
110. *G. S. Ladde et al.,* Oscillation Theory of Differential Equations with Deviating Arguments (1987)
111. *L. Dudkin et al.,* Iterative Aggregation Theory (1987)
112. *T. Okubo,* Differential Geometry (1987)

113. *D. L. Stancl and M. L. Stancl,* Real Analysis with Point-Set Topology (1987)
114. *T. C. Gard,* Introduction to Stochastic Differential Equations (1988)
115. *S. S. Abhyankar,* Enumerative Combinatorics of Young Tableaux (1988)
116. *H. Strade and R. Farnsteiner,* Modular Lie Algebras and Their Representations (1988)
117. *J. A. Huckaba,* Commutative Rings with Zero Divisors (1988)
118. *W. D. Wallis,* Combinatorial Designs (1988)
119. *W. Więsław,* Topological Fields (1988)
120. *G. Karpilovsky,* Field Theory (1988)
121. *S. Caenepeel and F. Van Oystaeyen,* Brauer Groups and the Cohomology of Graded Rings (1989)
122. *W. Kozlowski,* Modular Function Spaces (1988)
123. *E. Lowen-Colebunders,* Function Classes of Cauchy Continuous Maps (1989)
124. *M. Pavel,* Fundamentals of Pattern Recognition (1989)
125. *V. Lakshmikantham et al.,* Stability Analysis of Nonlinear Systems (1989)
126. *R. Sivaramakrishnan,* The Classical Theory of Arithmetic Functions (1989)
127. *N. A. Watson,* Parabolic Equations on an Infinite Strip (1989)
128. *K. J. Hastings,* Introduction to the Mathematics of Operations Research (1989)
129. *B. Fine,* Algebraic Theory of the Bianchi Groups (1989)
130. *D. N. Dikranjan et al.,* Topological Groups (1989)
131. *J. C. Morgan II,* Point Set Theory (1990)
132. *P. Biler and A. Witkowski,* Problems in Mathematical Analysis (1990)
133. *H. J. Sussmann,* Nonlinear Controllability and Optimal Control (1990)
134. *J.-P. Florens et al.,* Elements of Bayesian Statistics (1990)
135. *N. Shell,* Topological Fields and Near Valuations (1990)
136. *B. F. Doolin and C. F. Martin,* Introduction to Differential Geometry for Engineers (1990)
137. *S. S. Holland, Jr.,* Applied Analysis by the Hilbert Space Method (1990)
138. *J. Okniński,* Semigroup Algebras (1990)
139. *K. Zhu,* Operator Theory in Function Spaces (1990)
140. *G. B. Price,* An Introduction to Multicomplex Spaces and Functions (1991)
141. *R. B. Darst,* Introduction to Linear Programming (1991)
142. *P. L. Sachdev,* Nonlinear Ordinary Differential Equations and Their Applications (1991)
143. *T. Husain,* Orthogonal Schauder Bases (1991)
144. *J. Foran,* Fundamentals of Real Analysis (1991)
145. *W. C. Brown,* Matrices and Vector Spaces (1991)
146. *M. M. Rao and Z. D. Ren,* Theory of Orlicz Spaces (1991)
147. *J. S. Golan and T. Head,* Modules and the Structures of Rings (1991)
148. *C. Small,* Arithmetic of Finite Fields (1991)
149. *K. Yang,* Complex Algebraic Geometry (1991)
150. *D. G. Hoffman et al.,* Coding Theory (1991)
151. *M. O. González,* Classical Complex Analysis (1992)
152. *M. O. González,* Complex Analysis (1992)
153. *L. W. Baggett,* Functional Analysis (1992)
154. *M. Sniedovich,* Dynamic Programming (1992)
155. *R. P. Agarwal,* Difference Equations and Inequalities (1992)
156. *C. Brezinski,* Biorthogonality and Its Applications to Numerical Analysis (1992)
157. *C. Swartz,* An Introduction to Functional Analysis (1992)
158. *S. B. Nadler, Jr.,* Continuum Theory (1992)
159. *M. A. Al-Gwaiz,* Theory of Distributions (1992)
160. *E. Perry,* Geometry: Axiomatic Developments with Problem Solving (1992)
161. *E. Castillo and M. R. Ruiz-Cobo,* Functional Equations and Modelling in Science and Engineering (1992)
162. *A. J. Jerri,* Integral and Discrete Transforms with Applications and Error Analysis (1992)
163. *A. Charlier et al.,* Tensors and the Clifford Algebra (1992)
164. *P. Biler and T. Nadzieja,* Problems and Examples in Differential Equations (1992)
165. *E. Hansen,* Global Optimization Using Interval Analysis (1992)
166. *S. Guerre-Delabrière,* Classical Sequences in Banach Spaces (1992)
167. *Y. C. Wong,* Introductory Theory of Topological Vector Spaces (1992)
168. *S. H. Kulkarni and B. V. Limaye,* Real Function Algebras (1992)
169. *W. C. Brown,* Matrices Over Commutative Rings (1993)
170. *J. Loustau and M. Dillon,* Linear Geometry with Computer Graphics (1993)
171. *W. V. Petryshyn,* Approximation-Solvability of Nonlinear Functional and Differential Equations (1993)

172. *E. C. Young*, Vector and Tensor Analysis: Second Edition (1993)
173. *T. A. Bick*, Elementary Boundary Value Problems (1993)
174. *M. Pavel*, Fundamentals of Pattern Recognition: Second Edition (1993)
175. *S. A. Albeverio et al.*, Noncommutative Distributions (1993)
176. *W. Fulks*, Complex Variables (1993)
177. *M. M. Rao*, Conditional Measures and Applications (1993)
178. *A. Janicki and A. Weron*, Simulation and Chaotic Behavior of α-Stable Stochastic Processes (1994)
179. *P. Neittaanmäki and D. Tiba*, Optimal Control of Nonlinear Parabolic Systems (1994)
180. *J. Cronin*, Differential Equations: Introduction and Qualitative Theory, Second Edition (1994)
181. *S. Heikkilä and V. Lakshmikantham*, Monotone Iterative Techniques for Discontinuous Nonlinear Differential Equations (1994)
182. *X. Mao*, Exponential Stability of Stochastic Differential Equations (1994)
183. *B. S. Thomson*, Symmetric Properties of Real Functions (1994)
184. *J. E. Rubio*, Optimization and Nonstandard Analysis (1994)
185. *J. L. Bueso et al.*, Compatibility, Stability, and Sheaves (1995)
186. *A. N. Michel and K. Wang*, Qualitative Theory of Dynamical Systems (1995)
187. *M. R. Darnel*, Theory of Lattice-Ordered Groups (1995)
188. *Z. Naniewicz and P. D. Panagiotopoulos*, Mathematical Theory of Hemivariational Inequalities and Applications (1995)
189. *L. J. Corwin and R. H. Szczarba*, Calculus in Vector Spaces: Second Edition (1995)
190. *L. H. Erbe et al.*, Oscillation Theory for Functional Differential Equations (1995)
191. *S. Agaian et al.*, Binary Polynomial Transforms and Nonlinear Digital Filters (1995)
192. *M. I. Gil'*, Norm Estimations for Operation-Valued Functions and Applications (1995)
193. *P. A. Grillet*, Semigroups: An Introduction to the Structure Theory (1995)
194. *S. Kichenassamy*, Nonlinear Wave Equations (1996)
195. *V. F. Krotov*, Global Methods in Optimal Control Theory (1996)
196. *K. I. Beidar et al.*, Rings with Generalized Identities (1996)
197. *V. I. Arnautov et al.*, Introduction to the Theory of Topological Rings and Modules (1996)
198. *G. Sierksma*, Linear and Integer Programming (1996)
199. *R. Lasser*, Introduction to Fourier Series (1996)
200. *V. Sima*, Algorithms for Linear-Quadratic Optimization (1996)
201. *D. Redmond*, Number Theory (1996)
202. *J. K. Beem et al.*, Global Lorentzian Geometry: Second Edition (1996)
203. *M. Fontana et al.*, Prüfer Domains (1997)
204. *H. Tanabe*, Functional Analytic Methods for Partial Differential Equations (1997)
205. *C. Q. Zhang*, Integer Flows and Cycle Covers of Graphs (1997)
206. *E. Spiegel and C. J. O'Donnell*, Incidence Algebras (1997)
207. *B. Jakubczyk and W. Respondek*, Geometry of Feedback and Optimal Control (1998)
208. *T. W. Haynes et al.*, Fundamentals of Domination in Graphs (1998)
209. *T. W. Haynes et al.*, Domination in Graphs: Advanced Topics (1998)
210. *L. A. D'Alotto et al.*, A Unified Signal Algebra Approach to Two-Dimensional Parallel Digital Signal Processing (1998)
211. *F. Halter-Koch*, Ideal Systems (1998)
212. *N. K. Govil et al.*, Approximation Theory (1998)
213. *R. Cross*, Multivalued Linear Operators (1998)
214. *A. A. Martynyuk*, Stability by Liapunov's Matrix Function Method with Applications (1998)
215. *A. Favini and A. Yagi*, Degenerate Differential Equations in Banach Spaces (1999)
216. *A. Illanes and S. Nadler, Jr.*, Hyperspaces: Fundamentals and Recent Advances (1999)
217. *G. Kato and D. Struppa*, Fundamentals of Algebraic Microlocal Analysis (1999)
218. *G. X.-Z. Yuan*, KKM Theory and Applications in Nonlinear Analysis (1999)
219. *D. Motreanu and N. H. Pavel*, Tangency, Flow Invariance for Differential Equations, and Optimization Problems (1999)
220. *K. Hrbacek and T. Jech*, Introduction to Set Theory, Third Edition (1999)
221. *G. E. Kolosov*, Optimal Design of Control Systems (1999)
222. *N. L. Johnson*, Subplane Covered Nets (2000)
223. *B. Fine and G. Rosenberger*, Algebraic Generalizations of Discrete Groups (1999)
224. *M. Väth*, Volterra and Integral Equations of Vector Functions (2000)
225. *S. S. Miller and P. T. Mocanu*, Differential Subordinations (2000)

226. *R. Li et al.*, Generalized Difference Methods for Differential Equations: Numerical Analysis of Finite Volume Methods (2000)
227. *H. Li and F. Van Oystaeyen*, A Primer of Algebraic Geometry (2000)
228. *R. P. Agarwal*, Difference Equations and Inequalities: Theory, Methods, and Applications, Second Edition (2000)
229. *A. B. Kharazishvili*, Strange Functions in Real Analysis (2000)
230. *J. M. Appell et al.*, Partial Integral Operators and Integro-Differential Equations (2000)
231. *A. I. Prilepko et al.*, Methods for Solving Inverse Problems in Mathematical Physics (2000)
232. *F. Van Oystaeyen*, Algebraic Geometry for Associative Algebras (2000)
233. *D. L. Jagerman*, Difference Equations with Applications to Queues (2000)
234. *D. R. Hankerson et al.*, Coding Theory and Cryptography: The Essentials, Second Edition, Revised and Expanded (2000)
235. *S. Dăscălescu et al.*, Hopf Algebras: An Introduction (2001)
236. *R. Hagen et al.*, C*-Algebras and Numerical Analysis (2001)
237. *Y. Talpaert*, Differential Geometry: With Applications to Mechanics and Physics (2001)
238. *R. H. Villarreal*, Monomial Algebras (2001)
239. *A. N. Michel et al.*, Qualitative Theory of Dynamical Systems, Second Edition (2001)
240. *A. A. Samarskii*, The Theory of Difference Schemes (2001)
241. *J. Knopfmacher and W.-B. Zhang*, Number Theory Arising from Finite Fields (2001)
242. *S. Leader*, The Kurzweil-Henstock Integral and Its Differentials (2001)
243. *M. Biliotti et al.*, Foundations of Translation Planes (2001)
244. *A. N. Kochubei*, Pseudo-Differential Equations and Stochastics over Non-Archimedean Fields (2001)

Additional Volumes in Preparation

PSEUDO-DIFFERENTIAL EQUATIONS AND STOCHASTICS OVER NON-ARCHIMEDEAN FIELDS

Anatoly N. Kochubei
Ukrainian National Academy of Sciences
Kiev, Ukraine

MARCEL DEKKER
MARCEL DEKKER, INC. NEW YORK • BASEL

ISBN: 0-8247-0655-2

This book is printed on acid-free paper.

Headquarters
Marcel Dekker, Inc.
270 Madison Avenue, New York, NY 10016
tel: 212-696-9000; fax: 212-685-4540

Eastern Hemisphere Distribution
Marcel Dekker AG
Hutgasse 4, Postfach 812, CH-4001 Basel, Switzerland
tel: 41-61-261-8482; fax: 41-61-261-8896

World Wide Web
http://www.dekker.com

The publisher offers discounts on this book when ordered in bulk quantities. For more information, write to Special Sales/Professional Marketing at the headquarters address above.

Current printing (last digit):
10 9 8 7 6 5 4 3 2 1

PRINTED IN THE UNITED STATES OF AMERICA

Preface

A remarkable feature of contemporary mathematical analysis and probability theory is the occurrence of analytic and stochastic objects related to various structures appearing in other branches of mathematics, such as algebra, geometry, or topology. The interaction of different trends of mathematical thought becomes even more fruitful under the influence of mathematical physics. The study of analytic, probabilistic, and physical structures related to p-adic numbers (and general non-Archimedean local fields) is a good example of such an interaction.

In the hundred years since their discovery in 1899 by Hensel, p-adic numbers became very important in number theory (see e.g. [18, 71]) and representation theory [39]. Non-Archimedean counterparts were found and investigated for the main notions of the classical calculus [134], Fourier analysis [143], functional analysis [108, 127], complex analysis [28], analytic theory of differential equations [24, 25, 125], and geometry [19, 42, 11].

A strong impetus to the development of non-Archimedean analysis was given by the hypothesis about a possible p-adic structure of physical space-time at sub-Planck distances ($\leq 10^{-33}$ cm). This idea, expounded in a series of papers by Volovich (see especially [155]), gave rise to many publications by physicists. Whatever form the non-Archimedean models may take in the future (not only in sub-Planck physics, but also in such different areas as spin glasses, turbulence, quantum computing etc.), it has become clear that some non-Archimedean counterparts should be found for all basic structures of mathematical physics, such as the Laplace, Schrödinger, and heat equations, basic classes of random variables, and stochastic processes.

Note that there are essentially two kinds of p-adic analysis and, correspondingly, two kinds of p-adic models in physics. One of them deals typically with functions whose arguments and values are both non-Archimedean (e.g. polynomials and power series). The second considers real or complex-valued functions over non-Archimedean structures (characters, probability densities etc.). In this book we shall not touch the first approach; for a description of recent activity in this field see [68, 69].

For the second approach, most of the results obtained before 1992 were summarized in an important book by Vladimirov, Volovich, and Zelenov [154], which also contains a good exposition of some introductory material for beginners. However, a rapid development of the theory during the last decade made a new book necessary in order to describe the current state of this subject. Such a book is proposed now.

Writing a book on an interdisciplinary subject like non-Archimedean analysis one is faced with the problem of making the book accessible to a wide readership. In fact this book is intended primarily for people interested in analysis, partial differential equations, and stochastic processes. At the same time, the preparatory material needed for understanding the main part of the book is usually known mostly to number theorists (some of whom can of course find it interesting to see familiar objects in a completely new environment). On the other hand, a complete exposition of all preliminaries would require at least another book, or books, which are already available.

To resolve this dilemma, the author chose a compromise option. Chapter 1 contains most of the preliminaries given in a concise form, usually without proofs, which can be found in well-known textbooks and monographs on number theory and non-Archimedean analysis. Thus, at least for the first reading, the book is more or less self-contained and does not require any preliminary knowledge in non-Archimedean analysis (the only exception is Chapter 6, where some results from p-adic functional analysis are used; however, they are quite similar to the ones from the functional analysis over $\mathbb{C}$). On the other hand, the author tried to avoid repeating (with full proofs) the results available from other books, especially from [154].

As we stressed above, in this book we consider primarily operators on complex-valued functions defined over non-Archimedean spaces. Of course, one cannot differentiate such a function, but an appropriate harmonic analysis is available, so it is natural to consider pseudo-differential operators. The simplest of them, the fractional differentiation operator D^α ($\alpha > 0$), was introduced by Vladimirov (see [154]). In Chapter 2 we give further information about D^α including its remarkable connection (found by Haran) with zeroes of the Riemann zeta function. The next step is the study of elliptic and hyperbolic partial differential equations of the second order. The theory is quite different from the classical one. For example, elliptic operators over $\mathbb{Q}_p^n$ ($p \neq 2$) exist only for $n \leq 4$, and in the "physical" dimension $n = 4$ such an operator is unique up to an isomorphism! We compute and study their fundamental solutions and Green functions. Note that we concentrate on the non-Archimedean counterparts of classical equations of mathematical physics; for the general calculus of non-Archimedean pseudo-differential operators see [129, 54, 8].

The symbols of the above pseudo-differential operators over $\mathbb{Q}_p^n$ are of the form $|a(\xi_1, \dots, \xi_n)|_p^\alpha$, where a is a quadratic form of an appropriate kind. The analogy with real analysis need not work here. For example, the form $a(\xi_1, , \xi_2) = \xi_1^2 + \xi_2^2$ $(\xi_1, \xi_2 \in \mathbb{Q}_p)$ corresponds to an elliptic operator only if $p \equiv 3 \pmod 4$; otherwise $a(\xi_1, , \xi_2) = 0$ for some non-zero values of ξ_1, ξ_2. On the other hand, considering the symbol $|\xi_1 - b(\xi_2, \dots, \xi_n)|_p^\alpha$, where b generates an elliptic symbol, one is tempted to see the corresponding equation as a parabolic one. In fact, that is wrong, and the equation appears as a non-Archimedean analog of the nonstationary Schrödinger equation. Moreover, this equation is satisfied by a free wave function of p-adic quantum mechanics, and that is the first example of a pseudo-differential equation with a p-adic time parameter, for which the Cauchy problem is well-posed.

In Chapter 3 we turn to spectral theory. This chapter can be seen as a sequel to [154], where the spectral theory of the operator D^α over $\mathbb{Q}_p$, balls and spheres in $\mathbb{Q}_p$, and of the Schrödinger-like operator $D^\alpha + V$ with a discrete spectrum was given. We consider D^α over general local fields, and $D^\alpha + V$ for various classes of "potentials" V leading to non-empty essential spectra and a non-trivial spectral behavior like the massive singular spectrum. We also study the operator D^α over an arbitrary open subset G of a local field, and investigate relations between its spectral properties and the geometry of G. It appears that the shape of a p-adic "drum" can be heard more clearly than in the real geometry. For the case when G is the group of units, the operator D^α, and its eigenfunctions and eigenvalues are found to be related to the multiplicative structure of the local field, though initially D^α is defined in terms of its additive structure.

Chapter 4 is devoted to a class of equations with properties similar to those of classical parabolic equations. These are equations with positive real time variable, and spatial variable from a local field. The simplest example (a counterpart to the heat equation) is

$$\frac{\partial u(x,t)}{\partial t} + (D^\alpha u)(x,t) = f(x,t).$$

For this and much more general equations a fundamental solution of the Cauchy problem is constructed and investigated, the uniqueness theorem is proved, and the conditions for the stabilization of solutions as $t \to \infty$ are found. The non-Archimedean case exhibits some features quite different from the Euclidean case. For example, a fundamental solution may be non-negative for an equation of an arbitrary order, even greater than 2. On the other hand, in the non-Archimedean case there is no essential difference between the properties of fundamental solutions for integer and fractional order equations.

Recall that in the Euclidean case, a fundamental solution of a second order parabolic equation has an exponential asymptotic behavior as $x \to \infty$, while a fundamental solution of an equation of a fractional order $\alpha \in (1, 2)$ has a power-like decrease [74]. Expressed in terms of the Markov processes associated to these equations, this difference corresponds to the difference between a diffusion process, whose paths are continuous, and a process whose paths have discontinuities of the first kind. In the non-Archimedean case, there are no non-trivial processes with continuous paths, since a local field is totally disconnected, and carries no continuous curves. Thus, this peculiarity of non-Archimedean parabolic equations is related to the topological properties of local fields, although it is established by purely analytic methods.

The above connection with Markov processes is in part a motivation for investigating such equations. Chapter 5 deals with the processes themselves. The starting point is the process $\xi_\alpha(t)$ generated by the operator $-D^\alpha$, an analog of the symmetric stable process. We study stochastic integrals with respect to the Poisson random measure corresponding to ξ_α. A natural next step is the theory of stochastic differential equations, which gives a large class of Markov processes on $\mathbb{Q}_p$ and other local fields. On the other hand, we present (following Albeverio, Karwowski, and Yasuda) a description and study of all temporally and spatially homogeneous rotation-invariant processes with independent increments on a local field. A method is explained for gluing together processes defined on $\mathbb{Q}_p$ for each prime p, into a process on the ring of adeles. The chapter ends with a description of non-Archimedean analogs of general stable distributions (in terms of limits of normalized sums of independent identically distributed random variables). The description is similar in spirit to the classical results of Lévy and Khinchin, though it possesses some completely new features.

In Chapter 6 we turn to the non-Archimedean infinite-dimensional analysis. We start with a general framework (due to Evans and Mądrecki) of Gaussian measures on measurable vector spaces over local fields, and then concentrate on a class of infinite-dimensional spaces of purely arithmetical nature. These are infinite extensions of local fields, such as the maximal unramified or maximal abelian extensions, well known in algebraic number theory. We construct and study a Gaussian measure on the space conjugate to an infinite extension (considered with a natural topology), a Fourier transform, a fractional differentiation operator, and an analog of the symmetric stable process.

Up to this point all the stochastic processes we mentioned were parametrized, as usual, by a real positive time parameter. In the non-Archimedean setting it is natural also to consider processes with the non-Archimedean "time", and that is the subject of Chapter 7. This chapter is an introduction to the theory (developed by Evans) of the fully non-Archimedean Brownian motion, with both the time parameter and values from a local field, and to the theory of the complex-valued Wiener process with p-adic time (introduced by Bikulov and Volovich).

Since the book touches a rather large variety of topics from various branches of mathematics, it was difficult to maintain throughout the book a consistent system of notations. However, some notations are kept. In particular, unless stated otherwise, $\|\cdot\|$ always denotes the normalized absolute value on a local field K (other norms are distinguished by appropriate subscripts), O is the ring of integers, P a prime ideal, q and p are respectively the cardinality and characteristic of the residue field O/P, $S \subset O$ is a complete set of representatives of classes from O/P. Various positive constants will often be denoted by the same letter C.

Anatoly N. Kochubei

Contents

Chapter 1

Preliminaries

In this chapter we give a brief overview of various aspects of the theory of non-Archimedean local fields. Most of the results are well-known and included in order to introduce necessary notions making the exposition in subsequent chapters self- contained. Proofs, which are omitted, can be found in standard textbooks on algebra, number theory, and non-Archimedean analysis [99, 18, 138, 156, 104, 63, 39, 71, 134, 143, 154].

1.1 Field extensions

If a field k is a subfield of a field K, then K is called an extension of k. In order to emphasize that K is considered as an extension of k, we often denote the extension by K/k. If a field K_0 is a subfield of K containing k, that is $k \subset K_0 \subset K$, then K_0 is called an intermediate field of the extension K/k.

An extension K/k can be considered as a vector space over k.

Definition 1.1. *An extension K/k is called finite if K is a finite-dimensional vector space over k. Its dimension $(K : k)$ is called the degree of the extension K/k. Any basis of K over k is called a basis of the extension K/k.*

If an extension K/k is finite, then for any intermediate field K_0 the extensions K_0/k and K/K_0 are also finite. Conversely, the finiteness of the extensions K_0/k and K/K_0 implies the finiteness of K/k, with the chain rule

$$(K : k) = (K : K_0)(K_0 : k).$$

Definition 1.2. *Let k be a subfield of a field K, and $\theta \in K$. If θ is a root of a non-zero polynomial with coefficients in k, the element θ is called algebraic over k. An extension K/k is called algebraic if every element of K is algebraic over k.*

Theorem 1.1. *Every finite extension K/k is algebraic. For any polynomial $f \in k[x]$, $\deg f \geq 1$, there exists a finite extension K/k, in which f splits into a product of linear factors:*

$$f(x) = a(x - \alpha_1) \dots (x - \alpha_n), \quad a \in k, \alpha_1, \dots, \alpha_n \in K.$$

If the field K from Theorem 1.1 can be obtained by adjoining elements $\alpha_1, \dots, \alpha_n$ to the field k, then k is called *a splitting field* of the polynomial f. Such a field always exists; it is unique up to an isomorphism.

Let K/k be a finite extension of the degree n. For an arbitrary $\alpha \in K$ the mapping $\xi \mapsto \alpha\xi$ ($\xi \in K$) is a linear operator on K (as a vector space over k). Let $\omega_1, \dots, \omega_n$ be a basis of the extension K/k. Then

$$\alpha\omega_i = \sum_{j=1}^{n} a_{ij}\omega_j, \quad a_{ij} \in k.$$

The determinant and the trace of the matrix (a_{ij}) belong to k and do not depend on the choice of the basis $\{\omega_j\}$.

Definition 1.3. *The determinant $N_{K/k}(\alpha) = \det(a_{ij})$ and the trace $\mathrm{Tr}_{K/k}(\alpha) = \mathrm{tr}(a_{ij})$ are called the norm and the trace of the element $\alpha \in K$ with respect to the extension K/k.*

For any $\alpha, \beta \in K$, $a \in k$

$$\begin{gathered} N_{K/k}(\alpha\beta) = N_{K/k}(\alpha)N_{K/k}(\beta), \\ \mathrm{Tr}_{K/k}(\alpha + \beta) = \mathrm{Tr}_{K/k}(\alpha) + \mathrm{Tr}_{K/k}(\beta), \\ \mathrm{Tr}_{K/k}(a\alpha) = a\,\mathrm{Tr}_{K/k}(\alpha), \\ N_{K/k}(a) = a^n, \quad \mathrm{Tr}_{K/k}(a) = na. \end{gathered}$$

These identities show that the mapping $\alpha \mapsto N_{K/k}(\alpha)$ is a homomorphism of the multiplicative group K^* of the field K into the multiplicative group k^* of k; the mapping $\alpha \mapsto \mathrm{Tr}_{K/k}(\alpha)$ is a linear function on K with values in k.

An automorphism σ of the field K is called an automorphism of the extension K/k if $\sigma(a) = a$ for all $a \in k$. The set of all such automorphisms is a group (with natural operations). If the extension K/k is finite, the its group of automorphisms is finite. its order does not exceed $(K : k)$.

Definition 1.4. *A finite extension K/k is called a Galois extension, if the order of its group of automorphisms G coincides with $(K : k)$. In this case G is called the Galois group of the extension K/k.*

The class of Galois extension is important since for them there exists a one-to-one correspondence between intermediate fields and subgroups of the Galois group. On the other hand, the Galois group is a natural "symmetry group" for analytic objects related to an extension. Note that the norm and trace mappings are invariant with respect to the Galois group.

A finite extension K/k is called *cyclic* if it is a Galois extension, and its Galois group is cyclic.

Theorem 1.2 (Hilbert). *Let K/k be a cyclic extension with the Galois group G, σ be a generator of G, and β be an element from K. The trace $\mathrm{Tr}_{K/k}(\beta)$ equals 0 if and only if there exists such an element $\alpha \in K$ that $\beta = \alpha - \sigma(\alpha)$.*

A similar property holds for the norm mapping (see e.g. [99]).

1.2 Finite fields

The first example of a finite field is the field $\mathbb{F}_p$ (p is a prime number) consisting of the numbers $\{0, 1, \ldots, p-1\}$ with algebraic operations defined mod p. Equivalently, one can define $\mathbb{F}_p$ as the quotient ring $\mathbb{Z}/(p)$, where (p) is the principal ideal in $\mathbb{Z}$ generated by p, and verify that $\mathbb{F}_p$ is actually a field. It is easy to see that the characteristic $\mathrm{char}\,\mathbb{F}_p$ equals p.

Every finite field of characteristic p is a finite extension of $\mathbb{F}_p$. More precisely, we have the following result.

Theorem 1.3. *For any prime number p and any natural number n, the splitting field $\mathbb{F}_q$ of the polynomial $x^q - x$, $q = p^n$, over $\mathbb{F}_p$, is a finite field consisting of q elements. Any finite field of q elements is isomorphic to $\mathbb{F}_q$. The field $\mathbb{F}_q$ has characteristic p, $(\mathbb{F}_q : \mathbb{F}_p) = n$. Each subfield of $\mathbb{F}_q$ has the order p^m, where m is a positive divisor of the number n. Conversely, if m is a positive divisor of n, there exists only one subfield of $\mathbb{F}_q$ of the order p^m. This subfield $\mathbb{F}_{p^m}$ consists of the roots of the polynomial $x^{p^m} - x$ in $\mathbb{F}_q$.*

Consider an extension $\mathbb{F}_{q^m}/\mathbb{F}_q$. Its Galois group consists of the automorphisms $\sigma_0, \sigma_1, \ldots, \sigma_{m-1}$ defined as

$$\sigma_j(\alpha) = \alpha^{q^j}, \quad \alpha \in \mathbb{F}_{q^m}, \quad 0 \le j \le m-1.$$

Therefore this group is cyclic, with a generator σ_1.

The multiplicative group $\mathbb{F}_q^*$ of non-zero elements of the field $\mathbb{F}_q$ is also cyclic. Its generator is called a *primitive element* of the field $\mathbb{F}_q$.

One of the traditional (but far from completed) topics in the theory of finite fields is the study of various bases of an extension $\mathbb{F}_{q^m}/\mathbb{F}_q$. We shall mention three types of bases.

A basis $\{\varepsilon_1, \ldots, \varepsilon_m\}$ of the extension $\mathbb{F}_{q^m}/\mathbb{F}_q$ is called *self-dual* if $\mathrm{Tr}_{\mathbb{F}_{q^m}/\mathbb{F}_q}(\varepsilon_i \varepsilon_j) = \delta_{ij}$ (the Kronecker symbol), *normal* if $\varepsilon_2 = \varepsilon_1^q, \ldots \varepsilon_m = \varepsilon_1^{q^{m-1}}$, and *primitive* if the elements $\varepsilon_1, \ldots, \varepsilon_m$ are primitive elements of the field $\mathbb{F}_{q^m}$.

Theorem 1.4. *The extension $\mathbb{F}_{q^m}/\mathbb{F}_q$ possesses a normal primitive basis. If m is odd, or $m \equiv 2 \pmod 4$ and q is even, then the extension $\mathbb{F}_{q^m}/\mathbb{F}_q$ has a normal self-dual basis.*

For the proof see [63, 102, 103].

At least for some situations (for example, if $m = 3$, $q = 3$ or 5) there exists a basis, which is simultaneously normal, primitive, and self-dual. That follows from the explicit formulas (for the above values of m and q) for the polynomials, whose roots are elements of a self-dual normal basis [101], and the table of primitive polynomials [104]. It seems probable that at least for the extension $\mathbb{F}_{p^3}/\mathbb{F}_p$ such a basis exists for any prime number p.

Let us consider characters of the additive and multiplicative groups of the field $\mathbb{F}_q$ (we shall often call them additive and multiplicative characters of the field).

Let, as before, p be the characteristic of the field $\mathbb{F}_q$. Then we have an extension $\mathbb{F}_q/\mathbb{F}_p$. An additive character

$$\eta_1(\alpha) = \exp\left(\frac{2\pi i}{p} \mathrm{Tr}_{\mathbb{F}_q/\mathbb{F}_p}(\alpha)\right), \quad \alpha \in \mathbb{F}_q,$$

is called the *canonical additive character* of the field $\mathbb{F}_q$. Any other additive character has the form $\eta_\gamma(\alpha) = \eta_1(\gamma\alpha)$, $\gamma \in \mathbb{F}_q$. Thus the additive group $\mathbb{F}_q^+$ of $\mathbb{F}_q$ has the self-duality property – its character group is isomorphic to $\mathbb{F}_q^+$ itself.

Let ρ be a fixed primitive element of the field $\mathbb{F}_q$. Then for any j, $0 \le j \le q-2$, a function ψ_j of the form

$$\psi_j(\rho^k) = \exp\left(\frac{2\pi i j k}{q-1}\right), \quad 0 \le k \le q-2,$$

defines a multiplicative character of the field $\mathbb{F}_q$. Any multiplicative character of $\mathbb{F}_q$ coincides with the character ψ_j for some j, $0 \le j \le q-2$.

We shall use one important example of a multiplicative character. Suppose that $p \ne 2$. Define a function $\theta^{(2)} : \mathbb{F}_q^* \to \mathbb{R}$ setting $\theta^{(2)}(\zeta) = 1$ if $\zeta \in \mathbb{F}_q^*$ is a square of some element, and $\theta^{(2)}(\zeta) = -1$ otherwise. It is easy to see that $\theta^{(2)} = \psi_j$ for $j = \frac{q-1}{2}$. The character $\theta^{(2)}$ is called *the quadratic character* of the field $\mathbb{F}_q$.

The interrelations of the additive and multiplicative properties of a finite field are reflected in identities for sums of characters. In particular, if $q = p^n$, $p \ne 2$, then

$$\sum_{\zeta \in \mathbb{F}_q^*} \theta^{(2)}(\zeta)\eta_1(\zeta) = \begin{cases} (-1)^{n-1}q^{1/2}, & \text{if } p \equiv 1 \pmod 4, \\ (-1)^{n-1}i^n q^{1/2}, & \text{if } p \equiv 3 \pmod 4. \end{cases} \tag{1.1}$$

Some other frequently used identities are actually specializations of the orthogonality relations for characters of an arbitrary finite abelian group. We have

$$\sum_{\alpha \in \mathbb{F}_q} \eta_{\gamma_1}(\alpha)\overline{\eta_{\gamma_2}(\alpha)} = \begin{cases} 0 & , \text{if } \gamma_1 \ne \gamma_2, \\ q & , \text{if } \gamma_1 = \gamma_2. \end{cases} \tag{1.2}$$

In particular,

$$\sum_{\alpha \in \mathbb{F}_q} \eta_\gamma(\alpha) = 0 \quad \text{if } \gamma \ne 0. \tag{1.3}$$

1.3 Local fields

Let K be a non-discrete locally compact topological field. If K is connected, then either $K = \mathbb{R}$ or $K = \mathbb{C}$. We shall be interested in the opposite case of disconnected fields.

Definition 1.5. *A local field is a non-discrete disconnected locally compact field.*

There are two major examples of a local field. Let p be a prime number. The field of p-adic numbers is the completion $\mathbb{Q}_p$ of the field $\mathbb{Q}$ of rational numbers, with respect to the absolute value $|x|_p$ defined by setting $|0|_p = 0$,

$$|x|_p = p^{-\nu} \text{ if } x = p^{\nu}\frac{m}{n},$$

where $\nu, m, n \in \mathbb{Z}$, and m, n are prime to p.

Note that by Ostrowski's theorem there are no absolute values on $\mathbb{Q}$, which are not equivalent to the "Euclidean" one, or one of $|\cdot|_p$.

The absolute value $|x|_p$, $x \in \mathbb{Q}_p$, has the following properties:

$$|x|_p = 0 \text{ if and only if } x = 0;$$
$$|xy|_p = |x|_p \cdot |y|_p;$$
$$|x+y|_p \le \max(|x|_p, |y|_p).$$

The latter property called *the ultra-metric inequality* (or *the non-Archimedean property*) leads to substantial differences from the properties of the Archimedean fields ($\mathbb{R}$ and $\mathbb{C}$). In particular, $\mathbb{Q}_p$ is totally disconnected in the topology determined by the metric $|x-y|_p$. Standard subsets of $\mathbb{Q}_p$ (with respect to this metric), like balls or spheres, are simultaneously open and closed (they are actually compact), two balls either do not intersect, or one of them is contained in the other, etc. Proofs of these (and many other similar properties) are based on the following equality, which is a consequence of the ultra-metric inequality:

$$|x+y|_p = \max(|x|_p, |y|_p) \quad \text{if } |x|_p \ne |y|_p. \tag{1.4}$$

The absolute value $|x|_p$ takes the discrete set of non-zero values p^N, $N \in \mathbb{Z}$. If $|x|_p = p^N$, then x admits a (unique) canonical representation

$$x = p^{-N}\left(x_0 + x_1 p + x_2 p^2 + \cdots\right), \tag{1.5}$$

where $x_0, x_1, x_2, \ldots \in \{0, 1, \ldots, p-1\}$, $x_0 \ne 0$. The series converges in the topology of $\mathbb{Q}_p$. For example,

$$-1 = (p-1) + (p-1)p + (p-1)p^2 + \cdots, \quad |-1|_p = 1.$$

If K is a finite extension of $\mathbb{Q}_p$, an absolute value on K extending $|x|_p$ from $\mathbb{Q}_p$ can be obtained by setting

$$|x|_K = \left|N_{K/\mathbb{Q}_p}(x)\right|_p^{1/m}, \quad m = (K : \mathbb{Q}_p).$$

We shall often use also the absolute value

$$\|x\|_K = \left|N_{K/\mathbb{Q}_p}(x)\right|_p,$$

which is called *a normalized absolute value* (the meaning of this term will become clear a little later). Both of them generate the same topology on K. Properties of these absolute values are similar to the ones of $|x|_p$. Note that $\operatorname{char} K = \operatorname{char} \mathbb{Q}_p = 0$.

The second important example of a local field is the field $\mathbb{F}_q((t))$ of formal Laurent series

$$x = \sum_{i=n}^{\infty} a_i t^i, \quad n \in \mathbb{Z}, \tag{1.6}$$

with coefficients from a finite field $\mathbb{F}_q$, $q = p^f$, $f \geq 1$. The (ultrametric) absolute value $\|x\|$ of the element (1.6) equals q^{-n}. Note that $\operatorname{char} \mathbb{F}_q((t)) = p$. This field is also complete and totally disconnected.

Theorem 1.5. *Every local field of characteristic 0 is a finite extension of the field $\mathbb{Q}_p$. Every local field of a positive characteristic p is isomorphic to $\mathbb{F}_q((t))$ for some $q = p^f$, $f \geq 1$.*

Let K be an arbitrary local field. Denote

$$O = \{x \in K : \|x\| \leq 1\}, \; P = \{x \in K : \|x\| < 1\}, \; U = O \setminus P.$$

O is a subring of K called *the ring of integers*, P is an ideal in O called *the prime ideal.* The multiplicative subgroup U is called *the group of units.* The ideal P contains an element β such that $P = \beta O$ (*a prime element*). For example, if $K = \mathbb{Q}_p$ then one may take $\beta = p$; the elements of O are called p-adic integers; if $K = \mathbb{F}_q((t))$, one can choose $\beta = t$.

The canonical representation (1.5) can be generalized as follows. Let us consider the quotient ring O/P. It is easy to show that O/P is actually a finite field (it coincides with $\mathbb{F}_p$ for $K = \mathbb{Q}_p$, and with $\mathbb{F}_q$ for $K = \mathbb{F}_q((t))$). Let p be the characteristic of the field O/P, q be its cardinality. Then the normalized absolute value $\|\cdot\|$ takes the values q^N, $N \in \mathbb{Z}$, and $\|\beta\| = q^{-1}$ ("normalization"). Let $S \subset O$ be a complete system of representatives of the residue classes from O/P. Then any nonzero element $x \in K$ is uniquely representable in the form of the convergent series

$$x = \beta^{-n} \left(x_0 + x_1 \beta + x_2 \beta^2 + \cdots \right), \tag{1.7}$$

where $n \in \mathbb{Z}$, $\|x\| = q^n$, $x_j \in S$, $x_0 \notin P$.

Let $\operatorname{char} K = 0$. As we know, in this case K is a finite extension of $\mathbb{Q}_p$. Then $\operatorname{Tr}_{K/\mathbb{Q}_p} : O \to \mathbb{Z}_p$, where $\mathbb{Z}_p$ is the ring of integers of $\mathbb{Q}_p$. In addition, there exists an integer $d \geq 0$ such that $\operatorname{Tr}_{K/\mathbb{Q}_p}(x) \in \mathbb{Z}_p$ for

$\|x\| \le q^d$ but $\mathrm{Tr}_{K/\mathbb{Q}_p}(x_0) \notin \mathbb{Z}_p$ for some x_0 with $\|x_0\| = q^{d+1}$. The integer d is called *the exponent of the different* of the extension $K/\mathbb{Q}_p$. It is known that $d \ge e - 1$, where e is *the ramification index* defined by the relation $\|p\| = q^{-e}$; it is also known that $d = 0$ if and only if $e = 1$ (in this case K is called *unramified.* Note that the field $\mathbb{Q}_p$ (as every local field) has a unique unramified extension of any given degree.

Another important number attached to the extension $K/\mathbb{Q}_p$ is its *index of inertia.* Let as before $p = \mathrm{char}\, O/P$, $q = \mathrm{card}\, O/P$. The index of inertia $f \ge 1$ is defined by the relation $q = p^f$. It is known that $ef = (K : \mathbb{Q}_p)$. If $f = 1$, the extension is called *totally ramified.*

Note that the above notions and relations carry over (virtually unchanged) to arbitrary finite extensions K/L of local fields (where not necessarily $L = \mathbb{Q}_p$).

Let K/L be a finite extension of local fields. A basis $x_1, \dots, x_n$, $n = (K : L)$, of the field K over L is called *fundamental* if all x_j belong to the ring of integers O_K, and for any $x \in O_K$ all the coefficients in the expansion

$$x = \xi_1 x_1 + \cdots + \xi_n x_n, \quad \xi_j \in L,$$

belong to the ring of integers O_L of the field L.

A fundamental basis of an extension can be constructed as follows. Let P_K, P_L be the prime ideals of K and L respectively, β_K be a prime element of K. As before, we denote by e the ramification index of the extension K/L, and by f the index of inertia. Let $\omega_1, \dots, \omega_f \in O_K$ be such a system of representatives of residue classes from O_K/P_K that their classes form a basis of O_K/P_K over O_L/P_L. Then the system

$$\omega_i \beta_K^j, \quad i = 1, \dots, f;\ j = 0, 1, \dots, e-1,$$

is a fundamental basis of the extension K/L.

If the extension is unramified, that is $e = 1$, $f = n$, the above fundamental basis consists solely of the representatives $\omega_1, \dots, \omega_n$. This basis will be called *canonical.*

Any prime element of L is simultaneously a prime element of the unramified extension K. Note also that the Galois group of an unramified extension is cyclic. Its generator is called the *Frobenius automorphism* of the extension K/L.

Properties of the norm and trace mappings for an unramified extension are much simpler than in the general case. We mention here two results regarding trace, which are important for harmonic analysis.

(i) Let K/L be an unramified extension of local fields. Denote by ρ_K and ρ_L the canonical homomorphisms

$$\rho_K : O_K \longrightarrow O_K/P_K, \quad \rho_L : O_L \longrightarrow O_L/P_L.$$

Then for any $x \in O_K$

$$\rho_L\left(\mathrm{Tr}_{K/L}(x)\right) = \mathrm{Tr}_{(O_K/P_K)/(O_L/P_L)}(\rho_K(x)). \tag{1.8}$$

(ii) Let $P_K^\nu = \beta_K^\nu O_K$, $P_L^\nu = \beta_L^\nu O_L$, $\nu \in \mathbb{Z}$. If K/L is unramified, then $\mathrm{Tr}_{K/L}$ is a surjection of P_K^ν onto P_L^ν for every $\nu \in \mathbb{Z}$.

Let us consider additive characters of local fields. Below we shall always deal with continuous characters.

Theorem 1.6. *The additive group of any local field K is self-dual, that is if χ is any non-constant additive character of K, then any other additive character can be written as $\chi_{(a)}(x) = \chi(ax)$, $x \in K$, for some $a \in K$.*

Therefore, the description of the set of all additive characters is reduced to finding one non-trivial character. If $\mathrm{char}\, K = p$, that is $K = \mathbb{F}_q((t))$, this can be done by setting

$$\chi(x) = \eta_1(a_{-1}), \quad x \in K,$$

where a_{-1} is a coefficient of the representation of $x \in K$ in the form (1.6), η_1 is the canonical additive character of the finite field $\mathbb{F}_q$.

Let $\mathrm{char}\, K = 0$. We begin with the case $K = \mathbb{Q}_p$. Proceeding from the canonical representation (1.5) of an element $x \in \mathbb{Q}_p$, we define *the fractional part* of x as the rational number

$$\{x\}_p = \begin{cases} 0, & \text{if } N \le 0 \text{ or } x = 0; \\ p^{-N}\left(x_0 + x_1 p + \cdots + x_{N-1}p^{N-1}\right), & \text{if } N > 0. \end{cases}$$

The function $\chi_p(x) = \exp(2\pi i \{x\}_p)$ is an additive character of $\mathbb{Q}_p$ (*the canonical additive character*). It is clear that $\chi_p(x) = 1$ if $|x|_p \le 1$.

If K is a finite extension of $\mathbb{Q}_p$, we can obtain a non-trivial additive character of K taking the composition $\chi_p \circ \mathrm{Tr}_{K/\mathbb{Q}_p}$.

Additive characters on K are usually classified by their ranks. The rank of a character χ is the smallest integer n such that $\chi(x) \equiv 1$ if $\|x\| \le q^{-n}$. If a character χ has the rank n, then $\chi(\beta^k x)$ has the rank $n - k$. It follows from the continuity of a character that every character has a finite rank.

By Theorem 1.6, in order to describe all additive characters, one has only to specify one of them. Sometimes the above composition with the trace is used, but it is often more convenient to fix a rank zero character. That can be done by setting

$$\chi(x) = \chi_p\left(\mathrm{Tr}_{K/\mathbb{Q}_p}(\beta^{-d}x)\right).$$

Let us introduce some more objects related to additive characters. Consider the mapping $\varphi: O \to \mathbb{C}$ defined by

$$\varphi(x) = \chi_p\left(\mathrm{Tr}_{K/\mathbb{Q}_p}(\beta^{-d-1}x)\right).$$

It induces a mapping $\Phi: O/P \to \mathbb{C}$, which is actually a non-trivial additive character on the finite field O/P. Due to the latter's self-duality, there exists a nonzero element $\sigma \in O/P$ such that

$$\Phi(\xi) = \eta_1(\sigma\xi), \quad \xi \in O/P. \tag{1.9}$$

The element σ reflects the ramification properties of K (see [156], Chapter VIII, Proposition 2): if K is unramified over $\mathbb{Q}_p$, then $\sigma = 1$.

Using the representation (1.9) and the identities (1.1), (1.3) for trigonometric sums over finite fields, we find that

$$\sum_{s\in S}\varphi(as) = 0, \quad a \in U;$$

$$\sum_{s\in S}\varphi(as^2) = \theta^{(2)}(\widehat{a})G_K, \quad a \in U, p \neq 2$$

(the *Gauss sum* identity), where $a \mapsto \widehat{a}$ is the canonical mapping of O onto O/P, $\theta^{(2)}$ is the quadratic multiplicative character on O/P, and

$$G_K = \begin{cases} (-1)^{f-1}q^{1/2}\theta^{(2)}(\sigma), & \text{if } p \equiv 1 \pmod 4; \\ (-1)^{f-1}i^f q^{1/2}\theta^{(2)}(\sigma), & \text{if } p \equiv 3 \pmod 4. \end{cases}$$

A notion of rank (or *ramification degree*) is available also for multiplicative characters of a local field. In fact we shall use more frequently not multiplicative characters on K, but characters of the (multiplicative) group of units U.

A character θ of the group U is said to have the ramification degree $k \geq 1$ if $\theta(1+x) = 1$ as $\|x\| \leq q^{-k}$ and $\theta(1+x_0) \neq 1$ for some $x_0 \in U$ with $\|x_0\| = q^{-k+1}$. The only character of the ramification degree 0 is $\theta \equiv 1$. On the other hand, each character has a finite ramification degree. There exist $q-2$ characters of the ramification degree 1 and $q^{j-2}(q-1)^2$ characters of the ramification degree j for each $j \geq 2$.

An arbitrary multiplicative character ζ on K is determined completely by a character on U (the restriction of ζ to U) and the value of ζ on a prime element β.

A homomorphism of the multiplicative group K^* of a local field K to the multiplicative group of the field $\mathbb{C}$ is called *a multiplicative quasicharacter* on K. Each multiplicative quasicharacter ζ can be presented in the form

$$\zeta(x) = \|x\|^{s-1}\theta(x), \tag{1.10}$$

where s is a complex number, θ is a multiplicative character, such that $\theta(\beta) = 1$. The latter condition identifies θ with a character of the group U.

1.4 Polynomials and quadratic forms

Let $f_1(x_1, \dots, x_n), \dots, f_n(x_1, \dots, x_n)$ be polynomials with coefficients from the ring of integers O of a local field K. We are interested in conditions, under which the system of equations

$$f_j(x_1, \dots, x_n) = 0, \quad j = 1, \dots, n, \tag{1.11}$$

has a solution $(x_1, \dots, x_n) \in O^n$. Denote $P_N = \{x \in K : \|x\| \le q^{-N}\}$, $N \ge 1$ (so that $P_1 = P$). Let $M(x_1, \dots, x_n)$ be the Jacobi matrix $(\partial f_i / \partial x_j)_{i,j=1}^n$.

Theorem 1.7 (Hensel's lemma). *Let $(a_1, \dots, a_n) \in O^n$ be such a vector that $\det M(a_1, \dots, a_n)$ belongs to the group of units U, and $f_j(a_1, \dots, a_n) \in P_N$ for all $j = 1, \dots, n$. Then there exists a unique solution $(x_1, \dots, x_n) \in O^n$ of the system (1.11), such that $x_i - a_i \in P_N$ for all $i = 1, \dots, n$.*

There exists a lot of refinements and generalizations of Theorem 1.7 (see e.g. [20, 99, 138]). We shall need one of them, which deals with one polynomial instead of a quadratic system.

Theorem 1.7′. *Let $f(x_1, \dots, x_n)$ be a polynomial with coefficients from O. Suppose that $(a_1, \dots, a_n) \in O$ is such that*

$$f((a_1, \dots, a_n) \in P_N, \quad \left\| \frac{\partial f}{\partial x_j}(a_1, \dots, a_n) \right\| = q^k$$

for some j, where $0 \le 2k < N$. Then the equation

$$f(x_1, \dots, x_n) = 0$$

possesses such a solution that $x_i - a_i \in P_{N-k}$ for all $i = 1, \dots, n$.

A very important class of polynomials over local fields is the class of quadratic forms. We shall need a subclass of *anisotropic* (or *elliptic*) quadratic forms over $\mathbb{Q}_p$, $p \ne 2$, consisting of forms $h(x_1, \dots, x_n)$ satisfying the condition

$$h(x_1, \dots, x_n) \ne 0 \text{ if } |x_1|_p + \dots + |x_n|_p \ne 0. \tag{1.12}$$

Other quadratic forms on $\mathbb{Q}_p$ are called *isotropic*.

Theorem 1.8. *If $n \geq 5$, then there exist no anisotropic quadratic forms.*

Therefore, in order to describe anisotropic quadratic forms, we shall consider separately the cases, when $n = 2, 3$ or 4.

The case of *binary* forms (that is, $n = 2$), is connected with quadratic extensions (extensions of the degree 2) of the field $\mathbb{Q}_p$.

Let $h(x_1, x_2)$ satisfy (1.12). Then up to a linear isomorphism

$$h(x_1, x_2) = x_1^2 - \tau x_2^2,$$

where $\tau \in \mathbb{Q}_p$ is not a square of an element of $\mathbb{Q}_p$. In order to obtain all non-equivalent forms, one has to take a full set of representatives of residue classes from $\mathbb{Q}_p^*/\mathbb{Q}_p^{*2}$ (here $\mathbb{Q}_p^*$ is the multiplicative group of the field $\mathbb{Q}_p$, $\mathbb{Q}_p^{*2}$ is the subgroup of all squares):

$$\tau = \varepsilon; \quad \tau = p; \quad \tau = \varepsilon p,$$

where ε is an arbitrary element with $|\varepsilon|_p = 1$ (a choice of a different ε leads to an equivalent form). Note that we have excluded the case $p = 2$, for which the situation is more complicated.

Let us consider (for each of the above cases) an extension K of the field $\mathbb{Q}_p$ obtained by adjoining an element $\sqrt{\tau}$, $K = \mathbb{Q}_p(\sqrt{\tau})$, with natural algebraic operations. Then the normalized absolute value on K can be expressed in terms of the form h:

$$\|x_1 + \sqrt{\tau} x_2\| = |x_1^2 - \tau x_2^2|_p, \quad x_1, x_2 \in Q_p. \tag{1.13}$$

The extension $K/\mathbb{Q}_p$ is unramified if $\tau = \varepsilon$; otherwise it is totally ramified.

A *ternary* form $h(x_1, x_2, x_3)$ satisfying (1.12) equals (up to a linear isomorphism)

$$h(x_1, x_2, x_3) = p\varepsilon_1 x_1^2 + \varepsilon_2 x_2^2 + \varepsilon_3 x_3^2,$$

where $|\varepsilon_1|_p = |\varepsilon_2|_p = |\varepsilon_3|_p = 1$, $\varepsilon_2 x_2^2 + \varepsilon_3 x_3^2 \neq 0$ if $|x_2|_p + |x_3|_p \neq 0$.

Finally, a *quaternary* form $h(x_1, x_2, x_3, x_4)$ satisfying (1.12) is unique (up to a linear isomorphism) and can be represented as

$$h(x_1, x_2, x_3, x_4) = x_1^2 - s x_2^2 - p x_3^2 + sp x_4^2, \tag{1.14}$$

where $s \in \mathbb{Z}$ is a quadratic nonresidue $\mod p$.

It is possible to obtain for the form (1.14) a representation similar to (1.13). However in this case one has to deal not with extensions of $\mathbb{Q}_p$ but with a non-commutative object called *the quaternion algebra K* over $\mathbb{Q}_p$, which is defined as a four-dimensional vector space over $\mathbb{Q}_p$ with a basis $1, i, j, k$ and the multiplication table

$$i^2 = s,\ j^2 = p,\ ij = -ji = k,\ k^2 = sp,$$
$$ik = -ki = js,\ jk = -kj = -ip.$$

K is a division algebra, its centre coincides with $\mathbb{Q}_p$. Many of the notions defined above for local fields make sense also for the division algebra K. In particular, one can define a non-Archimedean absolute value on K as

$$\|x\| = |h(x_1, x_2, x_3, x_4)|_p^2, \quad x = x_1 + x_2 i + x_3 j + x_4 k. \tag{1.15}$$

The absolute value $\|\cdot\|$ takes the values q^N, $N \in \mathbb{Z}$, where $q = p^2$, and the vector j plays the role of a prime element, so that the absolute value is normalized: $\|j\| = q^{-1}$.

Let $\mathbb{Q}_p(\sqrt{s})$ be the quadratic extension of the field $\mathbb{Q}_p$ obtained by adjoining the square root $\sqrt{s}$. If $p \neq 2$, the quaternion algebra K is isomorphic to the algebra of 2×2 matrices with elements from $\mathbb{Q}_p(\sqrt{s})$. The isomorphism θ can be given as

$$\theta(x_1 + x_2 i + x_3 j + x_4 k) = \begin{pmatrix} x_1 + \sqrt{s}x_2 & x_3 + \sqrt{s}x_4 \\ p(x_3 - \sqrt{s}x_4) & x_1 - \sqrt{s}x_2 \end{pmatrix}. \tag{1.16}$$

Now the analogs of the norm and trace mappings are defined as follows. If $x = x_1 + x_2 i + x_3 j + x_4 k \in K$, then

$$N(x) = \det \theta(x), \quad \operatorname{Tr}(x) = \operatorname{tr} \theta(x).$$

It is easy to see that $N(x) = h(x_1, x_2, x_3, x_4)$, $\operatorname{Tr}(x) = 2x_1$. The mappings $N: K \to \mathbb{Q}_p$ and $\operatorname{Tr}: K \to \mathbb{Q}_p$ are specific cases of the reduced norm and the reduced trace defined for general division algebras [115].

Lemma 1.1. *The inequality $\|x\| \le 1$, $x = x_1 + x_2 i + x_3 j + x_4 k \in K$, is equivalent to the system of inequalities*

$$|x_l|_p \le 1, \quad l = 1, 2, 3, 4.$$

Proof. Suppose that $\|x\| \le 1$ but $|x_l|_p > 1$ for some l, that is

$$x_l = p^{-N}(\xi_{0l} + \xi_{1l} p + \cdots); \quad \xi_{0l}, \xi_{1l}, \ldots \in \{0, 1, \ldots p - 1\}, N > 0,$$

and not all the numbers ξ_{0l} $(l = 1, 2, 3, 4)$ are equal to zero. Since $|h(x_1, x_2, x_3, x_4)|_p \le 1$, it follows that

$$\xi_{01}^2 - s\xi_{02}^2 \equiv 0 \pmod{p},$$
$$2\xi_{01}\xi_{11} - 2s\xi_{02}\xi_{12} - \xi_{03}^2 + s\xi_{04}^2 \equiv 0 \pmod{p}.$$

The first congruence implies (see Chapter 1 in [18]) that $\xi_{01} = \xi_{02} = 0$ due to the assumption made with regard to s. Now it follows in a similar way from the second congruence that $\xi_{03} = \xi_{04} = 0$. ■

In order to obtain a character of the additive group of the algebra K, it is sufficient to take $\chi = \chi_p \circ \mathrm{Tr}$. It follows from Lemma 1.1 that χ is a rank zero character.

1.5 Integration and harmonic analysis

Let K be a local field. Denote by dx the Haar measure on the additive group of K normalized by the equality

$$\int_{O} dx = 1. \tag{1.17}$$

A remarkable connection between the additive and multiplicative structures of a local field is expressed in the relation

$$d(ax) = \|a\|\, dx, \quad a \in K, \tag{1.18}$$

which means, in particular, that dx is also the Haar measure on the (multiplicative) group of units U.

The relation (1.18) can be seen as a special case of the general formula for the change of variables in (multiple) integrals. We shall need the formula for linear transformations: if $f \in L_1(K^n)$, T is a non-degenerate $n \times n$ matrix with entries from K, then

$$\int_{K^n} f(Tx)\, d^n x = \|\det T\|^{-1} \int_{K^n} f(y)\, d^n y. \tag{1.19}$$

It follows from (1.17) and (1.18) that

$$\int_{\|x\| \le q^{\nu}} dx = q^{\nu}; \tag{1.20}$$

$$\int_{\|x\| = q^{\nu}} dx = (1 - q^{-1}) q^{\nu}. \tag{1.21}$$

The notation in the next formula refers to the canonical representation (1.7). The formula itself is a consequence of the preceding one and the invariance of the Haar measure:

$$\int\limits_{\|x\|=q^{\nu},x_0=k} dx = q^{\nu-1} \quad (k \notin P). \tag{1.22}$$

Representing the ball $O = \{x \in K : \|x\| \le 1\}$ as the union of the sets $\{x \in K : \|x\| = q^N\}$, $-\infty < N \le -1$, we find also that

$$\int\limits_{\|x\|\le 1} \|x\|^{s-1}\, dx = \frac{1-q^{-1}}{1-q^{-s}}, \quad \operatorname{Re} s > 0. \tag{1.23}$$

Let χ be a character of rank 0. Then for any $a \in K$

$$\int\limits_{\|x\|\le q^{\nu}} \chi(ax)\, dx = \begin{cases} q^{\nu}, & \text{if } \|a\| \le q^{-\nu}, \\ 0, & \text{if } \|a\| > q^{-\nu}, \end{cases} \tag{1.24}$$

$$\int\limits_{\|x\|=q^{\nu}} \chi(ax)\, dx = \begin{cases} q^{\nu}(1-q^{-1}), & \text{if } \|a\| \le q^{-\nu}, \\ -q^{\nu-1}, & \text{if } \|a\| = q^{-\nu+1}, \\ 0, & \text{if } \|a\| \ge q^{-\nu+2}. \end{cases} \tag{1.25}$$

Similar formulas can be written for characters of an arbitrary rank. In particular, if K is a finite extension of $\mathbb{Q}_p$, $\chi = \chi_p \circ \mathrm{Tr}_{K/\mathbb{Q}_p}$, and d is the exponent of the different, we get

$$\int\limits_{\|x\|\le q^{\nu}} \chi(ax)\, dx = \begin{cases} q^{\nu}, & \text{if } \|a\| \le q^{d-\nu}, \\ 0, & \text{if } \|a\| > q^{d-\nu}, \end{cases} \tag{1.26}$$

$$\int\limits_{\|x\|=q^{\nu}} \chi(ax)\, dx = \begin{cases} q^{\nu}(1-q^{-1}), & \text{if } \|a\| \le q^{d-\nu}, \\ -q^{\nu-1}, & \text{if } \|a\| = q^{d-\nu+1}, \\ 0, & \text{if } \|a\| \ge q^{d-\nu+2}. \end{cases} \tag{1.27}$$

The formulas (1.24)–(1.27) with $a \neq 0$ are special cases of the following general formula, which is actually a consequence of (1.25). Let χ be a rank zero additive character, f a radial function, that is $f(x) = g(\|x\|)$. Then

$$\int\limits_K \chi(x\xi) f(x)\, dx = (1-q^{-1})\|\xi\|^{-1} \sum_{\nu=0}^{\infty} q^{-\nu} g(q^{-\nu}\|\xi\|^{-1}) - \|\xi\|^{-1} g(q\|\xi\|^{-1}), \quad \xi \neq 0, \tag{1.28}$$

provided $\sum_{\nu=0}^{\infty} q^{-\nu}|g(q^{-\nu})| < \infty$. The integral is understood as the sum of integrals over the spheres $\{\|x\| = q^N\}$.

Using (1.28), it is easy to obtain many other useful formulas. For example, if $f(x) = \log\|x\|$, after elementary transformations employing the identity

$$\sum_{n=0}^{\infty} nq^{-n} = \frac{1}{q(1-q^{-1})^2},$$

we come to the formula

$$\int_K \chi(x\xi)\log\|\xi\|\,d\xi = -\frac{q\log q}{q-1}\|x\|^{-1}, \quad x \neq 0. \tag{1.29}$$

We shall use also the following well-known formula for multiplicative characters. Let θ be a character of the group U having the ramification degree $k \geq 1$. Denote $V_0 = U$, $V_j = \{y \in U : \|1-y\| \leq q^{-j}\}$, $j \geq 1$. Then

$$\int_{V_0} \theta(y)\,dy = 0;$$

$$\int_{V_j\setminus V_{j+1}} \theta(y)\,dy = \begin{cases} 0, & \text{if } 0 \leq j < k-1; \\ -q^{-k}, & \text{if } j = k-1; \\ (1-q^{-1})q^{-j}, & \text{if } j \geq k. \end{cases} \tag{1.30}$$

Let us turn to Fourier transforms of complex-valued functions on a local field K. Since an additive group of K is self-dual, the Fourier transform of a function is again a function on K. We shall use two different normalizations of the Fourier transform. If K is an arbitrary local field, we fix a rank zero character χ and define, for any $f \in L_1(K)$,

$$(Ff)(\xi) = \int_K \chi(x\xi)f(x)\,dx.$$

We shall often write $Ff = \widetilde{f}$. It is possible to extend F from $L_1(K)\cap L_2(K)$ to a unitary operator on $L_2(K)$. If $\widetilde{f} \in L_1(K)$, we get the inversion formula

$$f(x) = \int_K \chi(-x\xi)\widetilde{f}(\xi)\,d\xi.$$

If K is a finite extension of $\mathbb{Q}_p$, and we employ explicitly some properties of the extension, it is sometimes more convenient to use not a rank zero character χ in the definition of the Fourier transform, but rather the character $\chi = \chi_p \circ \mathrm{Tr}_{K/\mathbb{Q}_p}$. In such cases we shall prefer the normalization

$$\widetilde{f}(\xi) = q^{-d/2} \int\limits_K \chi(x\xi) f(x)\, dx, \tag{1.31}$$

with the inversion formula

$$f(x) = q^{-d/2} \int\limits_K \chi(-x\xi) \widetilde{f}(\xi)\, d\xi, \tag{1.32}$$

where d is the exponent of the different. Of course, this change of a normalization does not influence qualitative properties of the transform (in particular, its unitarity on $L_2(K)$). However, unless stated otherwise, the Fourier transforms used in this book will be based on a rank zero character.

1.6 Distributions

As usual, in order to define some classes of distributions on a local field K, one has first to introduce an appropriate class of test functions.

Definition 1.6. *A function $f : K \to \mathbb{C}$ is called locally constant if there exists such an integer $l \geq 0$ that for any $x \in K$*

$$f(x + x') = f(x) \quad \text{if } \|x'\| \leq q^{-l}.$$

The smallest number l with this property is called the exponent of local constancy of the function f.

Note that the above definition of local constancy is more restrictive than the one given in [154].

Typical examples of locally constant functions are additive characters, and also cutoff functions like

$$\Omega(x) = \begin{cases} 1, & \text{if } \|x\| \leq 1; \\ 0, & \text{if } \|x\| > 1. \end{cases}$$

It is clear that a locally constant function is continuous. Of course, the continuity of Ω is an expression of the non-Archimedean property of a local field; it is related to the fact that K is totally disconnected, and its balls are both open and closed.

Denote by $\mathcal{D}(K)$ the set of all locally constant functions with compact supports. $\mathcal{D}(K)$ is a vector space over $\mathbb{C}$. In order to furnish it with a topology, let us consider a subspace $\mathcal{D}_N^l \subset \mathcal{D}(K)$ consisting of functions with supports in the ball

$$B_N = \{x \in K : \ \|x\| \le q^N\}, \quad N \in \mathbb{Z},$$

and the exponent of local constancy $\le l$.

It can be shown that $\mathcal{D}_N^l$ is a finite-dimensional space; thus it possesses a natural topology of a direct product of several copies of K. Then we set

$$\mathcal{D}_N = \varinjlim_{l\to\infty} \mathcal{D}_N^l,$$

and the topology in $\mathcal{D}(K)$ is defined as the inductive limit topology:

$$\mathcal{D}(K) = \varinjlim_{N\to\infty} \mathcal{D}_N,$$

The corresponding convergence in $\mathcal{D}(K)$ has the following meaning – $f_n \to 0$ in $\mathcal{D}(K)$ means that f_n all vanish outside a fixed compact set, they have a bounded set of the exponents of local constancy, and tend uniformly to zero.

The space $\mathcal{D}(K)$ is a complete locally convex topological vector space; it is rich enough to contain partitions of unity: for any compact set $\mathfrak{C} \subset K$ and any finite open covering $\{U_i\}$ of $\mathfrak{C}$ there exist functions $\varphi_i \in \mathcal{D}(K)$ such that $\varphi_i(x) \ge 0$, $\operatorname{supp}\varphi_i \subset U_i$, and $\sum \varphi_i(x) = 1$ for $x \in \mathfrak{C}$. On the other hand, $\mathcal{D}(K)$ is dense in $L_\alpha(K)$ for any $\alpha \in [1, \infty)$.

If $\mathfrak{O}$ is an open set in K, the space $\mathcal{D}(\mathfrak{O})$ of test functions on $\mathfrak{O}$ is defined as a subspace of $\mathcal{D}(K)$ consisting of functions with supports in $\mathfrak{O}$.

Definition 1.7. *The space $\mathcal{D}'(K)$ of Bruhat-Schwartz distributions on a local field K is the strong conjugate space to the space $\mathcal{D}(K)$.*

Note that (in contrast to distributions on $\mathbb{R}$) any linear functional $\mathcal{D}(K) \to \mathbb{C}$ is automatically continuous. Thus $\mathcal{D}'(K)$ consists of all linear functionals on $\mathcal{D}(K)$. The action of a functional f upon a test function φ will be denoted $\langle f, \varphi\rangle$.

The notions of restriction and support of a distribution $f \in \mathcal{D}'(K)$ are defined in a standard way (as for the space $\mathcal{D}'(\mathbb{R})$). However, the following property is purely non-Archimedean.

Theorem 1.9. *If $f \in \mathcal{D}'(K)$, and its support* $\operatorname{supp} f = \{0\}$, *then $f = C\delta$, where C is a constant, δ is the Dirac distribution, $\langle \delta, \varphi \rangle = \varphi(0)$, $\varphi \in \mathcal{D}(K)$.*

As in the classical case, every function $f \in L_1^{\mathrm{loc}}(K)$ determines a distribution $f \in \mathcal{D}'(K)$ by the formula

$$\langle f, \varphi \rangle = \int_K f(x)\varphi(x)\,dx, \quad \varphi \in \mathcal{D}(K).$$

The above notions and results carry over easily to the case of several variables, that is to the case of test functions and distributions on K^n, $n > 1$.

Let us turn now to operations over distributions. We start with the notion of a direct product.

Let $f \in \mathcal{D}'(K^n)$, $g \in \mathcal{D}'(K^m)$. The *direct product* $f \times g \in \mathcal{D}'(K^{n+m})$ is defined by the formula

$$\langle f(x) \times g(y), \varphi \rangle = \langle f(x), \langle g(y), \varphi(x,y) \rangle \rangle, \quad \varphi \in \mathcal{D}(K^{n+m}).$$

The operation of the direct product is commutative and jointly continuous.

Denote

$$\Omega_l(x) = \begin{cases} 1, & \text{if } \|x\| \le q^l; \\ 0, & \text{if } \|x\| > q^l. \end{cases}$$

Let $f, g \in \mathcal{D}'(K)$. *The convolution* $f * g$ is defined as

$$\langle f * g, \varphi \rangle = \lim_{l \to \infty} \langle f(x) \times g(y), \Omega_l(x)\varphi(x+y) \rangle \tag{1.33}$$

if the limit in the right-hand side exists for any $\varphi \in \mathcal{D}(K)$.

If $f * g$ exists, then also the convolution $g * f$ exists, and $f * g = g * f$.

The simplest case when a convolution of two distributions exists, is the case, in which one of them has a compact support (then the expression on the right in (1.33) becomes independent of l for $l \ge l_0$ with l_0 large enough). As usual, $f * \delta = f$ for any $f \in \mathcal{D}'(K)$. If $g \in \mathcal{D}(K)$, the formula (1.33) can be simplified – the convolution $f * g$ is actually a function on K, and

$$(f * g)(x) = \langle f(y), g(x-y) \rangle,$$

which is in agreement with the standard definition of the convolution of "ordinary" functions.

Before introducing the Fourier transform of a distribution we note a property of the Fourier transform of test functions, which is in a sharp contrast to the classical situation.

Theorem 1.10. *The Fourier transform is a linear automorphism of the space $\mathcal{D}(K)$.*

This result makes it possible to define the Fourier transform of an arbitrary distribution $f \in \mathcal{D}(K)$:

$$\left\langle \widetilde{f}, \varphi \right\rangle = \langle f, \widetilde{\varphi} \rangle, \quad \varphi \in \mathcal{D}(K).$$

If $f \in \mathcal{D}(K)$, this definition coincides with the one given before.

Theorem 1.11. *The Fourier transform is a linear automorphism of the space $\mathcal{D}'(K)$, with the inversion formula*

$$f - F\left[\widetilde{f}(-\xi)\right], \quad f \in \mathcal{D}'(K).$$

In order to formulate some results regarding the relations between the Fourier transform and convolution, we need a notion of a product of distributions. If $f \in \mathcal{D}'(K)$, $\psi \in \mathcal{D}(K)$, then the product ψf is defined as the functional

$$\langle \psi f, \varphi \rangle = \langle f, \psi \varphi \rangle, \quad \varphi \in \mathcal{D}(K).$$

This definition can be extended to more general functions ψ, e.g. locally constant functions.

Denote

$$\delta_k(x) = q^k \Omega_{-k}(x), \quad x \in K, \; k = 0, 1, 2, \ldots .$$

It is easy to see that $\delta_k \to \delta$ and $\widetilde{\delta_k} \to 1$ as $k \to \infty$.

Definition 1.8. *The product $f \cdot g$ of two distributions $f, g \in \mathcal{D}'(K)$ is the functional*

$$\langle f \cdot g, \varphi \rangle = \lim_{k \to \infty} \langle g, (f * \delta_k) \varphi \rangle$$

provided the limit exists for any $\varphi \in \mathcal{D}(K)$.

Note that this definition agrees with the above one, when the latter is applicable.

If f and g are continuous functions on K, then their product in the sense of Definition 1.8 exists and coincides with the pointwise product.

Theorem 1.12. *Let $f, g \in \mathcal{D}'(K)$. The product $f \cdot g$ exists if and only if the convolution $\widetilde{f} * \widetilde{g}$ exists in $\mathcal{D}'(K)$. In this case*

$$F(f \cdot g) = \widetilde{f} * \widetilde{g}, \quad F(f * g) = \widetilde{f} \cdot \widetilde{g}.$$

In particular, the above assertions are valid if at least one of the distributions f, g has a compact support.

It follows from the commutativity of convolution that $f \cdot g$ exists whenever $g \cdot f$ exists, and in that case $f \cdot g = g \cdot f$.

Note that the above definitions and results carry over in a straightforward way to the case of functions of several variables.

An important class of distributions consists of Borel measures on K (or K^n), which are finite on any compact subset. Moreover, if μ is a finite measure on K^n (it is sufficient to assume that K is a probability measure), then the above definition of the Fourier transform $\widetilde{\mu}$ agrees with the classical definition (for measures we shall often write $\widehat{\mu}(x)$ instead of $\widetilde{\mu}$ in order to stress that the Fourier transform is interpreted here as an ordinary function):

$$\widehat{\mu}(x) = \int_{K^n} \chi(\langle x, y \rangle)\mu(dy), \quad x \in K^n,$$

where for $x = (x_1, \ldots, x_n), y = (y_1, \ldots, y_n) \in K^n$ we write $\langle x, y \rangle = x_1 y_1 + \cdots + x_n y_n$.

The classical Bochner theorem carries over to the case of a local field – a function $f(x)$ is a Fourier transform of a probability measure if and only if it is continuous and positive definite, that is $f(0) = 1$ and

$$\sum_{i,j=1}^{m} f(x_i - x_j) c_i \overline{c_j} \geq 0$$

for any $x_1, \ldots, x_m \in K^n$, $c_1, \ldots, c_m \in \mathbb{C}$.

1.7 Homogeneous distributions

Let ζ be a multiplicative quasicharacter on a local field K.

Definition 1.9. *A distribution $f(x) \in \mathcal{D}'(K)$ is called homogeneous of the degree ζ if for any $\varphi \in \mathcal{D}(K)$, $t \in K \setminus \{0\}$*

$$\langle f, \varphi(t^{-1}x) \rangle = \zeta(t) \|t\| \langle f, \varphi(x) \rangle.$$

The representation of the quasicharacter ζ in the form (1.10) makes it possible to interpret ζ as a distribution

$$\langle \zeta, \varphi \rangle = \int_K \|x\|^{s-1}\theta(x)\varphi(x)\, dx. \tag{1.34}$$

The integral (1.34) is absolutely convergent if $\operatorname{Re} s > 0$. If $\operatorname{Re} s < 0$, it can be defined via analytic continuation as follows.

Without restricting generality, we may assume that $\operatorname{supp}\varphi \subset O$. Then we can write

$$\langle \zeta, \varphi \rangle = \int_O \|x\|^{s-1}\theta(x)[\varphi(x) - \varphi(0)]\, dx + \varphi(0) \int_O \|x\|^{s-1}\theta(x)\, dx. \tag{1.35}$$

The first integral converges for any s, since φ is locally constant, so that the difference $\varphi(x) - \varphi(0)$ vanishes on a neighbourhood of the origin. It follows from (1.30) that the second integral is equal to zero unless $\theta(x) \equiv 1$, when it equals $\varphi(0)\dfrac{1-q^{-1}}{1-q^{-s}}$. The only singularity in the last expression is $s = 0$, which is a simple pole with the residue $\dfrac{q-1}{q\log q}$.

Theorem 1.13. *Each multiplicative quasicharacter ζ on a local field K, except the quasicharacter $\zeta_0(x) = \|x\|^{-1}$, generates a homogeneous distribution by means of the formula (1.35) and analytic continuation. The Dirac δ is a homogeneous distribution of the degree ζ_0. There are no other homogeneous distributions on K.*

Let us consider the Fourier transforms (corresponding to a rank zero basic additive character χ) of homogeneous distributions. It is easy to see that the Fourier transform of a homogeneous distribution of the degree ζ is again a homogeneous distribution, but of the degree $\zeta^{-1}\zeta_0^{-1}$.

In particular, the Fourier transform of $\zeta(x)\|x\|^{-1}$ ($\zeta \not\equiv 1$) is, up to a factor, the homogeneous distribution ζ^{-1}. The above factor is denoted $\Gamma(\zeta)$ and called *the gamma function.* The gamma function admits an integral representation

$$\Gamma(\zeta) = \int_K \chi(x)\zeta(x)\|x\|^{-1}\, dx,$$

where the integral is understood (as above) in the sense of analytic continuation. It satisfies the relation

$$\Gamma(\zeta)\Gamma(\zeta_0\zeta^{-1}) = \zeta(-1). \tag{1.36}$$

Let $\lambda = \zeta(\beta)$, where β is a (fixed) prime element of the field K. Since every element $x \in K^*$ can be written as $x = \beta^k y$, $k \in \mathbb{Z}$, $y \in U$, we find that

$$\zeta(x) = \lambda^k \rho(y),$$

where ρ is a character of the group U (note that a quasicharacter of a compact group is always a character; see [156]). Thus the set of multiplicative quasicharacters is identified with the set of pairs (λ, ρ), where $\lambda \in \mathbb{C}^*$, ρ is a character of the group U. Correspondingly we shall write $\Gamma(\lambda, \rho)$ instead of $\Gamma(\zeta)$.

Theorem 1.14. *If ρ is a character of the ramification degree $m \geq 1$, then:*

$$\text{(i)} \quad \int_U \chi(xy)\rho(y)\,dy = 0 \quad \text{if } x \in K, \|x\| \neq q^m.$$

$$\text{(ii)} \quad \Gamma(\lambda, \rho) = \lambda^{-m} \int_U \chi(\beta^{-m} y)\rho(y)\,dy.$$

The function $\Gamma(\lambda, \rho)$ satisfies the identities

$$\Gamma(\lambda, \rho)\Gamma(q^{-1}\lambda^{-1}, \overline{\rho}) = \rho(-1);$$

$$\Gamma(\lambda, \rho)\overline{\Gamma(q^{-1}\overline{\lambda}^{-1}, \rho)} = 1.$$

(iii) *If $\rho \equiv 1$, then*

$$\Gamma(\lambda, 1) = \frac{1 - q^{-1}\lambda^{-1}}{1 - \lambda}.$$

In particular, if $\zeta(x) = \|x\|^s$, $s > 0$, then $\lambda = q^{-s}$, $\Gamma(\zeta) = \dfrac{1 - q^{s-1}}{1 - q^{-s}}$.

Later (Chapter 2) we shall use the notation $\Gamma_K(s) = \dfrac{1 - q^{s-1}}{1 - q^{-s}}$. It follows from (1.36) that

$$\Gamma_K(s)\Gamma_K(1 - s) = 1. \tag{1.37}$$

Let us consider now the exceptional case of the quasicharacter ζ_0. It is possible to interpret it as a distribution too, though the latter will not be homogeneous. The distribution $\mathcal{P}\|x\|^{-1}$ is defined by the formula

$$\langle \mathcal{P}\|x\|^{-1}, \varphi\rangle = \int\limits_{\|x\|\le 1} \|x\|^{-1}[\varphi(x) - \varphi(0)]\,dx + \int\limits_{\|x\|>1} \|x\|^{-1}\varphi(x)\,dx, \quad \varphi \in \mathcal{D}(K).$$

We shall use this distribution in Chapter 2.

The procedure of analytic continuation employed above for defining homogeneous distributions can be used in much more general situations. We shall need just one result in this direction.

Let $h(x_1, \dots, x_n)$ $(n > 2)$ be a nondegenerate quadratic form over $\mathbb{Q}_p$, $p \neq 2$. Consider a distribution $|h|_p^\lambda \in \mathcal{D}'(\mathbb{Q}_p^n)$ defined for $\operatorname{Re}\lambda > 0$ by the formula

$$\langle |h|_p^\lambda, \varphi\rangle = \int\limits_{\mathbb{Q}_p^n} |h(x_1, \dots, x_n)|_p^\lambda \varphi(x_1, \dots, x_n)\,dx_1 \dots dx_n, \quad \varphi \in \mathcal{D}(\mathbb{Q}_p^n),$$

and for other values of $\lambda \in \mathbb{C}$ (wherever possible) by analytic continuation. It is sufficient to consider λ with $|\operatorname{Im}\lambda| \le \pi(\log q)^{-1}$.

Theorem 1.15 ([48, 120]). *The distribution $|h|_p^\lambda$ can have singularities only for $\lambda = -1$ and $\lambda = -n/2$. Both singularities are actually simple poles if h is isotropic; otherwise $|h|_p^\lambda$ is holomorphic at $\lambda = -1$, and has a simple pole at $\lambda = -n/2$.*

1.8 Gaussian integrals

This section is devoted to the computation of some integrals over a local field K or its subsets, which contain expressions like $\chi(ax^2)$, where χ is an additive character. Such integrals often appear in various problems of p-adic quantum mechanics, spectral theory etc.

We start with some integrals for $K = \mathbb{Q}_p$, $p \neq 2$. Denote for $a \in \mathbb{Q}_p$, $a \neq 0$,

$$\lambda_p(a) = \begin{cases} 1, & \text{if } \nu \text{ is even,} \\ \left(\dfrac{a_0}{p}\right), & \text{if } \nu \text{ is odd, } p \equiv 1 \pmod 4, \\ i\left(\dfrac{a_0}{p}\right), & \text{if } \nu \text{ is odd, } p \equiv 3 \pmod 4, \end{cases}$$

where ν and a_0 are taken from the canonical representation of $a \in \mathbb{Q}_p$:

$$a = p^\nu(a_0 + a_1 p + \cdots), \quad \nu \in \mathbb{Z}, a_j \in \{0, 1, \ldots, p-1\}, a_0 \neq 0;$$

$\left(\frac{a_0}{p}\right)$ is the Legendre symbol.

The proof of the following integral formula can be found in [154]: if $a, b \in \mathbb{Q}_p$, $N \in \mathbb{Z}$, then

$$\int\limits_{|x|_p \le p^N} \chi_p(ax^2 + bx)\, dx$$

$$= \begin{cases} p^N \omega(p^N |b|_p), & \text{if } |a|_p p^{2N} \le 1, \\ \lambda_p(a)|a|_p^{-1/2} \chi_p\left(-\frac{b^2}{4a}\right) \omega\left(p^{-N} \left|\frac{b}{a}\right|_p\right), & \text{if } |a|_p p^{2N} > 1, \end{cases} \tag{1.38}$$

where $\omega(r) = \begin{cases} 1, & 0 \le r \le 1, \\ 0, & r > 1. \end{cases}$

In the limit $N \to \infty$ the formula (1.38) yields the identity

$$\int\limits_{\mathbb{Q}_p} \chi_p(ax^2 + bx)\, dx = \lambda_p(a)|a|_p^{-1/2} \chi_p\left(-\frac{b^2}{4a}\right), \tag{1.39}$$

where the integral in the left-hand side is understood as conditionally convergent, that is as a limit of integrals over the balls $\{|x|_p \le p^N\}$, $N \to \infty$.

Next we shall calculate [81] some Gaussian integrals for general local fields with the characteristic of the residue field different from 2. For $K = \mathbb{Q}_p$ (including the case $p = 2$) such calculations were made in [154].

Let K be a finite extension of $\mathbb{Q}_p$, $p \neq 2$. We shall follow the notations of Sect. 1.3.

Consider the integral

$$I_1 = \int\limits_{\|x\| = q^\nu} \chi(\varepsilon x^2)\, dx,$$

where $\chi = \chi_p \circ \mathrm{Tr}_{K/\mathbb{Q}_p}$, $\|\varepsilon\| = 1$.

Lemma 1.2. *The following equalities are valid:*

$$\begin{aligned} I_1 &= q^\nu(1 - q^{-1}), && \textit{if } \nu \le d/2; \\ &= q^{(d-1)/2}\left[\theta^{(2)}(\widehat{\varepsilon}) G_K - 1\right], && \textit{if } d \textit{ is odd}, \ \nu = \frac{d+1}{2}; \\ &= 0, && \textit{if } \nu > (d+1)/2. \end{aligned}$$

Proof. If $\nu \leq d/2$, then $\chi(\varepsilon x^2) = 1$, and it remains to use (1.21). Suppose that d is odd and $\nu = (d+1)/2$. Using the canonical representation (1.7) we obtain

$$\varepsilon x^2 \equiv \varepsilon x_0^2 \beta^{-d-1} \pmod{\beta^{-d} O}.$$

We have used the assumption $p \neq 2$, whence $\|2\| = 1$. Now

$$I_1 = \sum_{k \in S, k \notin P} \varphi(\varepsilon k^2) \int_{\|x\|=q^\nu, x_0=k} dx.$$

If $k \in P$, then $\varphi(\varepsilon k^2) = 1$. Therefore the required formula follows from the Gauss sum identity (see Sect. 1.3) and the integration formula (1.22).

If $\nu > (d+1)/2$, then

$$\varepsilon x^2 \equiv \sum_{0 \leq l \leq \nu-(d+1)/2} \varepsilon x_l^2 \beta^{2l-2\nu} + \sum_{\substack{0 < i+j \leq 2\nu-d-1 \\ i \neq j}} 2\varepsilon x_i x_j \beta^{i+j-2\nu} \pmod{\beta^{-d} O},$$

whence

$$\chi(\varepsilon x^2) = L(x_0, x_1, \dots, x_{2\nu-d-2}) \varphi(2\varepsilon x_0 x_{2\nu-d-1}),$$

where L is a certain function. Due to the invariance of the measure, the integral

$$\int_{\|x\|=q^\nu, x_0=k_0, \dots, x_{2\nu-d-1}=k_{2\nu-d-1}} dx$$

does not depend on $k_0, \dots, k_{2\nu-d-1}$. Since $\|2\varepsilon x_0\| = 1$, the summation through all possible values of $x_{2\nu-d-1}$ can be expressed as the summation of the values of a nontrivial additive character of a finite field through all the elements. Therefore $I_1 = 0$. ■

The following lemma is proved in a similar way.

Lemma 1.3. *If*

$$I_2 = \int_{\|x\|=q^\nu} \chi(\varepsilon \beta x^2)\, dx, \quad \|\varepsilon\| = 1,$$

then

$$\begin{aligned} I_2 &= q^\nu(1 - q^{-1}), && \text{if } \nu \leq (d+1)/2; \\ &= q^{d/2}\left[\theta^{(2)}(\widehat{\varepsilon}) G_K - 1\right], && \text{if } d \text{ is even, } \nu = 1 + \frac{d}{2}; \\ &= 0, && \text{if } \nu > 1 + d/2. \end{aligned}$$

Next we shall consider a more complicated integral

$$I_3(x) = \int\limits_{\|y\|=q^\nu} \chi(\varepsilon(x-y)^2)\,dy, \quad \|\varepsilon\| = 1.$$

Lemma 1.4. *The following equalities are valid:*

$$\begin{aligned}
I_3(x) &= q^\nu(1-q^{-1})\chi(\varepsilon x^2), && \text{if } \|x\| \le q^{d-\nu},\ \nu \le d/2;\\
&= -q^{\nu-1}\chi(\varepsilon x^2), && \text{if } \|x\| = q^{d-\nu+1},\ \nu \le d/2;\\
&= 0, && \text{if } \|x\| \ge q^{d-\nu+2},\ \nu \le d/2;\\
&= q^{(d-1)/2}\left[\theta^{(2)}(\widehat{\varepsilon})G_K - 1\right], && \text{if } d \text{ is odd},\ \|x\| \le q^{d-\nu},\ \nu = \frac{d+1}{2};\\
&= q^{(d-1)/2}\left[\theta^{(2)}(\widehat{\varepsilon})G_K - \chi(\varepsilon x^2)\right], && \text{if } d \text{ is odd},\ \|x\| = q^{d-\nu+1},\ \nu = \frac{d+1}{2};\\
&= 0, && \text{if } d \text{ is odd},\ \|x\| \ge q^{d-\nu+2},\ \nu = \frac{d+1}{2};\\
&= 0, && \text{if } \|x\| \ne q^\nu,\ \nu > (d+1)/2;\\
&= q^{d/2}, && \text{if } d \text{ is even},\ \|x\| = q^\nu,\ \nu > (d+1)/2;\\
&= q^{(d-1)/2}\theta^{(2)}(\widehat{\varepsilon})G_K, && \text{if } d \text{ is odd},\ \|x\| = q^\nu,\ \nu > \frac{d+1}{2}.
\end{aligned}$$

Proof. If $\|x\| \le q^{d-\nu}$, $\|y\| = q^\nu$, then $\chi(2\varepsilon xy) = 1$, so that the computation is reduced to the application of Lemma 1.2. If $\nu \le d/2$, $\|y\| = q^\nu$, then $\chi(\varepsilon y^2) = 1$, and we can use (1.25).

Suppose that d is odd, $\nu = (d+1)/2$, $\|x\| = q^{d-\nu+1} = q^\nu$. Considering the canonical representation (1.7) for an element x and the analogous representation for y, we see that

$$\chi(\varepsilon(x-y)^2) = \varphi(\varepsilon(x_0-y_0)^2),$$

whence

$$I_3(x) = \sum_{k\in S, k\notin P} \varphi(\varepsilon(x_0-k)^2) \int\limits_{\|y\|=q^\nu, y_0=k} dy$$
$$= q^{\nu-1}\left[\sum_{k\in S} \varphi(\varepsilon(x_0-k)^2) - \varphi(\varepsilon x_0^2)\right].$$

The difference $x_0 - k$ runs through a complete system of representatives of residue classes from O/P; thus it is possible to use the Gauss sum identity, which implies the required formula.

Next, let d be odd, $\nu = (d+1)/2$, $\|x\| \geq q^{\nu+1}$. As above, $\chi(\varepsilon y^2) = \varphi(\varepsilon y_0^2)$, so that

$$I_3(x) = \chi(\varepsilon x^2) \sum_{k\in S, k\notin P} \varphi(\varepsilon k^2) \int\limits_{\|y\|=q^\nu, y_0=k} \chi(2\varepsilon xy)\, dy.$$

By the definition of d it is possible to choose such $z \in K$ that $\|z\| = q^{\nu-1}$ and $\operatorname{Tr}_{K/\mathbb{Q}_p}(2\varepsilon xz) \notin \mathbb{Z}_p$, so that $\chi(2\varepsilon xz) \neq 1$. We have

$$\{y:\ \|y\| = q^\nu, y_0 = k\} = \{y:\ \|y+z\| = q^\nu, y_0 = k\}.$$

Therefore

$$\int\limits_{\|y\|=q^\nu, y_0=k} \chi(2\varepsilon xy)\, dy = \int\limits_{\|y+z\|=q^\nu, y_0=k} \chi(2\varepsilon xy + 2\varepsilon xz)\, dy$$
$$= \chi(2\varepsilon xz) \int\limits_{\|y\|=q^\nu, y_0=k} \chi(2\varepsilon xy)\, dy,$$

whence

$$\int\limits_{\|y\|=q^\nu, y_0=k} \chi(2\varepsilon xy)\, dy = 0,$$

so that $I_3(x) = 0$.

Finally, let $\nu > (d+1)/2$. The case $\|x\| \neq q^\nu$ is considered in exactly the same way as in the proof of Lemma 1.2. If $\|x\| = q^\nu$, the same arguments lead to the equality

$$\int\limits_{\|y\|=q^\nu, y_0\neq x_0} \chi(\varepsilon(x-y)^2)\, dy = 0,$$

so that

$$I_3(x) = \int\limits_{\|y\|=q^\nu, y_0=x_0} \chi(\varepsilon(x-y)^2)\,dy = \int\limits_{\|x-y\|\le q^{\nu-1}} \chi(\varepsilon(x-y)^2)\,dy$$

$$= \int\limits_{\|z\|\le q^{\nu-1}} \chi(\varepsilon z^2)\,dz = \sum_{j=-\infty}^{\nu-1} \int\limits_{\|z\|=q^j} \chi(\varepsilon z^2)\,dz,$$

and by Lemma 1.2

$$I_3(x) = \sum_{j=-\infty}^{d/2} (1-q^{-1})q^j = q^{d/2},$$

if d is even, or

$$I_3(x) = \sum_{j=-\infty}^{(d-1)/2} (1-q^{-1})q^j + \int\limits_{\|z\|=q^{(d+1)/2}} \chi(\varepsilon z^2)\,dz$$

$$= q^{(d-1)/2} + q^{(d-1)/2}\left[\theta^{(2)}(\widehat{\varepsilon})G_K - 1\right] = q^{(d-1)/2}\theta^{(2)}(\widehat{\varepsilon})G_K,$$

if d is odd. ■

The integral

$$I_4(x) = \int\limits_{\|y\|=q^\nu} \chi(\varepsilon\beta(x-y)^2)\,dy, \quad \|\varepsilon\| = 1,$$

is calculated in the following lemma.

Lemma 1.5. *The following equalities are valid:*

$$
\begin{aligned}
I_4(x) &= q^{\nu}(1-q^{-1})\chi(\varepsilon\beta x^2), && \textit{if } \|x\| \le q^{d-\nu+1},\ \nu \le (d+1)/2;\\
&= -q^{\nu-1}\chi(\varepsilon\beta x^2), && \textit{if } \|x\| = q^{d-\nu+2},\ \nu \le (d+1)/2;\\
&= 0, && \textit{if } \|x\| \ge q^{d-\nu+3},\ \nu \le (d+1)/2;\\
&= q^{d/2}\left[\theta^{(2)}(\widehat{\varepsilon})G_K - 1\right], && \textit{if } d \textit{ is even},\ \|x\| \le q^{d-\nu+1},\\
& && \nu = 1+\frac{d}{2};\\
&= q^{d/2}\left[\theta^{(2)}(\widehat{\varepsilon})G_K - \chi(\varepsilon\beta x^2)\right], && \textit{if } d \textit{ is even},\ \|x\| = q^{d-\nu+2},\\
& && \nu = 1+\frac{d}{2};\\
&= 0, && \textit{if } d \textit{ is even},\ \|x\| \ge q^{d-\nu+3},\\
& && \nu = 1+\frac{d}{2};\\
&= 0, && \textit{if } \|x\| \ne q^{\nu},\ \nu > 1+\frac{d}{2};\\
&= q^{(d+1)/2}, && \textit{if } d \textit{ is odd},\ \|x\| = q^{\nu},\\
& && \nu > 1+\frac{d}{2};\\
&= q^{d/2}\theta^{(2)}(\widehat{\varepsilon})G_K, && \textit{if } d \textit{ is even},\ \|x\| = q^{\nu},\\
& && \nu > 1+\frac{d}{2}.
\end{aligned}
$$

The proof is similar to that of Lemma 1.4.

In Chapter 3 we shall have to deal with the integrals

$$
\begin{aligned}
I_5(x) &= \int\limits_{\|y\|=q^{\nu},\,y_0=k} \chi(\varepsilon(x-y)^2)\,dy,\\
I_6(x) &= \int\limits_{\|x\|=q^{\nu},\,y_0=k} \chi(\varepsilon\beta(x-y)^2)\,dy,
\end{aligned}
\tag{1.40}
$$

where $k \in S$, $k \notin P$, $\|\varepsilon\| = 1$. Their explicit expressions are extremely cumbersome, and we give here only a method for their calculation. Consider

$I_5(x)$, for example. After the change of variables $z = y - k\beta^{-\nu}$ we obtain

$$I_5(x) = \int\limits_{\|z\| \le q^{\nu-1}} \chi(\varepsilon(x - k\beta^{-\nu} - z)^2)\, dz$$

$$= \sum_{j=-\infty}^{\nu-1} \int\limits_{\|z\|=q^j} \chi(\varepsilon(x - k\beta^{-\nu} - z)^2)\, dz,$$

and it remains to use various formulas from Lemma 1.4.

Note that Lemmas 1.2-1.5 remain valid if K is a local field of characteristic p. In this case one should set $d = 0$ in all the integration formulas.

1.9 Adeles

The idea of the adeles is to incorporate in a single algebraic structure all completions of the field $\mathbb{Q}$ of rational numbers (or of an arbitrary global field; however we shall not consider that general situation).

Let $\mathbb{A}$ be the set of all sequences

$$a = (a_\infty, a_2, \dots, a_p, \dots),$$

where $a_\infty \in \mathbb{R}$, $a_p \in \mathbb{Q}_p$ for each prime number p, and all a_p, starting from a certain p (depending on a) are p-adic integers. The set of all such sequences forms a ring, with the component-wise addition and multiplication operations. This ring is called *the adele ring*, and its additive group is called *the adele group*.

Invertible elements of the ring $\mathbb{A}$ are called *ideles*. The set $\mathbb{A}^*$ of all ideles is a (multiplicative) group called *the idele group*. Its elements are sequences $\lambda = (\lambda_\infty, \lambda_2, \dots, \lambda_p, \dots)$, for which $\lambda_p \neq 0$ for all p, and $|\lambda_p|_p = 1$ for all p except a finite set.

The topology in $\mathbb{A}$ is given by declaring that a fundamental system of neighbourhoods of zero is the family of all sets $\prod\limits_{p \le \infty} V_p$, where V_p is a neighbourhood of zero in $\mathbb{Q}_p$ $(p < \infty)$, or in $\mathbb{R}$ $(p = \infty)$, and V_p coincides with the ring of integers $\mathbb{Z}_p$ for all p except a finite set.

The topology in $\mathbb{A}^*$ is introduced similarly. Both $\mathbb{A}$ and $\mathbb{A}^*$ are locally compact.

The field $\mathbb{Q}$ of rational numbers can be imbedded in $\mathbb{A}$ as

$$r \mapsto (r, r, \dots, r, \dots).$$

Such sequences form a discrete subring in $\mathbb{A}$ called *the ring of principal adeles.*

The fact that the adele ring is related to the most fundamental structures of mathematics is reflected in the following result.

Theorem 1.16. *The character group of the quotient group $\mathbb{A}/\mathbb{Q}$, where $\mathbb{Q}$ is identified with the subgroup of principal adeles, is isomorphic to the additive group of rational numbers.*

By duality, this means that the character group of the additive group of rational numbers is isomorphic to $\mathbb{A}/\mathbb{Q}$.

Let us consider the structure of the idele group $\mathbb{A}^*$. The multiplicative group $\mathbb{Q}^*$ of nonzero rational numbers can be imbedded monomorphically into $\mathbb{A}^*$. Namely, for any $r \in \mathbb{Q}$, $r \neq 0$, consider the sequence

$$(r, r, \dots, r, \dots). \tag{1.41}$$

Such a sequence is an idele, since $|r|_p = 1$ for any prime not contained in the decomposition of r as a product of primes. The set of all sequences (1.41) is a discrete subgroup of $\mathbb{A}^*$ called *the group of principal ideles.*

Denote by Λ the subgroup of all the ideles $\lambda = (\lambda_\infty, \lambda_2, \dots, \lambda_p, \dots)$, for which the real number λ_∞ is positive, and $|\lambda|_p = 1$ for all p.

Theorem 1.17. *The idele group is a direct product of the subgroup of principal ideles and the subgroup Λ: $\mathbb{A}^* = \mathbb{Q}^* \times \Lambda$, so that the quotient group $\mathbb{A}^*/\mathbb{Q}^*$ is isomorphic to Λ.*

Let da be a Haar measure on the adele group. It is convenient to use the following normalization:

$$\int_F da = 1,$$

where F is the (compact) set of such adeles $a = (a_\infty, a_2, \dots, a_p, \dots)$ that $0 \leq a_\infty \leq 1$, $|a_p|_p \leq 1$, $p = 2, 3, \dots$.

This measure can be expressed via the Haar measures on $\mathbb{Q}_p$:

$$da = da_\infty \cdot da_2 \cdot \dots \cdot da_p \cdot \dots, \tag{1.42}$$

where

$$\int_0^1 da_\infty = 1, \qquad \int_{|a_p|_p \leq 1} da_p = 1.$$

The equality (1.42) means that for any integrable function on $\mathbb{A}$ of the form

$$\varphi(a) = \varphi_\infty(a_\infty)\varphi_2(a_2)\dots\varphi_p(a_p)\dots$$

we have

$$\int \varphi(a)\,da = \int \varphi_\infty(a_\infty)\,da_\infty \int \varphi_2(a_2)\,da_2 \dots \int \varphi_p(a_p)\,da_p \dots .$$

Similarly, normalizing a Haar measure $d^*\lambda$ on the idele group $\mathbb{A}^*$ by the condition

$$\int_{\Lambda'} d^*\lambda = 1,$$

where $\Lambda' = \{\lambda = (\lambda_\infty, \lambda_2, \dots, \lambda_p, \dots) : \ 1 \leq \lambda_\infty \leq e, |\lambda_p|_p = 1, p = 2, 3, \dots\}$, we find that

$$d^*\lambda = d^*\lambda_\infty, d^*\lambda_2, \dots, d^*\lambda_p, \dots,$$

where

$$d^*\lambda_\infty = \frac{d\lambda_\infty}{|\lambda_\infty|_\infty}, \qquad \int_{|\lambda_p|_p=1} d^*\lambda_p = 1,$$

and $|\lambda_\infty|_\infty$ is a usual absolute value on $\mathbb{R}$.

Any idele λ generates an automorphism $a \mapsto \lambda a$ of $\mathbb{A}$. Since $d(\lambda a)$ is an invariant measure on $\mathbb{A}$, it is proportional to da. Denoting the factor by $|\lambda|$ we have

$$d(\lambda a) = |\lambda| da.$$

It is clear that $|\lambda'\lambda''| = |\lambda'| \cdot |\lambda''|$ for any $\lambda', \lambda'' \in \mathbb{A}^*$.

Theorem 1.18. *For any* $\lambda = (\lambda_\infty, \lambda_2, \dots, \lambda_p, \dots) \in \mathbb{A}^*$

$$|\lambda| = |\lambda_\infty| \cdot |\lambda_2|_2 \dots \cdot |\lambda_p|_p \dots .$$

If λ *is a principal idele, then* $|\lambda| = 1$.

Chapter 2

Fundamental Solutions

2.1 The Riesz kernel

Let K be a local field. The function

$$f_\alpha(x) = \frac{||x||^{\alpha-1}}{\Gamma_K(\alpha)}, \quad x \in K, \ \mathrm{Re}\,\alpha > 0, \ \alpha \neq 1,$$

is called *the Riesz kernel.* As we know from Chapter 1, it possesses, as a distribution from $\mathcal{D}'(K)$, a meromorphic continuation to all $\alpha \not\equiv 1$ (mod $\frac{2\pi i}{\log q}\mathbb{Z}$) given by

$$< f_\alpha, \varphi >= \frac{1-q^{-1}}{1-q^{\alpha-1}}\varphi(0)$$

$$+ \frac{1-q^{-\alpha}}{1-q^{\alpha-1}}\left[\int\limits_{||x||>1} \varphi(x)\frac{dx}{||x||^{1-\alpha}} + \int\limits_{||x||\leq 1} (\varphi(x)-\varphi(0))\frac{dx}{||x||^{1-\alpha}}\right]. \quad (2.1)$$

In particular, for $\mathrm{Re}\,\alpha > 0$

$$< f_\alpha, \varphi > = \frac{1-q^{-\alpha}}{1-q^{\alpha-1}}\int\limits_K \varphi(x)||x||^{\alpha-1}\,dx, \quad \alpha \not\equiv 1 \quad (\mathrm{mod}\ \frac{2\pi i}{\log q}\mathbb{Z}),$$

$$< f_{-\alpha}, \varphi > = \frac{1-q^{\alpha}}{1-q^{-\alpha-1}}\int\limits_K (\varphi(x)-\varphi(0))\frac{dx}{||x||^{1+\alpha}}.$$

The distribution f_α has simple poles at the points

$$1+\alpha_k = 1 + \frac{2k\pi i}{\log q}, \quad k = 0, \pm 1, \dots,$$

with the residues $-\frac{q-1}{q\log q}\|x\|^{\alpha_k}$. It is seen from (2.1) that $f_{\alpha_k}=\delta$.

Proposition 2.1. *If $\alpha\neq 1+\alpha_k$, then the Fourier transform of f_α equals $\|x\|^{-\alpha}$, where both f_α and $\|x\|^{-\alpha}$ are understood as distributions.*

Proof. This is a consequence of the general results on the Fourier transforms of homogeneous distributions (see Chapter 1). ■

If φ is a locally constant function, such that

$$\int_{\|x\|>1}\|\varphi(x)\|\frac{dx}{\|x\|^{1-\operatorname{Re}\alpha}}<\infty,$$

we can form the convolution $f_\alpha*\varphi$, which is actually an ordinary function of $t\in K$:

$$(f_\alpha*\varphi)(t)=\frac{1-q^{-1}}{1-q^{\alpha-1}}\varphi(t)+\frac{1-q^{-\alpha}}{1-q^{\alpha-1}}\left[\int_{\|x\|>1}\varphi(t+x)\frac{dx}{\|x\|^{1-\alpha}}\right.$$
$$\left.+\int_{\|x\|\le 1}(\varphi(t+x)-\varphi(t))\frac{dx}{\|x\|^{1-\alpha}}\right]. \qquad (2.2)$$

For $\operatorname{Re}\alpha>0$ we have

$$(f_\alpha*\varphi)(t)=\frac{1-q^{-\alpha}}{1-q^{\alpha-1}}\int_K\varphi(x+t)\|x\|^{\alpha-1}\,dx,\quad \alpha\not\equiv 1 \pmod{\frac{2\pi i}{\log q}\mathbb{Z}} \qquad (2.3)$$

$$(f_{-\alpha}*\varphi)(t)=\frac{1-q^{\alpha}}{1-q^{-\alpha-1}}\int_K(\varphi(t+x)-\varphi(t))\frac{dx}{\|x\|^{1+\alpha}}. \qquad (2.4)$$

If $\varphi\in\mathcal{D}(K)$, then $f_\alpha*\varphi$ is again locally constant and $(f_\alpha*\varphi)(x)=O\left(\|x\|^{\operatorname{Re}\alpha-1}\right)$. Thus we can form (in the above sense) the convolution $f_\beta*(f_\alpha*\varphi)$ for $\operatorname{Re}(\alpha+\beta)<1$.

Proposition 2.2. *If $\varphi\in\mathcal{D}(K)$, $\alpha,\beta\not\equiv 1 \pmod{\frac{2\pi i}{\log q}\mathbb{Z}}$, $\operatorname{Re}(\alpha+\beta)<1$, then*

$$f_\beta*(f_\alpha*\varphi)=f_{\alpha+\beta}*\varphi. \qquad (2.5)$$

Proof. If $\alpha, \beta, \alpha+\beta \in (0,1)$, the equality $f_\alpha * f_\beta = f_{\alpha+\beta}$ follows from Proposition 2.1 and properties of the convolution of ordinary integrable functions. The general case follows by analytic continuation. ■

Note that the identity (2.5) is valid also without the restriction $\mathrm{Re}(\alpha+\beta) < 1$, and for more general functions φ (even for a certain class of distributions). However then one cannot interpret the convolutions as ordinary functions defined as in (2.2)–(2.4). Since this more general situation will not be encountered in this book, we omit the details referring to [154].

The duality from Proposition 2 can be expressed in a different form, which will be crucial for probabilistic applications.

Proposition 2.3. *If* $\alpha > 0$ *then*

$$\|x\|^\alpha = \frac{1-q^\alpha}{1-q^{-\alpha-1}} \int_K \|s\|^{-\alpha-1}[\chi(xs)-1]\,ds, \tag{2.6}$$

where χ *is an arbitrary rank zero additive character on* K.

Proof. Let $\varphi \in \mathcal{D}(K)$,

$$\psi(x) = \int_K \chi(-x\xi)\varphi(\xi)\,d\xi.$$

By Proposition 2.1,

$$\begin{aligned}\langle \|x\|^\alpha, \varphi\rangle &= \Gamma_K(1+\alpha)\,\langle f_{1+\alpha}, \varphi\rangle = \Gamma_K(1+\alpha)\left\langle \widetilde{f}_{1+\alpha}, \psi\right\rangle \\ &= \Gamma_K(1+\alpha)\,\langle \|x\|^{-\alpha-1}, \psi\rangle \\ &= \Gamma_K(1+\alpha)\int_K \|x\|^{-\alpha-1}\,dx \int_K [\chi(-x\xi)-1]\varphi(\xi)\,d\xi.\end{aligned}$$

Suppose that $\varphi(\xi) = 0$ for $\|\xi\| > q^N$. Since χ has the rank zero,

$$\int_K [\chi(-x\xi)-1]\varphi(\xi)\,d\xi = 0$$

if $\|x\| \le q^{-N}$. Thus we may apply the Fubini theorem. It remains to note that $\Gamma_K(1+\alpha) = \frac{1-q^\alpha}{1-q^{-\alpha-1}}$. ■

It follows from the identity (2.6) that the function $x \mapsto \exp(-\tau\|x\|^\alpha)$ $(\tau > 0, \alpha > 0)$ is positive definite on K [10] . This result (another proof of which can be found in [60]) contrasts a similar property in real analysis, where necessarily $\alpha \le 2$.

2.2 The fractional differentiation operator

Denote by D^α, $\alpha > 0$, the operator defined (initially) on $\mathcal{D}(K)$ by the formula

$$(D^\alpha u)(x) = F^{-1}_{\xi\to x}\left[\|\xi\|^\alpha F_{y\to\xi} u\right], \tag{2.7}$$

where as before,

$$F_{y\to\xi} u = \int_K \chi(y\xi)u(y)\,dy,$$

$$F^{-1}_{\xi\to x}\zeta = \int_K \chi(-x\xi)\zeta(\xi)\,d\xi,$$

and χ is a rank zero additive character on K. It follows from Theorem 1.10 that the pseudo-differential operator $D^\alpha u$ is well-defined for $u \in \mathcal{D}(K)$. Note that $D^\alpha u$ need not belong necessarily to $\mathcal{D}(K)$ (since the function $x \mapsto \|x\|$ is not locally constant); however $D^\alpha u \in L_2(K)$.

By Proposition 2.1, $D^\alpha u = f_{-\alpha} * u$, that is

$$(D^\alpha u)(x) = \frac{1-q^\alpha}{1-q^{-\alpha-1}} \int_K \|y\|^{-\alpha-1}[u(x-y) - u(x)]\,dy. \tag{2.8}$$

Note that $(D^\alpha u)(x) \to u(x)$ if $\alpha \to 0, x \in K$, for every $u \in \mathcal{D}(K)$.

The right-hand side of (2.8) makes sense for a wider class of functions, for example, a class $\mathcal{E}_\alpha(K)$ of such locally constant functions $u(x)$ that

$$\int_{\|x\|\geq 1} \|x\|^{-\alpha-1}|u(x)|\,dx < \infty.$$

It is an easy consequence of the Fubini theorem that the right-hand side of (2.8) with $u \in \mathcal{E}_\alpha(K)$ coincides with the convolution $f_{-\alpha} * u$ understood in the distribution sense. The hypersingular integral operator (2.8) can be seen as a natural extension of the pseudo-differential operator (2.7).

Consider the equation

$$D^\alpha u = \varphi, \quad \varphi \in \mathcal{D}(K). \tag{2.9}$$

As usual, we shall understand *a fundamental solution* of an equation as a function (in more general situations, a distribution), such that its convolution with the right-hand side of the equation is a solution.

Theorem 2.1. (i) *The function*

$$E(x) = \begin{cases} \dfrac{1-q^{-\alpha}}{1-q^{-\alpha-1}}\|x\|^{\alpha-1}, & \text{if } \alpha \neq 1, \\ \dfrac{1-q}{q\log q}\log\|x\|, & \text{if } \alpha = 1, \end{cases} \tag{2.10}$$

is a fundamental solution of the equation (2.9).

(ii) *If* $u \in \mathcal{E}_\alpha(K), D^\alpha u = 0$, *then* $u \equiv$ *const.*

Proof. (i) Note first of all that

$$|(E * \varphi)(x)| \leq \text{const} \cdot \begin{cases} \|x\|^{\alpha-1}, & \text{if } \alpha \neq 1; \\ \log\|x\|, & \text{if } \alpha = 1, \end{cases}$$

if $\|x\|$ is large enough, and $u = E * \varphi$ belongs to $\mathcal{E}_\alpha(K)$, so that $D^\alpha u$ is well defined.

The assertion of the theorem is a consequence of Proposition 2.2, if $\alpha \neq 1$.

Let us consider the case $\alpha = 1$. Consider the (locally integrable) function

$$E(x) = -\frac{q-1}{q\log q}\log\|x\|$$

interpreting it as a distribution $E \in \mathcal{D}'(K)$. Using (1.29), we find that

$$\widetilde{E}(\xi) = \|\xi\|^{-1}, \quad \xi \neq 0.$$

By Theorem 1.9, this means that

$$\widetilde{E}(\xi) - \mathcal{P}\|\xi\|^{-1} = C\delta(\xi), \quad C \in \mathbb{C}. \tag{2.11}$$

In order to find the constant C, we can apply both sides of (2.11) to the test function $\Omega = \widetilde{\Omega}$. We find that

$$C = \left\langle \widetilde{E}, \Omega \right\rangle - \langle \mathcal{P}\|\xi\|^{-1}, \Omega\rangle = \langle E, \Omega\rangle - \int\limits_{\|\xi\|\leq 1} \frac{\Omega(\xi)-\Omega(0)}{\|\xi\|}\,d\xi$$

$$- \int\limits_{\|\xi\|>1} \Omega(\xi)\|\xi\|^{-1}\,d\xi = -\frac{q-1}{q\log q}\int\limits_{\|\xi\|\leq 1}\log\|\xi\|\,d\xi = \frac{1}{q}.$$

Thus

$$\widetilde{E}(\xi) = \mathcal{P}\|\xi\|^{-1} + q^{-1}\delta(\xi).$$

It follows from (2.8) and the Fubini theorem that $D^1 u = E * D^1\varphi$ whence $\widetilde{D^1 u} = \widetilde{E}\cdot\|\xi\|\widetilde{\varphi}$. If we prove that $\widetilde{E}\cdot\|\xi\| = 1$, we will get that $D^1 u = \varphi$, so that E is a fundamental solution.

By the definition, for any $\psi \in \mathcal{D}(K)$

$$\langle(\mathcal{P}\|\xi\|^{-1})\cdot\|\xi\|, \psi\rangle = \lim_{k\to\infty}\langle\mathcal{P}\|\xi\|^{-1}, (\|\xi\| * \delta_k)\psi\rangle$$

(we use the notation of Sect. 1.6). We have

$$(\|\xi\| * \delta_k)(z) = q^k \int\limits_{\|x\|\le q^{-k}} \|z-x\|\,dx. \tag{2.12}$$

If $\|z\| > q^{-k}$ then $(\|\xi\| * \delta_k)(z) = \|z\|$. Next,

$$(\|\xi\| * \delta_k)(0) = q^k \int\limits_{\|x\|\le q^{-k}} \|x\|\,dx,$$

and making a substitution $x = \beta^k y$ we find that $(\|\xi\| * \delta_k)(0) = q^{-k}$. On the other hand, the function (2.12) is locally constant, with the exponent of local constancy $\ge k$. For large k, we can assume the same with regard to ψ. Therefore

$$\begin{aligned}
&\langle\mathcal{P}\|\xi\|^{-1}, (\|\xi\| * \delta_k)\psi\rangle \\
&= \int\limits_{q^{-k}<\|z\|\le 1} \|z\|^{-1}[(\|\xi\| * \delta_k)(z)\psi(z) - (\|\xi\| * \delta_k)(0)\psi(0)]\,dz \\
&\qquad + \int\limits_{\|z\|>1} \|z\|^{-1}(\|\xi\| * \delta_k)(z)\psi(z)\,dz \\
&= \int\limits_{\|z\|>q^{-k}} \psi(z)\,dz - q^{-k}\psi(0) \int\limits_{q^{-k}<\|z\|\le 1} \|z\|^{-1}\,dz = \int\limits_{\|z\|>q^{-k}} \psi(z)\,dz \\
&- q^{-k}\psi(0)\sum_{l=-k+1}^{0}\int\limits_{\|z\|=q^l} \|z\|^{-1}\,dz = \int\limits_{\|z\|>q^{-k}} \psi(z) - (1-q^{-1})\psi(0)q^{-k}k,
\end{aligned}$$

so that

$$\langle\mathcal{P}\|\xi\|^{-1}, (\|\xi\| * \delta_k)\psi\rangle \longrightarrow \int\limits_K \psi(z)\,dz$$

as $k\to\infty$. Since

$$\langle\delta, (\|\xi\| * \delta_k)\psi\rangle = (\|\xi\| * \delta_k)(0)\psi(0) = q^{-k}\psi(0) \longrightarrow 0, \quad k\to\infty,$$

we come to the conclusion that $\widetilde{E} \cdot \|\xi\| = 1$.

(ii) It is easy to see that $D^\alpha u = f_{-\alpha} * u$, where the convolution is understood in the distribution sense. Since the convolution exists (equal to zero), we may apply the Fourier transform finding that

$$\|\xi\| \cdot \widetilde{u}(\xi) = 0$$

whence the distribution $\widetilde{u}$ is concentrated at 0. Therefore $\widetilde{u} = \text{const} \cdot \delta$ and $u = \text{const}$. ■

Note that Theorem 2.1 carries over to the case when K is a non-discrete totally disconnected locally compact division algebra. All the notions and constructions used in the proof are valid in this case too. In particular, below we shall use the formula (2.10) for the quaternion algebra over $\mathbb{Q}_p$.

2.3 The family $\{D^\alpha, \alpha > 0\}$ as a semigroup of operators

2.3.1. Local generator. It follows from Proposition 2.2 that

$$D^{\alpha_1}(D^{\alpha_2}\varphi) = D^{\alpha_1+\alpha_2}\varphi$$

for any $\alpha_1, \alpha_2 > 0$, $\varphi \in \mathcal{D}(K)$. This "semigroup property" prompts us to consider the infinitesimal generator $\frac{d}{d\alpha} D^\alpha|_{\alpha=0}$. In fact, with subsequent applications in mind (see the next section), we shall study only its action upon a *radial* function $g_f(x) = f(\|x\|)$, where f is a smooth compactly supported function on $(0, \infty)$.

Denote by $\mathcal{M}^s(f)$ the Mellin transform of a function f, that is

$$\mathcal{M}^s(f) = \int_0^\infty t^{s-1} f(t)\, dt.$$

Proposition 2.4. *For any smooth compactly supported function f on $(0, \infty)$*

$$\frac{d}{d\alpha}(D^\alpha g_f)(1)\bigg|_{\alpha=0} = -\frac{1}{2\pi i} \int_{\frac{1}{2}-i\infty}^{\frac{1}{2}+i\infty} \mathcal{M}^s(f)\, d\log\Gamma_K(s). \tag{2.13}$$

Proof. As in (2.2), we can write

$$(D^\alpha g_f)(1) = \frac{1-q^{-1}}{1-q^{-\alpha-1}} g_f(1) + \frac{1-q^{\alpha}}{1-q^{-\alpha-1}} \left\{ \int_{\|x\|>1} g_f(1-x)\|x\|^{-1-\alpha}\, dx + \int_{\|x\|\le 1} [g_f(1-x) - g_f(1)]\|x\|^{-1-\alpha}\, dx \right\}.$$

Differentiating with respect to α we find that the left-hand side of (2.13) is equal to

$$\begin{aligned} &-(1-q^{-1})^{-1}\log q \left\{ q^{-1} g_f(1) + \int_{\|x\|>1} g_f(1-x)\|x\|^{-1}\, dx + \int_{\|x\|\le 1} [g_f(1-x) - g_f(1)]\|x\|^{-1}\, dx \right\} \\ &= -(1-q^{-1})^{-1}\log q \left\{ q^{-1} g_f(1) + \int_{\|x\|>1} g_f(x)\|1-x\|^{-1}\, dx + \int_{\|x\|\le 1} [g_f(x) - g_f(1)]\|1-x\|^{-1}\, dx \right\} \\ &= -(1-q^{-1})^{-1}\log q \int_{\|x\|\ne 1} g_f(x)\|1-x\|^{-1}\, dx. \end{aligned}$$

We have used the fact that $g_f(x) - g_f(1) = 0$ for $\|x\| = 1$, and also (repeatedly) the ultrametric property of the absolute value.

Further, using the inversion formula for the Mellin transform, we get

$$\frac{d}{d\alpha}\left(D^{\alpha}g_f\right)(1)\bigg|_{\alpha=0} = -\log q \sum_{n\neq 0} f(q^n)\min(1,q^n)$$

$$= -\log q \sum_{n\neq 0} f(q^n)\min(1,q^n)\frac{1}{2\pi i}\int_{c-i\infty}^{c+i\infty} \mathcal{M}^s(f)q^{-ns}\,ds$$

$$= \frac{1}{2\pi i}\int_{c-i\infty}^{c+i\infty} \mathcal{M}^s(f)\left[-\log q\sum_{n\geq 1} q^{-ns} - \log q \sum_{n\geq 1} q^{-n(1-s)}\right] ds$$

for any $c \in (0,1)$. Fixing $c = 1/2$ and performing the summation we come to (2.13). ■

2.3.2. The case $p = \infty$. Planning to glue the formulas (2.13) for $K = \mathbb{Q}_p$ with all the different primes p into a single adelic picture, we have to consider also the case $p = \infty$, that is the operators corresponding to the Riesz kernels on $\mathbb{R}$.

Note that the Riesz kernel f_α on $\mathbb{Q}_p$ can be written as

$$f_\alpha(x) = \frac{\zeta_p(1-\alpha)}{\zeta_p(\alpha)}|x|_p^{\alpha-1}, \quad \operatorname{Re}\alpha > 0,$$

where

$$\zeta_p(s) = \int_{\mathbb{Q}_p^*} \Omega(x)|x|_p^{\alpha-1}\,d*x = \left(1-p^{-\alpha}\right)^{-1}$$

is called *a local zeta function.* Here $\mathbb{Q}_p^*(= \mathbb{Q}_p \setminus \{0\})$ is the multiplicative group of $\mathbb{Q}_p$, d^*x is a Haar measure on $\mathbb{Q}_p^*$ normalized in such a way that $\int_{U_p} d^*x = 1$, so that $d^*x = \left(1-p^{-1}\right)^{-1}|x|_p^{-1}\,dx$. The factor in the formula (2.8) for the operator D^α now equals $\frac{\zeta_p(1+\alpha)}{\zeta_p(-\alpha)}$.

Looking for an analogue for $p = \infty$, we recall that Ω is preserved by the Fourier transform. A function on $\mathbb{R}$ with properties similar to those of Ω, is $\omega_\infty(t) = \exp(-\pi t^2)$, $t \in \mathbb{R}$. Then the local zeta function ζ_∞ is defined as

$$\zeta_\infty(\alpha) = \int_{\mathbb{R}} \omega_\infty(t)|t|^{\alpha-1}\,dt, \quad \alpha > 0,$$

so that

$$\zeta_\infty(\alpha) = \pi^{-\alpha/2}\Gamma(\alpha/2),$$

and we define a renormalized fractional differentiation operator D_∞^α, $0 < \alpha < 1$, on $\mathbb{R}$ as the convolution with the distribution $\frac{\zeta_\infty(1+\alpha)}{\zeta_\infty(-\alpha)}|x|^{-\alpha-1}$:

$$(D_\infty^\alpha u)(x) = \frac{\zeta_\infty(1+\alpha)}{\zeta_\infty(-\alpha)} \int_{-\infty}^{\infty} |t|^{-\alpha-1}[u(x+t) - u(x)]\, dt, \quad u \in \mathcal{D}(\mathbb{R}),$$

or, which is equivalent,

$$\begin{aligned}(D_\infty^\alpha u)(x) &= \pi^{-(\frac{1}{2}+\alpha)} \frac{\Gamma\left(\frac{1}{2}+\frac{\alpha}{2}\right)}{\Gamma\left(1-\frac{\alpha}{2}\right)} u(x) \\ &+ \pi^{-(\frac{1}{2}+\alpha)} \frac{\Gamma\left(\frac{1}{2}+\frac{\alpha}{2}\right)}{\Gamma\left(-\frac{\alpha}{2}\right)} \left\{ \int_{|t|>1} |t|^{-\alpha-1} u(x+t)\, dt \right. \\ &\left. + \int_{|t|\le 1} |t|^{-\alpha-1}[u(x+t) - u(x)]\, dt \right\}. \qquad (2.14)\end{aligned}$$

Denote, as above, $g_f(x) = f(|x|)$, where f is a function on $[0, \infty)$.

Proposition 2.5. *For any smooth compactly supported function f on $(0, \infty)$*

$$\left.\frac{d}{d\alpha}(D_\infty^\alpha g_f)(1)\right|_{\alpha=0} = \frac{1}{2\pi i} \int_{\frac{1}{2}-i\infty}^{\frac{1}{2}+i\infty} \mathcal{M}^s(f)\, d\log \frac{\zeta_\infty(s)}{\zeta_\infty(1-s)}.$$

Proof. Using (2.14), we find that

$$\begin{aligned}\left.\frac{d}{d\alpha}(D_\infty^\alpha g_f)(1)\right|_{\alpha=0} &= \left.\frac{d}{d\alpha}\right|_{\alpha=0} \left\{ \pi^{-(\frac{1}{2}+\alpha)} \frac{\Gamma\left(\frac{1}{2}+\frac{\alpha}{2}\right)}{\Gamma\left(1-\frac{\alpha}{2}\right)} g_f(1) \right. \\ &+ \pi^{-(\frac{1}{2}+\alpha)} \frac{\Gamma\left(\frac{1}{2}+\frac{\alpha}{2}\right)}{\Gamma\left(-\frac{\alpha}{2}\right)} \left[\int_{|t|>1} |t|^{-\alpha-1} g_f(1-t)\, dt \right. \\ &\left.\left. + \int_{|t|\le 1} |t|^{-\alpha-1}(g_f(1-t) - g_f(1))\, dt \right] \right\}.\end{aligned}$$

Since $\Gamma(1) = 1, \Gamma(\frac{1}{2}) = \sqrt{\pi}, \Gamma'(1) = -\gamma$ (γ is the Euler constant), $\Gamma'(\frac{1}{2}) = -\sqrt{\pi}(\gamma + 2\log 2)$, $\Gamma(0) = \infty$, we obtain after differentiating that

$$\frac{d}{d\alpha}\left(D_\infty^\alpha g_f\right)(1)\Big|_{\alpha=0} = -(\gamma + \log 2\pi) g_f(1)$$

$$-\frac{1}{2}\left[\int\limits_{|t|>1} |t|^{-1} g_f(1-t)\,dt + \int\limits_{|t|\le 1} |t|^{-1}(g_f(1-t) - g_f(1))\,dt\right]$$

$$= -\left[(\gamma + \log\pi) f(1) + f(1)\int_1^2 \frac{dt}{t} + \frac{1}{2}\int\limits_{|1-t|>1} |1-t|^{-1} f(|t|)\,dt\right.$$

$$\left.+\frac{1}{2}\int\limits_{|1-t|\le 1} (f(|t|) - f(1))|1-t|^{-1}\,dt\right] = -\left[(\gamma + \log\pi) f(1)\right.$$

$$\left.+\int_1^\infty \frac{t^2 f(t) - f(1)}{t^2 - 1}\frac{dt}{t} + \int_0^1 \frac{f(t) - t f(1)}{1 - t^2}\,dt\right]$$

$$= -\left[(\gamma + \log\pi) f(1) + \int_1^\infty f(t)\frac{dt}{t} + \sum_{n=1}^\infty \int_1^\infty (f(t) - f(1)) t^{-2n}\frac{dt}{t}\right.$$

$$\left.+\int_0^1 f(t)\,dt + \sum_{n=1}^\infty \int_0^1 (t f(t) - f(1)) t^{2n}\frac{dt}{t}\right].$$

Now by the inversion formula for the Mellin transform we rewrite the last expression as follows:

$$\frac{d}{d\alpha}\left(D_\infty^\alpha g_f\right)(1)\Big|_{\alpha=0} = \frac{1}{2\pi i}\int_{\frac{1}{2}-i\infty}^{\frac{1}{2}+i\infty} \mathcal{M}^s(f)\left[-\frac{1}{2}(\gamma + \log\pi) - \frac{1}{s}\right.$$

$$\left.+\sum_{n=1}^\infty \left(\frac{1}{2n} - \frac{1}{2n+s}\right) - \frac{1}{2}(\gamma + \log\pi) - \frac{1}{1-s} + \sum_{n=1}^\infty \left(\frac{1}{2n} - \frac{1}{2n+1-s}\right)\right]$$

$$= \frac{1}{2\pi i}\int_{\frac{1}{2}-i\infty}^{\frac{1}{2}+i\infty} \mathcal{M}^s(f)\left[d\log\zeta_\infty(s) - d\log\zeta_\infty(1-s)\right],$$

where we have used the series representation of the function $\frac{\Gamma'(z)}{\Gamma(z)}$ (see [7]). ■

2.4 The fractional differential operator on ideles and zeroes of Riemann's zeta function

The aim of this section is to define D^α on ideles and to prove a remarkable identity connecting the infinitesimal generator of the semigroup $\{D^\alpha\}_{\alpha>0}$ with zeroes of the zeta function, which can be defined as

$$\zeta(s) = \prod_{p<\infty} \zeta_p(s), \quad \operatorname{Re} s > 1,$$

and continued analytically to a meromorphic function on $\mathbb{C}$ (with a single pole at $s = 1$, see [146]). We shall use the function

$$Z(s) = \zeta_\infty(s)\zeta(s),$$

which has the same non-real zeroes as $\zeta(s)$ and satisfies the relation

$$Z(s) = Z(1-s) \tag{2.15}$$

(see [146]).

Below we shall write $\mathbb{Q}_\infty = \mathbb{R}, \mathbb{Q}_\infty^* = \mathbb{R} \setminus \{0\}$. The set $\mathcal{D}(\mathbb{A}^*)$ of test functions on the idele group is defined as

$$\bigotimes_{p\le\infty}^{(\text{restr})} \mathcal{D}\left(\mathbb{Q}_p^*\right),$$

that is the space of complex-valued functions on $\mathbb{A}^*$ representable as finite linear combinations of "elementary" functions – infinite products $\varphi = \bigotimes_p \varphi_p$, where $\varphi_p \in \mathcal{D}\left(\mathbb{Q}_p^*\right)$, and for all p except a finite set $\varphi_p = \omega_p^*$ (the indicator function of the group of units U_p).

The fractional differentiation operator $\Delta_{\mathbb{A}^*}^\alpha \varphi$, $\varphi \in \mathcal{D}(\mathbb{A}^*)$, $\varphi = \bigotimes_{p\le\infty} \varphi_p$, is defined as

$$\Delta_{\mathbb{A}^*}^\alpha \varphi = \bigotimes_{p\le\infty} \frac{1}{c_p^{(\alpha)}} D_p^\alpha \varphi_p,$$

with further extension by linearity.

Here D_p^α is the operator D^α over $\mathbb{Q}_p$ defined above, if $p < \infty$, and D_∞^α, if $p = \infty$; the renormalization constants $c_p^{(\alpha)}$ are defined as follows: $c_\infty^{(\alpha)} = 1$,

$$c_p^{(\alpha)} = \left(D_p^\alpha \omega_p^*\right)(1), \quad \text{if } p < \infty.$$

It is easy to find that for $p < \infty$

$$c_p^{(\alpha)} = 1 + \frac{(p^\alpha - 1)(1 - p^{-\alpha})}{p - p^{-\alpha}} = 1 + O(\alpha^2), \quad \alpha \to 0.$$

A function $\varphi \in \mathcal{D}(\mathbb{A}^*)$, $\varphi = \bigotimes\limits_{p \le \infty} \varphi_p$, is called *radial* if all the functions φ_p are radial. For a radial φ denote

$$\widehat{\varphi}(\lambda) = \sum_{q \in \mathbb{Q}^*} \varphi(q\lambda), \quad \lambda \in \mathbb{A}^*. \tag{2.16}$$

Note that for any $x \in \mathbb{A}^*$ the sum in the right-hand side of (2.16) is taken over a finite set of rational numbers q. Indeed, otherwise we would have an infinite set $T \subset \mathbb{Q}^*$ of such non-zero rational numbers (= principal ideles) q that $|q|_p = 1$ for all primes p except a finite set of those p, for which $|q|_p \le C$, $q \in T$. It is clear that T is compact in the topology of $\mathbb{A}^*$. Since $\mathbb{Q}^*$ is a discrete subgroup of $\mathbb{A}^*$, T is finite, and we get a contradiction.

For a radial function φ the function $\widehat{\varphi}(\lambda)$, $\lambda = (\lambda_\infty, \lambda_2, \dots, \lambda_p, \dots) \in \Lambda$, depends only on $\lambda_\infty \in \mathbb{R}$, and we may write $\widehat{\varphi}(\lambda_\infty)$ instead of $\widehat{\varphi}(\lambda)$. Thus by Theorem 1.17 the Mellin transform

$$\mathcal{M}^s(\varphi) = \int\limits_{\mathbb{A}^*} \varphi(\lambda)|\lambda|^s \, d^*\lambda$$

can be written as

$$\mathcal{M}^s(\varphi) = \int\limits_{\Lambda} \left(\sum_{q \in \mathbb{Q}^*} \varphi(q\lambda) \right) |\lambda|^s \, d^*\lambda = \int\limits_0^\infty \widehat{\varphi}(\lambda_\infty)|\lambda_\infty|_\infty^s \, d^*\lambda_\infty = \mathcal{M}^s(\widehat{\varphi}),$$

where $\mathcal{M}^s(\widehat{\varphi})$ is the usual Mellin transform of a function on $(0, \infty)$.

The function $\widehat{\varphi}$ is smooth and compactly supported; therefore the Mellin transform is an entire function of $s \in \mathbb{C}$, and $|\mathcal{M}^s(\widehat{\varphi})| = O\left(|\operatorname{Im} s|^{-N}\right)$ uniformly on any vertical strip $a \le \operatorname{Re} s \le b$, for any N.

Theorem 2.2. *If $\varphi \in \mathcal{D}(\mathbb{A}^*)$ is radial, then*

$$\sum_{q \in \mathbb{Q}^*} \frac{\partial}{\partial \alpha}\bigg|_{\alpha = 0} \left(\Delta_{\mathbb{A}^*}^\alpha \varphi\right)(q) = \sum_{\zeta(s) = 0} \mathcal{M}^s(\varphi) - \mathcal{M}^0(\varphi) - \mathcal{M}^1(\varphi), \tag{2.17}$$

where the sum on the right is extended over all the non-real zeroes of ζ.

Proof. Let $\varphi = \bigotimes_p \varphi_p$ be an elementary radial function from $\mathcal{D}(\mathbb{A}^*)$. Then

$$\Sigma \stackrel{\text{def}}{=} \sum_{q\in\mathbb{Q}^*} \frac{\partial}{\partial\alpha}\bigg|_{\alpha=0} (\Delta^\alpha_{\mathbb{A}^*}\varphi)(q) = \sum_{q\in\mathbb{Q}^*} \frac{\partial}{\partial\alpha}\bigg|_{\alpha=0} \prod_p (D^\alpha_p \varphi_p)(q)$$

$$= \sum_{q\in\mathbb{Q}^*} \frac{\partial}{\partial\alpha}\bigg|_{\alpha=0} \prod_p \Delta^\alpha_p(\varphi_p(qx_p))(1),$$

where $\Delta^\alpha_p = \left[c_p^{(\alpha)}\right]^{-1} D^\alpha_p$, since $(D^\alpha_p\varphi_p)(q) = |q|_p^{-\alpha}\Delta^\alpha_p(\varphi_p(qx_p))(1)$ and $\prod_p |q|_p^{-\alpha} = 1$ by Theorem 1.18.

Further,

$$\Sigma = \sum_{q\in\mathbb{Q}^*}\sum_p \frac{\partial}{\partial\alpha}\bigg|_{\alpha=0} \Delta^\alpha_p(\varphi_p(qx_p))(1)\cdot \prod_{p'\neq p}\varphi_{p'}(q)$$

$$= \sum_p \frac{\partial}{\partial\alpha}\bigg|_{\alpha=0} \left(\Delta^\alpha_p g^{(p)}_{\widehat{\varphi}}\right)(1),$$

where $g^{(p)}_{\widehat{\varphi}}(x_p) = \widehat{\varphi}(|x_p|_p)$, $x_p \in \mathbb{Q}_p$. Here we have used the fact that

$$\sum_{q\in\mathbb{Q}^*}\varphi_p(qx_p)\prod_{p'\neq p}\varphi_{p'}(q) = \sum_{q\in\mathbb{Q}^*}\varphi(q(1,1,\dots,x_p,1,\dots))$$

$$= \sum_{q\in\mathbb{Q}^*}\varphi(q(|x_p|_p,|x_p|_p,\dots,|x_p|_p x_p,|x_p|_p,\dots)) = \widehat{\varphi}(|x_p|_p),$$

since φ is radial, and $(|x_p|_p,|x_p|_p,\dots,|x_p|_p x_p,|x_p|_p,\dots)\in\Lambda$.

Next, since $c_p^{(0)} = 1$, $\frac{\partial}{\partial\alpha}\big|_{\alpha=0} c_p^{(\alpha)} = 0$, we obtain that

$$\Sigma = \sum_p \frac{\partial}{\partial\alpha}\bigg|_{\alpha=0}\left(D^\alpha_p g^{(p)}_{\widehat{\varphi}}\right)(1),$$

and by Propositions 2.4 and 2.5

$$\Sigma = \sum_p \frac{1}{2\pi i} \int\limits_{\frac{1}{2}-i\infty}^{\frac{1}{2}+i\infty} \mathcal{M}^s(\widehat{\varphi})\, d\log \frac{\zeta_p(s)}{\zeta_p(1-s)}$$

$$= \sum_p \left\{ \frac{1}{2\pi i} \int\limits_{c-i\infty}^{c+i\infty} \mathcal{M}^s(\widehat{\varphi})\, d\log \zeta_p(s) - \frac{1}{2\pi i} \int\limits_{c-i\infty}^{c+i\infty} \mathcal{M}^{1-s}(\widehat{\varphi})\, d\log \zeta_p(s) \right\}$$

$$= \frac{1}{2\pi i} \int\limits_{c-i\infty}^{c+i\infty} \mathcal{M}^s(\widehat{\varphi})\, d\log Z(s) - \frac{1}{2\pi i} \int\limits_{c-i\infty}^{c+i\infty} \mathcal{M}^{1-s}(\widehat{\varphi})\, d\log Z(s)$$

for any $c > 1$.

Using the functional equation (2.15) we get

$$\Sigma = \frac{1}{2\pi i} \left(\int\limits_{c-i\infty}^{c+i\infty} + \int\limits_{1-c+i\infty}^{1-c-i\infty} \right) \mathcal{M}^s(\widehat{\varphi})\, d\log Z(s). \tag{2.18}$$

Since all zeroes of $Z(s)$ (or non-real zeroes of $\zeta(s)$) lie in the "critical strip" $0 \le \operatorname{Re} s \le 1$ [146], by the residue theorem (2.18) implies (2.17). ■

2.5 Pseudo-differential operators over a local field

2.5.1. Hypersingular integral representation. Let K be a local field, O its ring of integers, P its prime ideal, and U its group of units. As we know (see Chapter 1), $P = \beta O$, $\|\beta\| = q^{-1}$, where β is a prime element, $q = \operatorname{card} O/P$. Each element $x \in K$ can be presented as $x = \beta^{-N} u_x$, where $\|x\| = q^N$, $\|u_x\| = 1$.

In this section we shall consider pseudo-differential operators

$$\mathfrak{A} = F^{-1} M_a F,$$

where F is the Fourier transform, M_a is the operator of multiplication by the function ("symbol")

$$a(\xi) = \|\xi\|^\nu \gamma(u_\xi),$$

where $\nu > 0$, γ is a continuous strictly positive function on U. It is clear that $\mathfrak{A} : \mathcal{D}(K) \to L_2(K)$.

We shall assume the following local constancy condition upon γ: there exists such a natural number m that the inequality $\|u\| \le q^{-m}$ implies the equality

$$\gamma(u+v) = \gamma(u) \text{ for all } u \in U. \tag{2.19}$$

Let $\{\theta_{jl}\}$ be the system of all the characters of the group U. It is assumed that θ_{jl} is a character of the ramification degree j; the quantity of such characters will be denoted L_j.

The function γ can be expanded in the uniformly convergent series

$$\gamma(u) = \sum_{j=0}^{\infty} \sum_{l=1}^{L_j} c_{jl}\theta_{jl}(u),$$

where

$$c_{jl} = (1-q^{-1})^{-1} \int_U \gamma(u)\overline{\theta_{jl}(u)}\, du.$$

The condition (2.19) implies that $c_{jl} = 0$ for $j > m$. Indeed, if $j > m$, there exists such an element $v_{jl} \in K$, $\|v_{jl}\| = q^{-j+1} \le q^{-m}$, that $\theta_{jl}(1+v_{jl}) \ne 1$. Then

$$c_{jl} = (1-q^{-1})^{-1} \int_U \gamma(u(1+v_{jl}))\overline{\theta_{jl}(u(1+v_{jl}))}\, du = \overline{\theta_{jl}(1+v_{jl})}c_{jl},$$

whence $c_{jl} = 0$.

Thus

$$\gamma(u) = \sum_{j=0}^{m} \sum_{l=1}^{L_j} c_{jl}\theta_{jl}(u),$$

The operator $\mathfrak{A}$ admits a hypersingular integral representation generalizing the representation (2.8) for the operator D^α.

Proposition 2.6. *Let*

$$G(u) = \sum_{j=0}^{m} \sum_{l=1}^{L_j} c_{jl} \left[\Gamma\left(q^\nu, \overline{\theta_{jl}}\right)\right]^{-1} \overline{\theta_{jl}(u)}, \quad u \in U.$$

Then for any $z \in \mathcal{D}(K)$

$$(\mathfrak{A}z)(x) = \int_K \|y\|^{-\nu-1} G(u_y)[z(x-y) - z(x)]\, dy. \tag{2.20}$$

Proof. It is sufficient to verify (2.20) for $\gamma = \theta_{jl}$ with arbitrary j, l. In this case

$$G(u) = \left[\Gamma\left(q^{\nu}, \overline{\theta_{jl}}\right)\right]^{-1} \overline{\theta_{jl}(u)}, \quad u \in U.$$

The function $g(\xi) = \|\xi\|^{\nu} \theta_{jl}(u_{\xi})(Fz)(\xi)$ is the Fourier transform of the convolution $G_1 * z$, where the homogeneous distribution

$$G_1(x) = \left[\Gamma\left(q^{\nu}, \overline{\theta_{jl}}\right)\right]^{-1} \|x\|^{-\nu-1} \overline{\theta_{jl}(u_x)}$$

acts upon a test function $\varphi \in \mathcal{D}(K)$ as

$$\langle G_1, \varphi \rangle = \left[\Gamma\left(q^{\nu}, \overline{\theta_{jl}}\right)\right]^{-1} \left\{ \int\limits_{\|x\| \le 1} \|x\|^{-\nu-1} \overline{\theta_{jl}(u_x)} [\varphi(x) - \varphi(0)]\, dx \right.$$

$$\left. + \int\limits_{\|x\| > 1} \|x\|^{-\nu-1} \overline{\theta_{jl}(u_x)} \varphi(x)\, dx + \varphi(0) \int\limits_{\|x\| \le 1} \|x\|^{-\nu-1} \overline{\theta_{jl}(u_x)}\, dx \right\};$$

The last integral on the right is defined via analytic continuation (see Sect. 1.7) and equals either 0, if $j \ge 1$, or $\varphi(0) \frac{1-q^{-1}}{1-q^{\nu}}$, if $j = 0$ (that is $\theta_{01}(u_x) \equiv 1$).

If $j \ge 1$, then it follows from (1.30) that

$$\int\limits_{\|x\| > 1} \|x\|^{-\nu-1} \overline{\theta_{jl}(u_x)}\, dx = 0,$$

and we come to (2.20). If $j = 0$, then we note that

$$\int\limits_{\|x\| > 1} \|x\|^{-\nu-1}\, dx = \sum_{N=1}^{\infty} \int\limits_{\|x\| = q^N} \|x\|^{-\nu-1}\, dx = \sum_{N=1}^{\infty} (1 - q^{-1}) q^{-N\nu}$$

$$= \frac{(1 - q^{-1}) q^{-\nu}}{1 - q^{-\nu}} = -\frac{1 - q^{-1}}{1 - q^{\nu}},$$

which implies (2.20) for this case too. ■

2.5.2. Fundamental solution. Let us consider the equation

$$\mathfrak{A} z = \varphi, \quad \varphi \in \mathcal{D}(K). \tag{2.21}$$

As in the case of the simplest operator $\mathfrak{A} = D^{\nu}$, the hypersingular integral representation (2.20) enables us to define solutions of (2.21) in the class

$\mathcal{E}_\nu(K)$, and the definition of $\mathfrak{A}z$ as a hypersingular integral can be interpreted also as the convolution (in the distribution sense) of $\|y\|^{-\nu-1}G(u_y)$ with z. Looking for the fundamental solution E, such that $E * \varphi \in \mathcal{E}_\nu(K)$ is a solution of (2.21), we see formally that $(FE)(\xi)$ must coincide with $\|\xi\|^{-\nu}[\gamma(u_\xi)]^{-1}$. The following proposition, which generalizes Theorem 2.1, gives a regularization of the inverse Fourier transform.

Since $\gamma(u) > 0$, the function $[\gamma(u)]^{-1}$ satisfies the condition similar to (2.19), so that

$$[\gamma(u)]^{-1} = \sum_{j=0}^{m}\sum_{l=1}^{L_j} c'_{jl}\theta_{jl}(u), \quad u \in U,$$

for certain $c'_{jl} \in \mathbb{C}$.

Proposition 2.7. *The function*

$$E(x) = \|x\|^{\nu-1}\sum_{j=0}^{m}\sum_{l=1}^{L_j} c'_{jl}\left[\Gamma\left(q^{-\nu}, \overline{\theta_{jl}}\right)\right]^{-1}\overline{\theta_{jl}(u_x)}, \quad \nu \neq 1,$$

or

$$E(x) = -c_{01}\frac{q-1}{q\log q}\log\|x\| + \sum_{j=1}^{m}\sum_{l=1}^{L_j} c'_{jl}\left[\Gamma\left(q^{-\nu}, \overline{\theta_{jl}}\right)\right]^{-1}\overline{\theta_{jl}(u_x)}, \quad \nu = 1,$$

is a fundamental solution for the equation (2.21).

Proof. As in the proof of Theorem 2.1, we see that $z = E * \varphi \in \mathcal{E}_\nu(K)$. Let us consider the case $\nu = 1$ (if $\nu \neq 1$, the reasoning is similar, though somewhat simpler).

We have

$$\widetilde{E}(\xi) = c'_{01}[\mathcal{P}\|\xi\|^{-1} + q^{-1}\delta(\xi)] + \sum_{j=1}^{m}\sum_{l=1}^{L_j} c'_{jl}\|\xi\|^{-1}\theta_{jl}(u_\xi),$$

where

$$\langle \mathcal{P}\|\xi\|^{-1}, \psi\rangle = \int\limits_{\|\xi\leq 1} [\psi(\xi) - \psi(0)]\|\xi\|^{-1}\,d\xi + \int\limits_{\|\xi\|>1} \|\xi\|^{-1}\psi(\xi)\,d\xi, \quad \psi \in \mathcal{D}(K).$$

On the other hand, for any $\psi \in \mathcal{D}(K)$

$$\langle F\mathfrak{A}z, \psi\rangle = \langle \mathfrak{A}z, \widetilde{\psi}\rangle = \int\limits_K \|y\|^{-2}G(u_y)\,dy \int\limits_K [z(x-y) - z(x)]\widetilde{\psi}(x)\,dx.$$

Denoting $\psi_y(t) = \chi(yt)\psi(t)$, we get

$$\int_K [z(x-y) - z(x)]\widetilde{\psi}(x)\,dx = \langle z, \psi_y - \psi\rangle = \langle \widetilde{E}, \widetilde{\varphi}\cdot(\psi_y - \psi)\rangle.$$

Since the function $\widetilde{\varphi}\cdot(\psi_y - \psi)$ vanishes near the origin, we find that

$$\langle \widetilde{E}, \widetilde{\varphi}\cdot(\psi_y - \psi)\rangle = \int_K \|\xi\|^{-1}[\gamma(u_\xi)]^{-1}\widetilde{\varphi}(\xi)[\chi(y\xi) - 1]\psi(\xi)\,d\xi.$$

Introducing the notations

$$R_l(\xi) = \begin{cases} \|\xi\|^{-1}[\gamma(u_\xi)]^{-1}\widetilde{\varphi}(\xi)\psi(\xi), & \|\xi\| > q^l, \\ 0, & \|\xi\| \le q^l, \end{cases}$$

$$I_l(y) = \int_{\|\xi\|\le q^l} \|\xi\|^{-1}[\gamma(u_\xi)]^{-1}\widetilde{\varphi}(\xi)[\chi(y\xi) - 1]\psi(\xi)\,d\xi,$$

we obtain that

$$\langle F\mathfrak{A}z, \psi\rangle = \int_K \|y\|^{-2}G(u_y)\left[\widetilde{R_l}(y) - \widetilde{R_l}(0)\right]dy + \int_K \|y\|^{-2}G(u_y)I_l(y)\,dy \tag{2.22}$$

for any $l \in \mathbb{Z}$.

The distribution $\|y\|^{-2}G(u_y)$ is the Fourier transform of the function $\|\xi\|\gamma(u_\xi)$ (see Sect. 1.7). Hence

$$\int_K \|y\|^{-2}G(u_y)\left[\widetilde{R_l}(y) - \widetilde{R_l}(0)\right]dy = \int_K \|\xi\|\gamma(u_\xi)R_l(\xi)\,d\xi$$

$$= \int_{\|\xi\|>q^l} \widetilde{\varphi}(\xi)\psi(\xi)\,d\xi \longrightarrow \langle\widetilde{\varphi}, \psi\rangle, \quad l \to -\infty.$$

We have used the fact that $R_l \in \mathcal{D}(K)$.

Let us consider the second integral in (2.22). For any $y \in K$ $I_l(y) \to 0$ as $l \to -\infty$. If $l \le 0$, then $I_l = 0$ for $\|y\| \le 1$,

$$I_l(y) \le C \int_{\|y\|^{-1}\le\|\xi\|\le 1} \|\xi\|^{-1}\,d\xi$$

$$= C \sum_{j=-\log_q\|y\|}^{0} \int_{\|\xi\|=q^j} \|\xi\|^{-1}\,d\xi \le C\left[\log_q\|y\| + 1\right]$$

for $||y|| > 1$. By the dominated convergence theorem the second summand in (2.22) tends to zero as $l \to -\infty$.

Therefore we obtain the inequality

$$\langle F\mathfrak{A}z, \psi\rangle = \langle\widetilde{\varphi}, \psi\rangle,$$

whence $\mathfrak{A}z = \varphi$. ■

2.5.3. Green function. Another equation of interest is

$$(\mathfrak{A} + \lambda)z = \varphi, \quad \lambda > 0. \tag{2.23}$$

Its fundamental solution $E_\lambda(x)$ (called *the Green function* of the operator $\mathfrak{A}$) can be interpreted as an integral kernel of the resolvent $(\mathfrak{A} + \lambda)^{-1}$, if $\mathfrak{A}$ is considered as a (self-adjoint positive) operator in $L_2(K)$. As before, $E_\lambda(x)$ (which is apriori a distribution from $\mathcal{D}'(K)$) is the inverse Fourier transform of the function $[a(\xi) + \lambda]^{-1}$.

Proposition 2.8. (i) *If* $\nu > 1$, *then the function* $E_\lambda(x)$ *is continuous, while for* $0 < \nu \leq 1$ *the function* $E_\lambda(x)$ *is continuous for* $x \neq 0$, *and*

$$|E_\lambda(x)| \leq \begin{cases} C\|x\|^{\nu-1}, & 0 < \nu < 1; \\ C|\log\|x\|| + C', & \nu = 1, \end{cases} \tag{2.24}$$

if $\|x\| \leq 1$.

(ii) *As* $\|x\| \to \infty$,

$$E_\lambda(x) = \lambda^{-2}\Bigg[c_{01}\frac{q^{\nu+1}(q^\nu - 1)}{q^{\nu+1} - 1} - \sum_{j=1}^{m} q^{j(\nu+1)} \sum_{l=1}^{L_j} c_{jl}\Gamma(1, \theta_{jl})\overline{\theta_{jl}(-u_x)}\Bigg] \|x\|^{\nu-1} + O\left(\|x\|^{-2\nu-1}\right). \tag{2.25}$$

(iii) *If* $G(u) \leq 0$ *for all* $u \in U$, *then* $E_\lambda \geq 0$ *for all* $x \neq 0$.

Proof. (i) It follows from the continuity of the Fourier transform F on $\mathcal{D}'(K)$ that

$$E_\lambda(x) = \sum_{N=-\infty}^{\infty} e_\lambda^{(N)}(x), \tag{2.26}$$

where

$$e_\lambda^{(N)}(x) = \int_{\|\xi\|=q^N} \frac{\chi(-x\xi)}{a(\xi) + \lambda}\, d\xi, \quad x \in K, \tag{2.27}$$

and the series (2.26) converges in $\mathcal{D}'(K)$.

Let $x \in K$, $\|x\| = q^k$. Suppose first that $N \geq m - k + 1$, where the number m appears in the local constancy condition (2.19). Since χ is a rank zero character, it is possible to choose such an element $\tau \in K$ that $\|\tau\| = q^{1-k-N}$, $\chi(-x\beta^{-N}\tau) \neq 1$. Making in (2.27) a substitution $\xi = \eta + \beta^{-N}\tau$ and remembering that $\|\beta^{-N}\tau\| = q^{1-k} \leq q^{N-m}$, we find that

$$e_\lambda^{(N)}(x) = \chi(-x\beta^{-N}\tau) \int\limits_{\|\xi\|=q^N} \frac{\chi(-x\eta)\,d\eta}{\|\eta + \beta^{-N}\tau\|^\nu \gamma(\beta^N\eta + \tau) + \lambda}$$

$$= \chi(-x\beta^{-N}\tau) e_\lambda^{(N)}(x),$$

whence $e_\lambda^{(N)}(x) = 0$, so that

$$E_\lambda(x) = \sum_{N=-\infty}^{m-k} e_\lambda^{(N)}(x), \quad \|x\| = q^k.$$

If $\|x\| = q^k$, $N \leq m - k$, then

$$\left| e_\lambda^{(N)}(x) \right| \leq (1 - q^{-1}) q^N (q^{\nu N}\gamma_0 + \lambda)^{-1}, \tag{2.28}$$

where $\gamma_0 = \inf\limits_{u \in U} \gamma(u) > 0$. This implies the uniform convergence of the series (2.26) on the sphere $\{\|x\| = q^k\}$ and, equivalently, its uniform convergence on compact subsets of $K \setminus \{0\}$. Hence $E_\lambda(x)$ is always continuous for $x \neq 0$. If $\nu > 1$, the estimate (2.28) yields also the uniform convergence on K and the continuity of $E_\lambda(x)$ on the whole K.

If $0 < \nu < 1$, then by (2.28) for $\|x\| < 1$

$$|E_\lambda(x)| \leq \gamma_0^{-1}(1 - q^{-1}) \sum_{N=-\infty}^{m-k} q^{(1-\nu)N} = \gamma_0^{-1} \frac{1 - q^{-1}}{1 - q^{\nu - 1}} q^{(1-\nu)(m-k)}$$

$$= \gamma_0^{-1} q^{(1-\nu)m} \frac{1 - q^{-1}}{1 - q^{\nu-1}} \|x\|^{\nu - 1}.$$

Finally, if $\nu = 1$, then for $\|x\| \leq 1$

$$|E_\lambda(x)| \leq \lambda^{-1}(1 - q^{-1}) \sum_{N=-\infty}^{0} q^N + (1 - q^{-1}) \sum_{N=1}^{m-k} \gamma_0^{-1}$$

$$= \lambda^{-1} + (1 - q^{-1})\gamma_0^{-1}(m - k) = C|\log\|x\|| + C',$$

and we have proved (2.24).

(ii) Let us turn to the study of the asymptotic behavior of $E_\lambda(x)$ as $\|x\| \to \infty$. Let $\|x\| = q^k$. If $\|\xi\| \le q^{-k}$, then $\chi(-x\xi) = 1$. Therefore $E_\lambda(x) = E_\lambda^{(1)}(x) + E_\lambda^{(2)}(x)$, where

$$E_\lambda^{(1)}(x) = \sum_{N=-\infty}^{-k} \int\limits_{\|\xi\|=q^N} \frac{d\xi}{q^{\nu N}\gamma(\beta^N\xi) + \lambda},$$

$$E_\lambda^{(2)}(x) = \sum_{N=-k+1}^{-k+m} \int\limits_{\|\xi\|=q^N} \frac{\chi(-x\xi)d\xi}{q^{\nu N}\gamma(\beta^N\xi) + \lambda}.$$

Using the relation

$$(\lambda + t)^{-1} = \lambda^{-1} - t\lambda^{-2} + O(t^2), \quad t \to 0, \tag{2.29}$$

we get

$$\int\limits_{\|\xi\|=q^N} \frac{d\xi}{q^{\nu N}\gamma(\beta^N\xi) + \lambda} = \lambda^{-1}(1 - q^{-1})q^N - \lambda^{-2}q^{\nu N} \int\limits_{\|\xi\|=q^N} \gamma(\beta^N\xi)\, d\xi$$
$$+ O\left(q^{(2\nu+1)N}\right), \quad N \to -\infty. \tag{2.30}$$

It follows from the expansion (2.30) and the integral formula (1.27) that

$$E_\lambda^{(1)}(x) = \lambda^{-1}\|x\|^{-1} - \lambda^{-2}c_{01}\frac{1 - q^{-1}}{1 - q^{-\nu-1}}\|x\|^{\nu-1}$$
$$+ O\left(\|x\|^{-2\nu-1}\right), \quad \|x\| \to \infty. \tag{2.31}$$

Let us consider the asymptotics of $E_\lambda^{(2)}(x)$. We shall use again the asymptotic relation (2.29). The emerging integrals are computed as follows (see Chapter 1):

$$\int\limits_{\|\xi\|=q^N} \chi(-x\xi)\, d\xi = \begin{cases} -q^k, & N = -k+1, \\ 0, & N > -k+1; \end{cases}$$

if $j \ge 1$, we perform the substitution $\xi = -x^{-1}\beta^{-N-k}\eta$, so that

$$\int\limits_{\|\xi\|=q^N} \theta_{jl}(\beta^N\xi)\chi(-x\xi)\, d\xi = q^N \int\limits_{\|\eta\|=1} \theta_{jl}(-\beta^{-k}x^{-1}\eta)\chi(\beta^{-N-k}\eta)\, d\eta$$
$$= q^N\overline{\theta_{jl}(-\beta^k x)} \int\limits_{\|\eta\|=1} \theta_{jl}(\eta)\chi(\beta^{-N-k}\eta)\, d\eta,$$

where

$$\int\limits_{\|\eta\|=1} \theta_{jl}(\eta)\chi(\beta^{-N-k}\eta)\,d\eta = \begin{cases} \Gamma(1,\theta_{jl}), & N+k=j, \\ 0, & N+k\neq j. \end{cases}$$

As a result, we obtain

$$E_\lambda^{(2)}(x) = -\lambda^{-1}\|x\|^{-1} + \lambda^{-2}\|x\|^{-\nu-1}\left[q^\nu c_{01} - \sum_{j=1}^{m} q^{j(\nu+1)} \sum_{l=1}^{L_j} c_{jl}\Gamma(1,\theta_{jl})\overline{\theta_{jl}(-u_x)}\right].$$

This formula, together with (2.31), implies (2.25).

(iii) Suppose that $G(u)\leq 0$, $u\in U$. We shall prove that the function $z=E_\lambda * \varphi$ is non-negative for an arbitrary non-negative function $\varphi\in\mathcal{D}(K)$. It is clear that z is locally constant and (in view of the asymptotics (2.25)), $z(x)\to 0$ for $\|x\|\to\infty$, $z\in\mathcal{E}_\nu(K)$.

The equality $(\mathfrak{A}+\lambda)z=\varphi$ means that

$$\varphi(x) = \int\limits_K \|y\|^{-\nu-1}G(u_y)[z(x-y)-z(x)]\,dy + \lambda z(x), \quad x\in K. \tag{2.32}$$

If $z(x)$ takes negative values at certain points, then (since $z(x)\to 0$ as $\|x\|\to\infty$) there exists a point x_0 of the global minimum for $z(x)$, and $z(x_0)<0$. Substituting $x=x_0$ in (2.32), we find that the integral in the right-hand side is non-positive and $\lambda z(x_0)<0$ whereas $\varphi(x_0)\geq 0$, and we come to a contradiction. ■

2.6 Coordinate representation

Let K be a field of zero characteristic. Then K is an extension of the field $\mathbb{Q}_p$ of a finite degree n. Suppose that $n\geq 2$. A rank zero character on the additive group of K can be given by the formula

$$\chi(x) = \chi_p(\mathrm{Tr}(\beta^{-d}x)), \tag{2.33}$$

where $\mathrm{Tr}\,:\,K\to\mathbb{Q}_p$ is the trace in K over $\mathbb{Q}_p$, d is an exponent of the different (see Chapter 1).

Consider a bilinear form

$$\rho(x,y) = \mathrm{Tr}(xy\beta^{-d})\,, \quad x,y\in K, \tag{2.34}$$

with values in $\mathbb{Q}_p$. Since ρ is symmetric and non-degenerate, the field K possesses a ρ-orthogonal basis $e_1, \dots, e_n$ over $\mathbb{Q}_p$ (see [6]):

$$\rho(e_i, e_j) = 0 \quad \text{if} \quad i \neq j, \quad \rho(e_i, e_i) = \mu_i \neq 0, \ i = 1, \dots, n.$$

If $x = x_1 e_1 + \cdots + x_n e_n$, $y = y_1 e_1 + \cdots + y_n e_n$, $x_i, y_i \in \mathbb{Q}_p$ then

$$\chi(xy) = \prod_{i=1}^{n} \chi_p(\mu_i x_i y_i).$$

The normalized additive Haar measure on K can be represented as

$$dx = c_K dx_1 \cdots dx_n, \tag{2.35}$$

where dx_j is the normalized additive Haar measure on $\mathbb{Q}_p$,

$$c_K^{-1} = \int_{\|x_1 e_1 + \cdots + x_n e_n\| \leq 1} dx_1 \dots dx_n.$$

Let $\mathfrak{A}$ be a pseudo-differential operator studied in Sect. 2.5. Consider an operator B over $\mathbb{Q}_p^n$ defined by the relation

$$(Bg)(x_1, \dots, x_n) = (\mathfrak{A} f)(x_1 e_1 + \cdots + x_n e_n)$$

where $g \in \mathcal{D}(\mathbb{Q}_p^n)$; $f \in \mathcal{D}(K)$ is such that

$$g(x_1, \dots, x_n) = f(x_1 e_1 + \cdots + x_n e_n) \ , \quad x_j \in \mathbb{Q}_p.$$

It is easy to see that B is a pseudo-differential operator $F^{-1} M_b F$ with the symbol

$$b(\xi_1, \dots, \xi_n) = c_K^2 \left(\prod_{j=1}^{n} |\mu_j|_p^{-1} \right) a \left(\sum_{k=1}^{n} \mu_k^{-1} \xi_k e_k \right). \tag{2.36}$$

Here F is the n-dimensional Fourier transform

$$(F\varphi)(\xi_1, \dots, \xi_n) = \int_{\mathbb{Q}_p^n} \chi_p(x_1 \xi_1 + \cdots + x_n \xi_n) \varphi(x_1, \dots, x_n) \, dx_1 \dots dx_n.$$

The basis $\{e_1, \dots, e_n\}$ is called a *basis of coordinate representation* for the operator $\mathfrak{A}$. Below such a basis will be frequently used as follows. Starting from a pseudo-differential operator A over $\mathbb{Q}_p^n$ with a symbol $|h(\xi_1, \dots, \xi_n)|_p^{\alpha}$, where h is a p-adic quadratic form, $\alpha > 0$, we find an extension K, a basis of coordinate representation,an appropriate symbol $a(\xi)$, $\xi \in K$, and a matrix T of order n over $\mathbb{Q}_p$ such that

$$|h(\xi_1, \dots, \xi_n)|_p^\alpha = b(T(\xi_1, \dots, \xi_n)'),$$

where b has the form (2.36). Now $A = W_T^{-1} B W_T$, $(W_T g)(x) = g(T'x)$, $x \in \mathbb{Q}_p^n$.The hypersingular integral representation (2.20) implies the representation

$$\begin{aligned}(Ag)&(x_1, \dots, x_n)\\ &= \int_K \|y\|^{-\nu-1} G(u_y) \left[g(x_1 - (T'Y)_1, \dots, x_n - (T'Y)_n) \right.\\ &\qquad\qquad\qquad\qquad \left. - g(x_1, \dots, x_n) \right] dy, \qquad (2.37)\end{aligned}$$

where $Y = (y_1, \dots, y_n) \in \mathbb{Q}_p^n$ is determined from the expansion $y = y_1 e_1 + \dots + y_n e_n$, $y \in K$. The relation between α and ν and the choice of K and a will be specified separately for different n and various classes of forms h.

The expression (2.37) makes sense, for example, for locally constant functions $g(x_1, \dots, x_n)$ growing at infinity not faster than $\max\limits_{1 \le j \le n} |x_j|_p^{\nu - \varepsilon}$, $\varepsilon > 0$ (or, as above, an integral growth restriction can be imposed).

If $E(x)$ is a fundamental solution for the operator $\mathfrak{A}$ constructed in Proposition 2.7, then a fundamental solution $\mathcal{E}(x_1, \dots, x_n)$ for the operator A (understood in just the same way as in Sect. 2.5) is given by the formula

$$\mathcal{E}(x_1, \dots, x_n) = |\det T|_p^{-1} E\left((T'^{-1} X)_1 e_1 + \dots + (T'^{-1} X)_n e_n \right), \qquad (2.38)$$

where $X = (x_1, \dots x_n)$. This formula is verified by a straightforward calculation.

It is clear that a similar connection is valid for the Green functions, that is fundamental solutions of "shifted" operators $\mathfrak{A} + \lambda$ and $A + \lambda$.

The relations between the operators A and $\mathfrak{A}$ including the formula (2.38) carry over to the case when K is a (non-commutative) division algebra. Namely, let K be a central division algebra over $\mathbb{Q}_p$, $\dim K = m^2$ (see [115, 156]). One should only substitute $m - 1$ for d, and the reduced trace t for Tr in (2.33) and (2.34). The hypersingular integral representation (2.20) for general symbols is possible only for a commutative K. However, for the operator D_K^ν the whole reasoning carries over to non-commutative algebras, for example, the quaternion algebra over $\mathbb{Q}_p$.

The role of the basis of coordinate representation is, of course, in expressing the factor $\chi(\xi x)$ in the definition of the Fourier transform on K, in terms of the corresponding factor for the multi-dimensional Fourier transform over $\mathbb{Q}_p$. Sometimes it is convenient to start from another basis having in mind a possibility to transform it by a linear transformation into a basis of coordinate representation, with subsequent use of the formula (2.38). As a first example of this approach, we shall consider the case, where the symbol of A is not defined via a quadratic form (this is postponed until the next section), but equals

$$\left[\max_{1\le j\le n} |\xi_j|_p\right]^\alpha, \quad \alpha > 0. \tag{2.39}$$

In a moment it will become clear that the best choice of an extension K, in order to deal with the symbol (2.39), is the unramified extension of the degree n.

Lemma 2.1. *Let K be an unramified extension of the field $\mathbb{Q}_p$. Then $\|x\| = \|x\|_{\max}^n$, where $n = (K : \mathbb{Q}_p)$,*

$$\|x\|_{\max} = \max_{1\le j\le n} |\xi_j|_p,$$

and $\xi_j \in \mathbb{Q}_p$ are the coefficients of the expansion of $x \in K$ with respect to a canonical basis.

Proof. For the unramified extension $\beta = p$, $q = p^n$. Let $x = p^N z$, $\|z\| = 1$, so that $\|x\| = q^{-N} = p^{-nN}$. On the other hand,

$$\|x\|_{\max} = p^{-N}\|z\|_{\max} = \|x\|^{1/n}\|z\|_{\max},$$

and it remains to show that $\|z\|_{\max} = 1$.

Consider the expansion of the element z with respect to the canonical basis $\{f_j\}$:

$$z = \sum_{j=1}^n \zeta_j f_j, \quad \zeta_j \in \mathbb{Q}_p.$$

Since the basis is fundamental, we have $|\zeta_j|_p \le 1$ for all j. Since $\|f_j\| = 1$, $j = 1, \dots, n$, we find that

$$1 = \|z\| \le \max |\zeta_j|_p \le 1$$

whence $\|z\|_{\max} = 1$. ■

It follows from Lemma 2.1 that the pseudo-differential operator A with the symbol (2.39) can be studied by means of the reduction to the operator $D_K^{\alpha/n}$, where K is the unramified extension of $\mathbb{Q}_p$ of the degree n. For example, it follows immediately from Proposition 2.8 that the Green function of A is continuous outside the origin (and at the origin, if $\alpha > n$), non-negative, and tends to zero at infinity.

2.7 Fundamental solutions of elliptic operators

Let $h(\xi_1, \dots, \xi_n)$ be a quadratic form on $\mathbb{Q}_p^n$, $p \neq 2$. We shall consider a pseudo-differential operator A with the symbol $|h(\xi_1, \dots, \xi_n)|_p^\alpha$, $\alpha > 0$.

Theorem 2.3. *If*

$$h(\xi_1, \dots, \xi_n) \neq 0 \quad \textit{when } |\xi_1|_p + \dots + |\xi_n|_p \neq 0, \tag{2.40}$$

then there exists a fundamental solution $\mathcal{E}(x_1, \dots, x_n)$ for the equation $Az = f$, which is locally integrable, continuous for $(\xi_1, \dots, \xi_n) \neq 0$, and has at most a power-like growth at infinity. If $f \in \mathcal{D}(\mathbb{Q}_p^n)$ then the function $z = \mathcal{E} * f$ satisfies the equation $Az = f$, where A is understood as a hyper-singular integral operator (2.37).

The proof given below consists of an explicit construction of $\mathcal{E}$ for all possible quadratic forms satisfying the ellipticity condition (2.40). We shall also write down the explicit forms of the hypersingular integral representations (2.37).

a) *Binary forms*

Let $h(\xi_1, \xi_2)$ satisfy (2.40). Then up to a linear isomorphism

$$h(\xi_1, \xi_2) = \xi_1^2 - \tau \xi_2^2$$

where $\tau \in \mathbb{Q}_p$ is not a square in $\mathbb{Q}_p$ (see Sect. 1.4). To obtain all non-equivalent forms one should take a complete set of representatives τ of cosets from $\mathbb{Q}_p^*/\mathbb{Q}_p^{*2}$ (here $\mathbb{Q}_p^*$ is the multiplicative group of the field $\mathbb{Q}_p$,and $\mathbb{Q}_p^{*2}$ is the subgroup of all squares):

$$\tau = \varepsilon, |\varepsilon|_p = 1\,; \quad \tau = p\,; \quad \tau = \varepsilon p.$$

In each of the three cases we set $K = \mathbb{Q}_p(\sqrt{\tau})$. The absolute value of an element $x = x_1 + \sqrt{\tau}\, x_2 \in K$ is given by

$$\|x\| = |x_1^2 - \tau x_2^2|_p.$$

The operator of multiplication by $x = x_1 + \sqrt{\tau}\, x_2$ is represented by the matrix $\begin{pmatrix} x_1 & x_2 \\ x_2\tau & x_1 \end{pmatrix}$ with respect to the basis $\{1, \sqrt{\tau}\}$. Thus $\operatorname{Tr} x = 2x_1$.

Let us consider three possible cases separately.

Let $\tau = \varepsilon$. Then K is unramified over $\mathbb{Q}_p$ whence $q = p^2, d = 0$. We can take $e_1 = 1, e_2 = \sqrt{\varepsilon}$ as a basis of coordinate representation.

It is easy to calculate that $c = 1, \mu_1 = 2, \mu_2 = 2\varepsilon, |\mu_1|_p = |\mu_2|_p = 1$. Taking $\nu = \alpha, \mathfrak{A} = D_K^\alpha, T = \begin{pmatrix} 2 & 0 \\ 0 & 2\varepsilon \end{pmatrix}$ we obtain a hypersingular integral representation

$$(Ag)(x_1, x_2) = \frac{1 - p^{2\alpha}}{1 - p^{-2\alpha-2}} \int_{\mathbb{Q}_p^2} |y_1^2 - \varepsilon y_2^2|_p^{-\alpha-1} [g(x_1 - 2y_1, x_2 - 2\varepsilon y_2) - g(x_1, x_2)]\, dy_1\, dy_2.$$

A fundamental solution is given by

$$\mathcal{E}(x_1, x_2) = E\left(\frac{x_1}{2} + \sqrt{\varepsilon}\,\frac{x_2}{2\varepsilon}\right), \quad x_1, x_2 \in \mathbb{Q}_p,$$

where for every $x \in K, x \neq 0$,

$$E(x) = \begin{cases} \dfrac{1 - p^{-2\alpha}}{1 - p^{2(\alpha-1)}} \|x\|^{\alpha-1}, & \text{if } \alpha \neq 1, \\ \dfrac{1 - p^2}{2p^2 \log p} \log \|x\|, & \text{if } \alpha = 1. \end{cases}$$

Let $\tau = p$. Then K is a totally ramified extension of $\mathbb{Q}_p$, $q = p, d = 1$. One may set $\beta = \sqrt{p}$ and take

$$e_1 = 1 - \sqrt{p} \quad , \quad e_2 = 1 + \sqrt{p}$$

as a basis of coordinate representation. Then $\mu_1 = -4$, $\mu_2 = 4, |\mu_1|_p = |\mu_2|_p = 1$.

Since K is totally ramified, we have a fundamental basis $\{1, \sqrt{p}\}$ (see Sect. 1.3). It follows that the basis $\{e_1, e_2\}$ is also fundamental. Indeed, $\|e_1\| = \|e_2\| = 1$, and if $\|x_1 e_1 + x_2 e_2\| \leq 1$, $x_1, x_2 \in \mathbb{Q}_p$, that is

$$\|(x_1 + x_2) + (x_2 - x_1)\sqrt{p}\| \leq 1,$$

then $|x_1 + x_2|_p \leq 1$, $|x_2 - x_1|_p \leq 1$, whence $|x_1|_p \leq 1$, $|x_2|_p \leq 1$. Therefore $c = 1$.

As before we take $\nu = \alpha, \mathfrak{A} = D_K^\alpha$. In this case $T = 1/2 \begin{pmatrix} 1 & 1 \\ -1 & 1 \end{pmatrix}$,

$$(Ag)(x_1, x_2) = \frac{1 - p^\alpha}{1 - p^{-\alpha-1}} \int_{\mathbb{Q}_p^2} |(y_1 + y_2)^2 - p(y_2 - y_1)^2|_p^{-\alpha-1} \times \left[g(x_1 - \frac{y_1}{2} + \frac{y_2}{2}, x_2 - \frac{y_1}{2} - \frac{y_2}{2}) - g(x_1, x_2)\right] dy_1 dy_2.$$

The expression for a fundamental solution is

$$\mathcal{E}(x_1, x_2) = E((x_1 + x_2)(1 - \sqrt{p}) + (x_2 - x_1)(1 + \sqrt{p})) = E(2x_2 - 2x_1\sqrt{p})$$

where

$$E(x) = \begin{cases} \dfrac{1 - p^{-\alpha}}{1 - p^{\alpha-1}} \|x\|^{\alpha-1}, & \text{if } \alpha \neq 1, \\ \dfrac{1 - p}{p \log p} \log \|x\|, & \text{if } \alpha = 1, \end{cases}$$

$x \in K, x \neq 0$.

The third case ($\tau = \varepsilon p$) is virtually identical to the second one.

b) *Ternary forms*

If the quadratic form $h(\xi_1, \xi_2, \xi_3)$ satisfies (2.40) then up to a linear isomorphism

$$h(\xi_1, \xi_2, \xi_3) = p\varepsilon_1\xi_1^2 + \varepsilon_2\xi_2^2 + \varepsilon_3\xi_3^2 \tag{2.41}$$

where $|\varepsilon_1|_p = |\varepsilon_2|_p = |\varepsilon_3|_p = 1$, $\varepsilon_2\xi_2^2 + \varepsilon_3\xi_3^2 \neq 0$ if $|\xi_2|_p + |\xi_3|_p \neq 0$. So it is sufficient to consider only the forms (2.41).

Let K be the unramified extension of $\mathbb{Q}_p$ with degree 3. Its residue field $\mathbb{F}_q = O/P$ consisting of $q = p^3$ elements is an extension of a simple finite field $\mathbb{F}_p = \mathbb{Z}_p/p\mathbb{Z}_p$. Let λ be a primitive element of the field $\mathbb{F}_q$. Then $O/P = \{0, 1, \lambda, \dots, \lambda^{q-2}\}$, the elements $1, \lambda, \lambda^2$ form a basis in $\mathbb{F}_q$ over $\mathbb{F}_p$. Let $\{f_1, f_2, f_3\}$ be a canonical basis in K over $\mathbb{Q}_p$ corresponding to the basis $\{1, \lambda, \lambda^2\}$; this means that f_1, f_2 and f_3 are arbitrary representatives of the residue classes $1, \lambda$ and λ^2 respectively. The absolute value in K has a simple expression in terms of coefficients with respect to a canonical basis (see Lemma 2.1): if $x \in K$, $x = \eta_1 f_1 + \eta_2 f_2 + \eta_3 f_3$, then $\|x\| = [\max\{|\eta_1|_p, |\eta_2|_p, |\eta_3|_p\}]^3$.

An interpretation of the operator A associated with the form (2.41) as a coordinate representation of a pseudo-differential operator over K can be given as follows. Consider the following function γ on the group of units U of the field K. Let $u \in U$, $u = \eta_1 f_1 + \eta_2 f_2 + \eta_3 f_3$. Then $\max\limits_{1 \leq j \leq 3} |\eta_j|_p = 1$. Set

$$\gamma(u) = \begin{cases} p^{-\alpha}, & \text{if } |\eta_1|_p = 1, |\eta_2|_p < 1, |\eta_3|_p < 1, \\ 1, & \text{otherwise.} \end{cases}$$

The operator $\mathfrak{A}$ is,by definition,a pseudo-differential operator over K with the symbol

$$a(x) = \rho\|x\|^{2\alpha/3}\gamma(u_x), \quad x \in K,\ \rho = \text{const} > 0. \tag{2.42}$$

If $\{e_1, e_2, e_3\}$ is a basis of coordinate representation then the matrix T appearing in relations between $\mathfrak{A}$ and A equals $M\Theta'$, where

$$M = \operatorname{diag}(\mu_1, \mu_2, \mu_3),$$

and the matrix $\Theta = (\theta_{ij})$ transforms the bases:

$$f_i = \sum_{j=1}^{3} \theta_{ij} e_j \ , \quad i = 1, 2, 3.$$

Indeed, it is sufficient to show that

$$|h(\eta_1, \eta_2, \eta_3)|_p^\alpha = \|x\|^{2\alpha/3} \gamma(u_x), \tag{2.43}$$

where $x = \eta_1 f_1 + \eta_2 f_2 + \eta_3 f_3$.

If

$$\max_{j=1,2,3} |\eta_j|_p = p^N.$$

then

$$h(\eta_1, \eta_2, \eta_3) = p^{-2N} h(\zeta_1, \zeta_2, \zeta_3), \tag{2.44}$$

where

$$\zeta_j = p^N \eta_j, \quad |\zeta_j|_p \leq 1, \quad \max |\zeta_j|_p = 1.$$

In particular, if

$$|\zeta_1|_p = 1, \quad |\zeta_2|_p < 1, \quad |\zeta_3|_p < 1,$$

then

$$|\zeta_2|_p \leq p^{-1}, \ |\zeta_3|_p \leq p^{-1}, \ |\varepsilon_2 \zeta_2^2 + \varepsilon_3 \zeta_3^2|_p \leq p^{-2}, \ |p\varepsilon_1 \zeta_1^2|_p = p^{-1},$$

whence

$$|h(\zeta_1, \zeta_2, \zeta_3)|_p = p^{-1}. \tag{2.45}$$

For the remaining values of $\zeta_1, \zeta_2, \zeta_3)$

$$|h(\zeta_1, \zeta_2, \zeta_3)|_p = 1. \tag{2.46}$$

In order to prove this, it is sufficient to verify that $|\varepsilon_2 \zeta_2^2 + \varepsilon_3 \zeta_3^2|_p = 1$. Supposing otherwise, we have $\varepsilon_2 \zeta_2^2 + \varepsilon_3 \zeta_3^2 \equiv 0 \pmod{p}$ (a congruence in $\mathbb{Z}_p$ modulo $p\mathbb{Z}_p$), and either $2\varepsilon_2 \zeta_2 \not\equiv 0 \pmod{p}$, or $2\varepsilon_3 \zeta_3 \not\equiv 0 \pmod{p}$, which implies (see [18]) the nontrivial solvability of the equation $\varepsilon_2 \zeta_2^2 + \varepsilon_3 \zeta_3^2 = 0$. We have arrived at a contradiction.

Now the representation (2.43) is a consequence of (2.44)-(2.46).

Let us expand the function $\gamma(u)$, $u \in U$, with respect to the characters of the group U. Each element $u \in U$ can be uniquely represented in the form $u = f_2^m y$, where $0 \leq m \leq q-2$, $\|1-y\| < 1$. Let

$$\psi_j(u) = \exp\left(\frac{2\pi ijm}{q-1}\right), \quad 0 \leq j \leq q-2.$$

Obviously, ψ_j is a character of the group U. We have

$$\gamma(u) - 1 = \frac{p^{-\alpha}-1}{p^2+p+1} \sum_{l=0}^{p(p+1)} \psi_{l(p-1)}(u). \tag{2.47}$$

Indeed, let $u = \eta_1 f_1 + \eta_2 f_2 + \eta_3 f_3$. If $|\eta_1|_p = 1$, $|\eta_2|_p < 1$, $|\eta_3|_p < 1$, then in this (and only this) case the class of the element u in O/P is identical to the class of the element $\eta_1 \in \mathbb{Q}_p$. Let $\delta = \lambda^{p^2+p+1}$. Then $\delta^{p-1} = 1$, all the degrees δ^i, $0 \leq i \leq p-2$, are different and together with zero constitute a subfield of O/P consisting of p elements. Such a subfield is unique (see Theorem 1.3). Hence it coincides with the field $\mathbb{Z}_p/p\mathbb{Z}_p$ considered as a subfield in O/P. Therefore for the above values of u in the representation $u = f_2^m y$ we have $m = l_1(p^2+p+1)$, $0 \leq l_1 \leq p-2$, whence $\psi_{l(p-1)}(u) = 1$, so that the right-hand side of (2.47) is equal to $p^{-\alpha} - 1$, which equals the left-hand side.

Consider the remaining values of u. We have

$$\psi_{l(p-1)}(u) = \exp\left(\frac{2\pi ilm}{p^2+p+1}\right),$$

where $m \not\equiv 0 \pmod{p^2+p+1}$. The function

$$\tau(l) = \exp\left(\frac{2\pi ilm}{p^2+p+1}\right)$$

is a nontrivial character of the additive group of classes of residues $(\mathrm{mod}\ (p^2+p+1))$. Therefore

$$\sum_{l=0}^{p(p+1)} \tau(l) = 0,$$

and the right-hand side of (2.47) equals zero, as we needed to prove.

It follows from (2.47) that the function γ satisfies the local constancy condition (2.19). Let us calculate the function $G(u)$ using the expression given in Proposition 2.5. In our case the characters contained in the expansion of γ are $\theta_{00} \equiv 1$ and some of the characters of the ramification degree 1, namely $\psi_{l(p-1)}$ with $l > 0$.

Since

$$\Gamma\left(q^{2\alpha/3}, 1\right) = \frac{1 - q^{-2\alpha/3-1}}{1 - q^{2\alpha/3}},$$

$$\overline{\psi_{l(p-1)}(u)} = \psi_{(p^2+p+1-l)(p-1)}(u), \quad l \neq 0,$$

we get (changing the index of summation) that

$$\rho^{-1}G(u) = \frac{1-p^{2\alpha}}{1-p^{-2\alpha-3}}\left(1 + \frac{p^{-\alpha}-1}{p^2+p+1}\right) + \frac{p^{-\alpha}-1}{p^2+p+1}\sum_{l=1}^{p(p+1)} \frac{\psi_{l(p-1)}(u)}{\Gamma\left(q^{2\alpha/3}, \psi_{l(p-1)}\right)}. \tag{2.48}$$

By Theorem 1.14 we have for $l \neq 0$, $j \neq 0$

$$\Gamma\left(q^{2\alpha/3}, \psi_{l(p-1)}\right) = \left[\overline{\Gamma\left(q^{-1-2\alpha/3}, \psi_{l(p-1)}\right)}\right]^{-1},$$

$$\Gamma\left(q^{-1-2\alpha/3}, \psi_j\right) = q^{2\alpha/3+1}\int_{\|y\|=1} \chi(p^{-1}y)\psi_j(y)\,dy.$$

The last integral is equal to the sum of integrals over the sets

$$\{y = f_2^m z : \|1 - z\| < 1\}, \quad 0 \le m \le q-2.$$

By the invariance of the Haar measure, the measure of each of them is $(1-q^{-1})(q-1)^{-1} = q^{-1}$. Hence

$$\Gamma\left(q^{-1-2\alpha/3}, \psi_j\right) = q^{2\alpha/3}\sum_{m=0}^{q-2} \chi(p^{-1}f_2^m)\exp\left(\frac{2\pi ijm}{q-1}\right).$$

Substituting all this into (2.48) we find that for $u = f_2^N z$, $\|1 - z\| < 1$,

$$\rho^{-1}G(u) = \frac{1-p^{2\alpha}}{1-p^{-2\alpha-3}}\left(1 + \frac{p^{-\alpha}-1}{p^2+p+1}\right)$$
$$+ p^{2\alpha}(p^{-\alpha}-1)\sum_{m=0}^{q-2}\sum_{l=1}^{p(p+1)} \chi(-p^{-1}f_2^m)\exp\left(\frac{2\pi i l(N-m)}{p^2+p+1}\right)$$
$$= \frac{1-p^{2\alpha}}{1-p^{-2\alpha-3}}\left(1 + \frac{p^{-\alpha}-1}{p^2+p+1}\right)$$
$$+ p^{2\alpha}(p^{-\alpha}-1)\left(\Sigma - \frac{1}{p^2+p+1}\sum_{m=0}^{q-2}\chi(-p^{-1}f_2^m)\right),$$

where

$$\Sigma = \frac{1}{p^2+p+1}\sum_{m=0}^{q-2}\sum_{l=0}^{p(p+1)} \chi(-p^{-1}f_2^m)\exp\left(\frac{2\pi i l(N-m)}{p^2+p+1}\right).$$

Note that

$$\sum_{m=0}^{q-2}\chi(-p^{-1}f_2^m) = \sum_{m=0}^{q-2}\chi_p\left(-p^{-1}\operatorname{Tr}_{K/\mathbb{Q}_p} f_2^m\right). \tag{2.49}$$

The expression $\chi_p(p^{-1}x)$, $x \in \mathbb{Z}_p$, defines a non-trivial additive character $\varkappa$ on $\mathbb{Z}_p/p\mathbb{Z}_p$. Since K is unramified, the sum in (2.49) equals

$$\sum_{m=0}^{q-2}\varkappa(-\operatorname{tr}(\lambda^m)),$$

that is a sum of values of a non-trivial additive character $\varkappa(-\operatorname{tr}(\cdot))$ on all non-zero elements of a finite field O/P. Therefore

$$\sum_{m=0}^{q-2}\chi(-p^{-1}f_2^m) = -1.$$

Let us calculate Σ. We have, as above,

$$\sum_{l=0}^{p(p+1)}\exp\left(\frac{2\pi i l(N-m)}{p^2+p+1}\right)$$
$$= \begin{cases} p^2+p+1, & \text{if } m \equiv N \pmod{p^2+p+1}, \\ 0, & \text{otherwise,}\end{cases}$$

so that

$$\Sigma = \sum_{\substack{0\le m\le q-2 \\ m\equiv N \pmod{p^2+p+1}}} \chi\left(-p^{-1}f_2^m\right) = \sum_{j=0}^{p-2} \chi\left(-p^{-1}f_2^N \cdot f_2^{j(p^2+p+1)}\right)$$

$$= \sum_{j=0}^{p-2} \varkappa(-\operatorname{tr}(\lambda^N\delta^j)),$$

where $\delta = \lambda^{p^2+p+1}$. Since the elements $0, 1, \delta, \delta^2, \dots, \delta^{p-2}$ form a subfield in O/P, which coincides with $\mathbb{Z}_p/p\mathbb{Z}_p$, it is clear that

$$\Sigma = \begin{cases} p-1, & \text{if } \operatorname{tr}(\lambda^N) = 0; \\ -1, & \text{if } \operatorname{tr}(\lambda^N) \neq 0. \end{cases} \tag{2.50}$$

Thus we have proved that

$$\rho^{-1}G(u) = \frac{1-p^{2\alpha}}{1-p^{-2\alpha-3}}\left(1 + \frac{p^{-\alpha}-1}{p^2+p+1}\right) + p^{2\alpha}(p^{-\alpha}-1)(\Sigma + (p^2+p+1)^{-1}), \tag{2.51}$$

where Σ is determined by (2.50). Now the hypersingular integral representation of the operator A follows from (2.37).

Virtually the same calculations,together with Proposition 2.7,lead to the formula for the fundamental solution $E(x)$ of the operator $\mathfrak{A}$. If $\alpha \neq 3/2$ then

$$E(x) = \rho^{-1}\|x\|^{2\alpha/3-1}g(u_x)\ , \quad x \in K, \tag{2.52}$$

and for $u_x = f_2^N y$, $\|1-y\| < 1$,

$$g(u_x) = \left(1 - \frac{1-p^\alpha}{p^2+p+1}\right)\frac{1-p^{-2\alpha}}{1-p^{2\alpha-3}} + \frac{(p^\alpha-1)p^{-2\alpha+3}}{p^2+p+1},$$

if $\operatorname{tr}(\lambda^N) = 0$ and

$$g(u_x) = \left(1 - \frac{1-p^\alpha}{p^2+p+1}\right)\frac{1-p^{-2\alpha}}{1-p^{2\alpha-3}} - \frac{(p^\alpha-1)(p+1)p^{-2\alpha+1}}{p^2+p+1},$$

if $\operatorname{tr}(\lambda^N) \neq 0$. If $\alpha = 3/2$ then

$$E(x) = \frac{(p+2)(1-p)}{3p^2\log p}\rho^{-1}\log\|x\| + \frac{(p-1)\rho^{-1}}{p^3(p^2+p+1)}g_1(u_x), \tag{2.53}$$

$$g_1(u_x) = \begin{cases} p^3, & \text{if } \operatorname{tr}(\lambda^N) = 0 \\ -p(p+1), & \text{if } \operatorname{tr}(\lambda^N) \neq 0. \end{cases}$$

As before, a fundamental solution $\mathcal{E}$ for the operator A is given by (2.38). However this time the expression for $\mathcal{E}$ is not completely explicit since we cannot explicitly find the matrix T. Now we shall present another approach giving $\mathcal{E}$ explicitly under an additional assumption (which can turn out to be superfluous).

Suppose that $\mathbb{F}_q$ has a self-dual normal primitive basis over $\mathbb{F}_p$ (see Sect. 1.2). Then the field K possesses a basis $\{e_1, e_2, e_3\}$ of coordinate representation which is a canonical basis: the classes $\widehat{e}_1$, $\widehat{e}_2$, $\widehat{e}_3$ of elements e_j in $\mathbb{F}_q = O/P$ form a basis in $\mathbb{F}_q$ over $\mathbb{F}_p$.

Indeed, the self-duality of the basis $\{\lambda_1, \lambda_2, \lambda_3\}$ and Theorem 1.2 imply the existence of such non-zero elements $\zeta_{12}, \zeta_{13}, \zeta_{23} \in \mathbb{F}_q$ that

$$\lambda_1\lambda_2 = \zeta_{12}^p - \zeta_{12}\ , \quad \lambda_1\lambda_3 = \zeta_{13}^p - \zeta_{13}\ , \quad \lambda_2\lambda_3 = \zeta_{23}^p - \zeta_{23}. \tag{2.54}$$

The group of units U contains such an element u_0 that $u_0^{q-1} = 1$, and the set $M = \{u_0, u_0^2, \dots, u_0^{q-1}\}$ is a complete systems of representatives for residue classes from O/P. Let $s_{12}, s_{13}, s_{23} \in M$ be representatives of the classes $\zeta_{12}, \zeta_{13}, \zeta_{23}$ respectively. Then $s_{ij}^p = \sigma_K(s_{ij})$ where σ_K is the Frobenius automorphism.

Let $e_1^0, e_2^0, e_3^0 \in U$ be arbitrary representatives of the classes λ_1, λ_2, λ_3. The identities (2.54) imply that

$$e_i^0 e_j^0 - (s_{ij}^p - s_{ij}) \equiv 0 \pmod{P}\ , \quad 1 \le i < j \le 3. \tag{2.55}$$

Consider the system of equations

$$e_i e_j - (s_{ij}^p - s_{ij}) = 0\ , \quad 1 \le i < j \le 3, \tag{2.56}$$

with respect to $e_1, e_2, e_3 \in K$. The determinant of the corresponding Jacobi matrix equals $-2e_1e_2e_3$; for $e_j = e_j^0$ it is an invertible element of the ring O (since $p \neq 2$). It follows from Theorem 1.7 that there exists a solution $\{e_j\}$ of (2.56) with $e_j - e_j^0 \in P$, $j = 1, 2, 3$. Since $s_{ij}^p = \sigma_K(s_{ij})$ we have

$$\operatorname{Tr}(e_1e_2) = \operatorname{Tr}(e_1e_3) = \operatorname{Tr}(e_2e_3) = 0.$$

On the other hand,$\widehat{e}_j = \lambda_j$, $j = 1, 2, 3$.

It follows from the construction of the basis $\{e_1, e_2, e_3\}$ (and from the fact that K is not ramified over $\mathbb{Q}_p$) that

$$|\operatorname{Tr}(e_j^2)|_p = 1\ , \quad j = 1, 2, 3.$$

Let us construct the operator $\mathfrak{A}$ just as it was done above but with the basis $\{e_1, e_2, e_3\}$ instead of $\{f_1, f_2, f_3\}$. In this case $\rho = 1$, T equals the identity matrix whence

$$\mathcal{E}(x_1, x_2, x_3) = E(x_1 e_1 + x_2 e_2 + x_3 e_3).$$

Repeating the above arguments we obtain for G and E the formulas identical to (2.51), (2.52), (2.53), with $\rho = 1$, and $\mathrm{tr}\left(\lambda_1^{N+1}\right)$ instead of $\mathrm{tr}\left(\lambda^N\right)$.

The formulas for the fundamental solution will become more transparent if we express the condition $\mathrm{tr}(\lambda_1^{N+1}) = 0$ for N related to $x \in K, x = x_1 e_1 + x_2 e_2 + x_3 e_3$, $x_j \in \mathbb{Q}_p$, in terms of the coefficients x_j.

Lemma 2.2. *Let* $u_x = e_1^N y$, $\|1 - y\| < 1$. *The equality* $\mathrm{tr}(\lambda_1^{N+1}) = 0$ *is equivalent to the inequality*

$$|x_1|_p < \max\{|x_2|_p, |x_3|_p\}. \tag{2.57}$$

Proof. If $N = p$ or $N = p^2$ then $\mathrm{tr}(\lambda_1^{N+1}) = 0$ because the basis $\{\lambda_1, \lambda_2, \lambda_3\}$ is self-dual, $\lambda_2 = \lambda_1^p, \lambda_3 = \lambda_1^{p^2}$. Consider the set $\mathfrak{N} = \ker \mathrm{tr} \subset \mathbb{F}_q$ which is a vector subspace in $\mathbb{F}_q$ over $\mathbb{F}_p$. Obviously $\dim \mathfrak{N} \leq 2$. On the other hand, the vectors $\lambda_1^{p+1} \in \mathfrak{N}$ and $\lambda_1^{p^2+1} \in \mathfrak{N}$ are linearly independent over $\mathbb{F}_p$: otherwise it would mean that $\lambda_1^{p^2-p} \in \mathbb{F}_p$ which contradicts the description of $\mathbb{F}_p \setminus \{0\}$ as the set of elements $\lambda_1^{\nu(p^2+p+1)}$, $\nu = 0, 1, \dots, p-2$.

Hence,the set $\{x \in K \ : \ \mathrm{tr}(\lambda_1^{N+1}) = 0\}$ consists of those x for which

$$\lambda_1^{N+1} = \sigma_1 \lambda_1 \lambda_2 + \sigma_2 \lambda_1 \lambda_3, \quad \sigma_1, \sigma_2 \in \mathbb{F}_p.$$

This is equivalent to

$$u_x \equiv s_2 e_2 + s_3 e_3 \pmod{P}$$

where $s_2, s_3 \in \mathbb{Q}_p$, $\max\{|s_2|_p, |s_3|_p\} = 1$. It means that

$$x = p^{-m}(s_2 e_2 + s_3 e_3) + x' , \quad \|x\| = q^m , \ \|x'\| < q^m$$

(for an unramified extension $\beta = p$). This equality means that the coefficients of the expansion $x = x_1 e_1 + x_2 e_2 + x_3 e_3$ satisfy the inequality (2.57). ■

c) *Quaternary forms*

As we know (see Sect. 1.4), a quadratic form $h(\xi_1, \xi_2, \xi_3, \xi_4)$ satisfying the ellipticity condition (2.40) is unique up to a linear isomorphism. It may be written as

$$h(\xi_1, \xi_2, \xi_3, \xi_4) = \xi_1^2 - s\xi_2^2 - p\xi_3^2 + sp\xi_4^2,$$

where $s \in \mathbb{Z}$ is a quadratic non-residue mod p.

Let $K = \left(\frac{s,p}{\mathbb{Q}_p}\right)$ be the quaternion algebra over $\mathbb{Q}_p$, with the basis $1, i, j, k$ and the multiplication table

$$i^2 = s \ , \ j^2 = p \ , \ ij = -ji = k \ , \ k^2 = -sp,$$
$$ik = -ki = js \ , \ jk = -kj = -ip.$$

K is a central division algebra over $\mathbb{Q}_p$, with the normalized absolute value

$$\|x\| = |h(x_1, x_2, x_3, x_4)|_p^2 \ , \quad x = x_1 + x_2 i + x_3 j + x_4 k.$$

In this case $q = p^2$, $\beta = j$. The reduced trace $t \ : \ K \to \mathbb{Q}_p$ is $t(x) = 2x_1$.

The basis of coordinate representation can be chosen as

$$e_1 = \frac{1}{2}(1 - j) \ , \ e_2 = \frac{1}{2}(1 + j) \ , \ e_3 = \frac{1}{2s}(i - k) \ , e_4 = \frac{1}{2}(i + k).$$

It is easy to show using Lemma 1.1 that $\mu_1 = -1$, $\mu_2 = 1$, $\mu_3 = -s^{-1}$, $\mu_4 = s$, $c = 1$. Taking $\mathfrak{A} = D_K^{\alpha/2}$,

$$T = \begin{pmatrix} -1 & 0 & 1 & 0 \\ 1 & 0 & 1 & 0 \\ 0 & -1 & 0 & 1 \\ 0 & s & 0 & s \end{pmatrix}, \quad |\det T|_p = 1,$$

we obtain a hypersingular integral representation of the operator A with $\Omega(u) \equiv \dfrac{1 - p^{\alpha}}{1 - p^{-\alpha-2}}$. The fundamental solution can be written as

$$\mathcal{E}(x_1, x_2, x_3, x_4) = E\left(\frac{x_3 - x_1}{2} e_1 + \frac{x_1 + x_3}{2} e_2 + \frac{x_4 - x_2}{2} e_3 + \frac{x_2 + x_4}{2s} e_4\right),$$

$$E(x) = \begin{cases} \dfrac{1 - p^{-\alpha}}{1 - p^{\alpha-2}} \|x\|^{\alpha/2-1}, & \text{if } \alpha \neq 2; \\ \dfrac{1 - p^2}{2p^2 \log p} \log \|x\|, & \text{if } \alpha = 2. \end{cases}$$

Since there are no quadratic forms over $\mathbb{Q}_p$ satisfying (2.40) with $n \geq 5$, the above results complete the proof of Theorem 2.3. ■

2.8 Green functions of elliptic operators

As in the previous section, we shall consider the pseudo-differential operator A with the symbol $|h(\xi_1, \dots \xi_n)|_p^\alpha$, where $\alpha > 0$ and h is a quadratic form satisfying (2.40). As in Sect. 2.5, we shall call a fundamental solution $\mathcal{E}_\lambda(x_1, \dots, x_n)$ of the equation

$$(A + \lambda)z = f, \quad \lambda > 0, \tag{2.58}$$

a *Green function* of the operator A.

The case $n = 1$ is of course contained in Proposition 2.8, and we have to study the cases when $n = 2, 3$ or 4. The method of coordinate representation developed above carries over virtually unchanged to the case of Green functions. In particular, the formula (2.38) remains valid with $\mathcal{E}_\lambda$ substituted for $\mathcal{E}$, and E_λ for E respectively. The quantities necessary in order to use Proposition 2.8 and the formula (2.38) have all been found in the course of proving Theorem 2.3.

We start with the following qualitative result. As before, we assume in this section that $p \neq 2$.

Theorem 2.4. *If the quadratic form h satisfies (2.40), then there exists a Green function $\mathcal{E}_\lambda(x_1, \dots, x_n)$ of the operator A, which is locally integrable, continuous for $(x_1, \dots, x_n) \neq 0$, non-negative, and tends to zero at infinity. If $f \in \mathcal{D}(\mathbb{Q}_p^n)$, then the function $x = \mathcal{E}_\lambda * f$ satisfies the equation (2.58), where A is understood as a hyper-singular integral operator (2.37).*

The *proof* follows from the above considerations. In particular, the non-negativity of $\mathcal{E}_\lambda$ is a consequence of Proposition 2.8 and the non-positivity of the function G appearing in (2.37) for each type of the quadratic form h. The inequality $G(u) \leq 0$ is evident from explicit formulas for G found in the proof of Theorem 2.3. ■

The results regarding the behaviour of $\mathcal{E}_\lambda$ near the origin and infinity are collected in the following theorems.

Theorem 2.5. *Let $n = 2$, $h(\xi_1, \xi_2) = \xi_1^2 - \tau\xi_2^2$. If $\tau = \varepsilon$, $|\varepsilon|_p = 1$, and ε is not a square, then $\mathcal{E}_\lambda(x_1, x_2)$ depends in fact on $< x >_\varepsilon = |x_1^2 - \varepsilon^{-1}x_2^2|_p$. $\mathcal{E}_\lambda$ is continuous at the origin if $\alpha > 1$; otherwise*

$$\mathcal{E}_\lambda(x_1, x_2) = \begin{cases} O\left(< x >_\varepsilon^{\alpha-1}\right), & 0 < \alpha < 1; \\ O\left(|\log < x >_\varepsilon|\right), & \alpha = 1, \end{cases}$$

$< x >_\varepsilon \to 0$. As $< x >_\varepsilon \to \infty$,

$$\mathcal{E}_\lambda(x_1, x_2) = \frac{p^{2(\alpha+1)}(p^{2\alpha} - 1)}{p^{2(\alpha+1)} - 1}\lambda^{-2} < x >_\varepsilon^{-\alpha-1} + O\left(< x >_\varepsilon^{-2\alpha-1}\right).$$

If $\tau = p$ then $\mathcal{E}_\lambda$ depends on $< x >_p = |x_2^2 - px_1^2|_p$. The behaviour of $\mathcal{E}_\lambda$ near the origin is as above, with $< x >_p$ substituted for $< x >_\varepsilon$. As $< x >_p \to \infty$,

$$\mathcal{E}_\lambda(x_1, x_2) = \frac{p^{\alpha+1}(p^\alpha - 1)}{p^{\alpha+1} - 1}\lambda^{-2} < x >_p^{-\alpha-1} + O\left(< x >_p^{-2\alpha-1}\right).$$

If $\tau = \varepsilon p$, the asymptotic behaviour is the same as for $\tau = p$, with $< x >_{\varepsilon p} = |x_2^2 - \varepsilon p x_1^2|_p$ substituted for $< x >_p$.

Theorem 2.6. *Let $n = 3$,*

$$h(\xi_1, \xi_2, \xi_3) = p\varepsilon_1\xi_1^2 + \varepsilon_2\xi_2^2 + \varepsilon_3\xi_3^2,$$

where $|\varepsilon_1|_p = |\varepsilon_2|_p = |\varepsilon_3|_p = 1$, $\varepsilon_2\xi_2^2 + \varepsilon_3\xi_3^2 \neq 0$ for $|\xi_2|_p + |\xi_3|_p \neq 0$. If $\alpha > 3/2$, then the function $\mathcal{E}_\lambda$ is continuous at the origin; otherwise

$$\mathcal{E}_\lambda(x_1, x_2, x_3) = \begin{cases} O\left(< x >^{2\alpha-3}\right), & 0 < \alpha < 3/2; \\ O\left(|\log < x > |\right), & \alpha = 3/2, \end{cases}$$

$< x > \equiv \max\limits_{j=1,2,3} |x_j|_p \to 0$.

For $< x > > 1$,

$$C_1 < x >^{-2\alpha-3} \leq \mathcal{E}_\lambda(x_1, x_2, x_3) \leq C_2 < x >^{-2\alpha-3}, \quad C_1, C_2 > 0. \quad (2.59)$$

If there exists a self-dual normal primitive basis of $\mathbb{F}_{p^3}$ over $\mathbb{F}_p$, then

$$\mathcal{E}_\lambda(x_1, x_2, x_3) = \lambda^{-2}\Big[p^{2\alpha} - 1 + \frac{(p^\alpha - 1)(p^{\alpha+1} - 1)}{p^{2\alpha+3} - 1} + (p^{2\alpha} - p^\alpha)\mu(x)\Big] < x >^{-2\alpha-3} + O\left(< x >^{-4\alpha-3}\right).$$

as $< x > \to \infty$, where

$$\mu(x) = \begin{cases} p - 1, & \text{if } < x > > |x_1|_p, \\ -1, & \text{if } < x > = |x_1|_p. \end{cases}$$

Theorem 2.7. *Let $n = 4$,*

$$h(\xi_1, \xi_2, \xi_3, \xi_4) = \xi_1^2 - s\xi_2^2 - p\xi_3^2 + sp\xi_4^2,$$

$s \in \mathbb{Z}$ is a quadratic nonresidue $\bmod p$. *The Green function $\mathcal{E}_\lambda(x_1,x_2,x_3,x_4)$ depends on $\ll x \gg = |spx_1^2 - px_2^2 - sx_3^2 + x_4^2|_p$. $\mathcal{E}_\lambda$ is continuous at the origin if $\alpha > 2$; otherwise*

$$\mathcal{E}_\lambda(x_1,x_2,x_3,x_4) = \begin{cases} O\left(\ll x \gg^{\alpha-2}\right), & 0 < \alpha < 2; \\ O\left(|\log \ll x \gg|\right), & \alpha = 2, \end{cases}$$

$\ll x \gg \to 0$. *As* $\ll x \gg \to \infty$,

$$\mathcal{E}_\lambda(x_1,x_2,x_3,x_4) = \frac{p^{\alpha+2}(p^\alpha - 1)}{p^{\alpha+2}-1}\lambda^{-2} \ll x \gg^{-\alpha-2} + O\left(\ll x \gg^{-2\alpha-2}\right).$$

Among all the assertions of Theorems 2.5-2.7 there is only one still requiring a *proof*, namely the lower bound (2.59) in Theorem 2.6. By virtue of (2.38) and (2.25), it is sufficient to show that the expression in square brackets in (2.25) with $q = p^3$, $\nu = 2\alpha/3$, $m = 1$,

$$c_{01} = 1 + \frac{p^{-\alpha}-1}{p^2+p+1},$$

$$c_{1l} = \begin{cases} \dfrac{p^{-\alpha}-1}{p^2+p+1}, & l = \nu(p-1),\ \nu = 1, \dots, p(p+1); \\ 0, & \text{for the remaining } l \le p^3 - 2, \end{cases}$$

is bounded from below by a positive constant independent of x. Thus we have to obtain a lower estimate for the expression

$$\begin{aligned} \Xi = {} & \left(1 + \frac{p^{-\alpha}-1}{p^2+p+1}\right) \frac{p^{2\alpha+3}(p^{2\alpha}-1)}{p^{-2\alpha+3}-1} \\ & - p^{2\alpha+3}\frac{p^{-\alpha}-1}{p^2+p+1} \sum_{\nu=1}^{p(p+1)} \Gamma\left(1, \psi_{\nu(p-1)}\right) \overline{\psi_{\nu(p-1)}(-u_x)}, \end{aligned} \tag{2.60}$$

where $\psi_{\nu(p-1)}(-u_x) = \exp\left(\frac{2\pi i \nu N}{p^2+p+1}\right)$, N depends on x, $0 \le N \le q-2$ (see the proof of Theorem 2.3),

$$\begin{aligned} \Gamma\left(1, \psi_{\nu(p-1)}\right) &= \int\limits_{\|y\|=1} \chi(p^{-1}y)\psi_{\nu(p-1)}(y)\, dy \\ &= q^{-1} \sum_{k=0}^{q-2} \chi\left(p^{-1} f_2^k\right) \exp\left(\frac{2\pi i \nu k}{p^2+p+1}\right). \end{aligned}$$

We find that

$$\Xi_1 = \sum_{\nu=1}^{p(p+1)} \Gamma\left(1, \psi_{\nu(p-1)}\right) \overline{\psi_{\nu(p-1)}(-u_x)}$$

$$= q^{-1} \sum_{k=0}^{q-2} \chi\left(p^{-1} f_2^k\right) \left[\sum_{\nu=0}^{p(p+1)} \exp\left(\frac{2\pi i \nu (k-N)}{p^2+p+1}\right) - 1 \right].$$

The sum with respect to ν is different from zero (and equal to p^2+p+1) only for $k \equiv N \pmod{(p^2+p+1)}$. On the other hand, $\sum\limits_{k=0}^{q-2} \chi\left(p^{-1} f_2^k\right) = -1$ as the sum of the values of a non-trivial addtitive character on the non-zero elements of a field. Thus

$$\Xi_1 = q^{-1}(p^2+p+1) \sum_{j=0}^{p-2} \chi\left(p^{-1} f_2^{j(p^2+p+1)+N}\right) + q^{-1}. \tag{2.61}$$

As we know (see (2.50)), the sum in (2.61) equals either $p-1$ or -1, depending on the value of N, whence

$$\Xi_1 \le p^{-3}(p^2+p+1)(p-1) + p^{-3} = 1.$$

Substituting into (2.60) we find that

$$\Xi \ge \frac{p^{2\alpha+3}(p^{2\alpha}-1)}{p^{2\alpha+3}-1} + \frac{(1-p^{-\alpha})p^{4\alpha+3}(p^3-1)}{(p^{2\alpha+3}-1)(p^2+p+1)} > 0. \quad \blacksquare$$

2.9 Hyperbolic equations

In this section we shall consider the equation $Az = f$, where A is a pseudo-differential operator over $\mathbb{Q}_p^n$ with the symbol $P_\alpha(\xi_1, \dots, \xi_n) = |h(\xi_1, \dots, \xi_n)|_p^\alpha$, corresponding to the quadratic form

$$h(\xi_1, \dots, \xi_n) = \xi_1^2 - g(\xi_2, \dots, \xi_n), \tag{2.62}$$

$$g(\xi_2, \dots, \xi_n) \neq 0 \quad \text{if } (\xi_2, \dots, \xi_n) \neq 0. \tag{2.63}$$

As we know, quadratic forms (2.62) satisfying (2.63) exist only for $n \le 5$.

The case when $n = 2$ is quite simple. For example, if $h(\xi_1, \xi_2) = \xi_1^2 - \xi_2^2$, then just as for the wave equation over $\mathbb{R}^2$, A can be transformed (by the obvious change of variables) into the product of two one-dimensional D^α-operators. This leads to a hyper-singular integral representation

$$(Au)(x,t) = \left(\frac{1-p^\alpha}{1-p^{-\alpha-1}}\right)^2 \int_{\mathbb{Q}_p^2} |y\theta|_p^{-\alpha-1} [u(x-y-\theta, t-y+\theta)$$

$$-u(x-\theta, t+\theta) - u(x-y, t-y) + u(x,t)]\, dy\, d\theta$$

and the formulas for a fundamental solution (understood just as in the elliptic case):

$$G(x,t) = \left(\frac{1-p^{-\alpha}}{1-p^{\alpha-1}}\right)^2 |x+t|_p^{\alpha-1} |x-t|_p^{\alpha-1} ,$$

if $\alpha \neq 1$,or

$$G(x,t) = \left(\frac{p-1}{p\log p}\right)^2 (\log|x+t|_p)\log|x-t|_p ,$$

if $\alpha = 1$.

The case when $n > 2$ is more difficult because a solution may be a true distribution, and the action of A on $\mathcal{D}'(\mathbb{Q}_p^n)$ is not defined.Denote

$$C = \{\xi \in \mathbb{Q}_p^n : \ \xi_1^2 = g(\xi_2, \dots, \xi_n)\},$$

$$\mathcal{D}_C(\mathbb{Q}_p^n) = \{\varphi \in \mathcal{D}(\mathbb{Q}_p^n) : \ \widetilde{\varphi}(\xi) = 0 \ , \ \xi \in C\}.$$

This space is a closed subspace of $\mathcal{D}(\mathbb{Q}_p^n)$. Of course, the Fourier transform does not preserve $\mathcal{D}_C(\mathbb{Q}_p^n)$. However for the operator $A = F^{-1}P_\alpha F$ the opposite is true.

Proposition 2.9. *The operator A acts continuously from $\mathcal{D}_C(\mathbb{Q}_p^n)$ to $\mathcal{D}_C(\mathbb{Q}_p^n)$.*

Proof. Let $\varphi \in \mathcal{D}_C(\mathbb{Q}_p^n)$. Then $\widetilde{\varphi} \in \mathcal{D}(\mathbb{Q}_p^n)$. Suppose that

$$\operatorname{supp}\widetilde{\varphi} \subset \{\xi \in \mathbb{Q}_p^n \ : \ \max_{1\le j\le n} |\xi_j|_p \le p^N\},$$

$$\widetilde{\varphi}(\xi+\xi') = \widetilde{\varphi}(\xi) \quad \text{if} \quad \max_{1\le j\le n} |\xi_j'|_p \le p^l,$$

$N, l \in \mathbb{Z}$. We may assume that $N \geq 0$, $l \leq N-1$.

Let us show that there exists such $\nu \in \mathbb{Z}$ depending only on N and l that

$$\widetilde{\varphi}(\xi_1, \dots, \xi_n) = 0 \quad \text{if} \quad |h(\xi_1, \dots, \xi_n)|_p < p^\nu.$$

It is sufficient to consider only $\xi_1, \dots, \xi_n$ with $\max_{1\le j\le n} |\xi_j|_p \le p^N$. Denote $\eta_j = p^N\xi_j$, $j = 1, \dots, n$. Then $\eta_j \in \mathbb{Z}_p$. If $|h(\xi_1, \dots, \xi_n)|_p < p^\nu$ then $|h(\eta_1, \dots, \eta_n)|_p < p^{\nu+2N}$.

There are two options: either

$$\max_{1\le j\le n} |\eta_j|_p \le p^{l-N}$$

implying that $\max_{1\le j\le n} |\xi_j|_p \le p^l$ and $\widetilde{\varphi}(\xi_1, \dots, \xi_n) = \widetilde{\varphi}(0, \dots, 0) = 0$,- or at least for some j $|\eta_j|_p \ge p^{l-N+1}$. Suppose for example that $|\eta_1|_p = p^k$, $l - N + 1 \le k \le 0$.

If $\nu < 2k - 2N$ we can use Theorem 1.7′, which guarantees the existence of such $\eta_j^0 \in \mathbb{Z}_p$, $j = 1, \dots, n$, that

$$h(\eta_1^0, \dots, \eta_n^0) = 0 \ , \quad \max |\eta_j^0 - \eta_j|_p \le p^{-k+\nu+2N}.$$

Set $\xi_j^0 = p^{-N}\eta_j^0$, $j = 1, \dots, n$. Then $(\xi_1^0, \dots, \xi_n^0) \in C$,

$$\max_{1\le j\le n} |\xi_j - \xi_j^0|_p \le p^{N+\nu-k} \le p^{2N+\nu-l-1}.$$

Now if we choose $\nu \le 2l - 4N + 1$ then we get $\nu < 2k - 2N$ (the condition for validity of the preceding construction),

$$\max_{1\le j\le n} |\xi_j - \xi_j^0|_p \le p^l$$

whence $\widetilde{\varphi}(\xi_1, \dots, \xi_n) = 0$.

The function P_α is locally constant on a neighbourhood of the support of $\widetilde{\varphi}$. This implies the inclusion $P_\alpha\widetilde{\varphi} \in \mathcal{D}(\mathbb{Q}_p^n)$. Hence $A\varphi \in \mathcal{D}_C(\mathbb{Q}_p^n)$. The continuity of A follows from the definition of topology in $\mathcal{D}_C(\mathbb{Q}_p^n)$ induced from $\mathcal{D}(\mathbb{Q}_p^n)$. ■

The conjugate $A' : \mathcal{D}'_C(\mathbb{Q}_p^n) \to \mathcal{D}'_C(\mathbb{Q}_p^n)$ is an extension of the operator A. Note that the Fourier transform is not defined on $\mathcal{D}'_C(\mathbb{Q}_p^n)$ since it does not preserve $\mathcal{D}_C(\mathbb{Q}_p^n)$.

The convolution of test functions and the inversion $\check{\varphi}(x) = \varphi(-x)$ are well-defined in $\mathcal{D}_C(\mathbb{Q}_p^n)$. In a standard way we can define the convolution of a distribution and a test function. Below we shall understand a fundamental solution as a solution $E \in \mathcal{D}'_C(\mathbb{Q}_p^n)$ of the equation $A'E = \delta$. For a fundamental solution E and a test function $f \in \mathcal{D}_C(\mathbb{Q}_p^n)$ the function $z = E * f$ considered as a distribution from $\mathcal{D}'_C(\mathbb{Q}_p^n)$ satisfies the equation $A'z = f$.

In order to construct a fundamental solution $E \in \mathcal{D}'_C(\mathbb{Q}_p^n)$ it is sufficient to consider the function

$$P_{-\alpha}(\xi_1, \dots, \xi_n) = |h(\xi_1, \dots, \xi_n)|_p^{-\alpha}$$

as a distribution from $\mathcal{D}'(\mathbb{Q}_p^n \setminus C)$ and to set

$$E = FP_{-\alpha}, \tag{2.64}$$

where the Fourier transform acting isomorphically from $\mathcal{D}_C(\mathbb{Q}_p^n)$ to $\mathcal{D}(\mathbb{Q}_p^n \setminus C)$ is extended to the action $\mathcal{D}'(\mathbb{Q}_p^n \setminus C) \to \mathcal{D}'(\mathbb{Q}_p^n)$ (note that the functions h and P_λ are even,so that here we need not distinguish between the direct and inverse transforms).

According to Theorem 1.15, the function P_λ, $\lambda \in (0, \infty)$, with values in $\mathcal{D}'(\mathbb{Q}_p^n)$, admits analytic continuation to a meromorphic function on $\mathbb{C}$. Its only real poles are at the points $\lambda = -1$ and $\lambda - -n/2$. This implies the following result.

Theorem 2.8. *Suppose that $\alpha \neq 1$ and $\alpha \neq n/2$. Then the fundamental solution (2.64) can be extended to a distribution from $\mathcal{D}'(\mathbb{Q}_p^n)$.*

2.10 Cauchy problem for an equation of the Schrödinger type

2.10.1. Classes of functions. In this section we consider the initial value problem

$$(Az)(x, t) = f(x, t), \quad x \in \mathbb{Q}_p^n, \ t \in \mathbb{Q}_p, \tag{2.65}$$

$$z(x, 0) = z_0(x), \tag{2.66}$$

where $n \geq 1$, $f \in \mathcal{D}(\mathbb{Q}_p^{n+1})$, $z_0 \in \mathcal{D}(\mathbb{Q}_p)$,

$$(Az)(x,t) = F^{-1}_{(\xi,\tau)\to(x,t)} \left[|\tau - \left(a_1\xi_1^2 + \dots + a_n\xi_n^2\right)|_p^\alpha F_{(y,\theta)\to(\xi,\tau)} z \right], \tag{2,67}$$

$\alpha > 0$, $0 \neq a_j \in \mathbb{Q}_p$ $(j = 1, \dots, n)$.

As in the cases of elliptic and hyperbolic operators, our first task is to find a hypersingular integral representation of the operator A, which would extend it to a wider class of functions. In the present case two such representations will be given. One of them maintains the Fourier transform with respect to the spatial variables (thus the operator remains pseudo-differential with respect to them). The second representation resembles the ones appearing in the preceding sections, being hypersingular integral with respect to all the variables.

Denote by $\mathcal{B}_\alpha$ the class of such measurable complex-valued functions $z(x,t)$, $x \in \mathbb{Q}_p^n$, $t \in \mathbb{Q}_p$, belonging to $L_2(\mathbb{Q}_p^n)$ for each t , that the partial Fourier transform $v(y,t) = F_{x\to y} z(x,t)$ possesses the following properties. There exist N , $l \in \mathbb{Z}$ and a positive function $c(t)$ (depending on u) such that

$$v(y,t) = 0 \quad \text{for all } t \in \mathbb{Q}_p \ \text{ if } \|y\| > p^N \text{ ; here } \|y\| = \max_{1\le j\le n} |y_j|_p \ ; \tag{2.68}$$

$$|v(y,t)| \le c(t) \ , \ y \in \mathbb{Q}_p^n \ , \ t \in \mathbb{Q}_p \ ; \ \int\limits_{|t|_p>1} c(t)|t|_p^{-\alpha-1} dt \ < \ \infty \ ; \tag{2.69}$$

$$v(y,t+t') = v(y,t) \text{ for all } y \in \mathbb{Q}_p^n \ , \ t \in \mathbb{Q}_p \ , \text{if } |t'|_p \ \le \ p^l \ . \tag{2.70}$$

Note that the conditions (2.68) , (2.70) imply that

$$z(x,t) = \int\limits_{\mathbb{Q}_p^n} \chi_p(-x\cdot y) v(y,t)dy, \quad x\cdot y = x_1y_1 + \cdots + x_ny_n,$$

is jointly locally constant: there exists such a number $\nu \in \mathbb{Z}$ that

$$u(x+x', t+t') = u(x,t)$$

for all $x \in \mathbb{Q}_p^n$, $t \in \mathbb{Q}_p$ if $\|x'\| \le p^\nu$, $|t'|_p \le p^\nu$.

Lemma 2.3. *If $z \in \mathcal{B}_\alpha$ then the expression*

$$(A_1 z)(x,t) = \frac{1-p^\alpha}{1-p^{-\alpha-1}}$$
$$\times F_{y\to x}^{-1} \int\limits_{\mathbb{Q}_p} |\tau|_p^{-\alpha-1} \left[\chi_p(-\tau(a_1y_1^2 + \cdots + a_ny_n^2))v(y,t-\tau) - v(y,t)\right] d\tau \tag{2.71}$$

exists in the sense of absolute convergence of the double integral. If $z \in \mathcal{D}(\mathbb{Q}_p^{n+1})$ then $A_1 z = Az$.

Proof. Due to the condition (2.68) the integration in the inverse Fourier transform is conducted over the set $\{ \|y\| \le p^N\}$. If $|\tau|_p \le p^{-N_1}$, $\|y\| \le p^N$, where $N_1 = 2N + \max_{1\le j\le n} \log_p |a_1|_p$, then $\chi_p(-\tau(a_1y_1^2+\cdots+a_ny_n^2)) = 1$. Now the condition (2.70) implies that the integral in (2.71) is actually taken over the set $\{|\tau|_p > p^{l_1}\}$, $l_1 = \min\{l, -N_1\}$. So the integral contains no singularity,and the absolute convergence is guaranteed by the condition (2.69).

If $z \in \mathcal{D}(\mathbb{Q}_p^{n+1})$ then

$$
\begin{aligned}
(A_1 z)(x,t) &\\
= F^{-1}_{y\to x} &\{\chi_p(-t(a_1y_1^2 + \cdots + a_ny_n^2))F^{-1}_{\tau\to t}[\, |\tau|_p^\alpha F_{\theta\to\tau}(\chi_p(\theta(a_1y_1^2 \\
&+ \cdots + a_ny_n^2))v(y,\theta))\,]\}
\end{aligned}
$$

in accordance with (2.8). Observe that

$$
F_{\theta\to\tau}\left[\chi_p(\theta(a_1y_1^2 + \cdots + a_ny_n^2))v(y,\theta)\right] = F_{\theta\to\tau+a_1y_1^2+\cdots+a_ny_n^2}\left[v(y,\theta)\right].
$$

After the change of variables we come to the expression (2.71). ■

In a sequel we shall omit the subscript identifying the operator A with its extension A_1 when the latter exists.

Let us introduce a more narrow class $\mathcal{C}_\alpha$ of functions for which the Fourier transform can be eliminated from the definition of the operator A.

The class $\mathcal{C}_\alpha$ consists of functions $z \in \mathcal{B}_\alpha$ possessing the following additional properties. For every $z \in \mathcal{C}_\alpha$ there exist such finite functions $a(t) \ge 1$, $b(t) \ge 0$ $(t \in \mathbb{Q}_p)$ depending only on $|t|_p$ that:

$$
z(x,t) = 0 \quad \text{if } \|x\| \ge a(t)\ ; \tag{2.72}
$$

$$
|z(x,t)| \le b(t) \quad \text{for all } x \in \mathbb{Q}_p^n\ ,\ t \in \mathbb{Q}_p; \tag{2.73}
$$

$$
\int\limits_{|t|_p>1} [a(t)]^n b(t) |t|_p^{-\alpha-n/2-1} dt \ < \ \infty. \tag{2.74}
$$

We shall say that the integral $\int\limits_{\mathbb{Q}_p} f(x)\,dx$ of a locally integrable complex-valued function f is conditionally convergent if the limit

$$
I = \lim_{N\to\infty} \int\limits_{\|x\|\le p^N} f(x)\,dx
$$

exists in $\mathbb{C}$. Then by definition

$$\int_{\mathbb{Q}_p} f(x)\,dx = I.$$

Lemma 2.4. *If $z \in \mathcal{C}_\alpha$ then*

$$(Az)(x,t)$$
$$= \frac{1-p^{\alpha}}{1-p^{-\alpha-1}} \prod_{i=1}^{n} |a_i|_p^{-1/2} \int_{\mathbb{Q}_p} \left\{ \prod_{j=1}^{n} \lambda_p(-\tau a_j) \right\} |\tau|_p^{-\alpha-n/2-1}\, d\tau$$
$$\times \int_{\mathbb{Q}_p^n} \left\{ \prod_{k=1}^{n} \chi_p\left(\frac{y_k^2}{4\tau a_k}\right) \right\} [z(x-y,t-\tau) - z(x,t)]\, dy \quad (2.75)$$

(the function λ_p was defined in Sect. 1.8). The integral is understood as an iterated one,with the conditionally convergent internal integral.

Proof. Denote the internal integral in (2.75) by $w(x,t,\tau)$. Then for $\tau \neq 0$ we have $w(x,t,\tau) = w_1(x,t,\tau) - w_2(x,t,\tau)$, where

$$w_1(x,t,\tau) = \int_{\mathbb{Q}_p^n} \left\{ \prod_{j=1}^{n} \chi_p\left(\frac{y_j^2}{4\tau a_j}\right) \right\} z(x-y,t-\tau)\, dy$$

(the integral is absolutely convergent due to the condition (2.72)),

$$w_2(x,t,\tau) = z(x,t) \prod_{j=1}^{n} \int_{\mathbb{Q}_p} \chi_p\left(\frac{y_j^2}{4\tau a_j}\right) dy_j$$
$$= z(x,t)|\tau|_p^{n/2} \prod_{j=1}^{n} \left(\lambda_p\left(\frac{1}{\tau a_j}\right) |a_j|_p^{1/2} \right).$$

We have used the formula (1.39) and the relation $\lambda_p(c^2 a) = \lambda_p(a)$.

Let us study some properties of $w(x,t,\tau)$. For fixed x,t and $|\tau|_p \geq p^k$ we have the estimates

$$|w_1(x,t,\tau)| \leq \int_{\|x-y\| \leq a(t-\tau)} |z(x-y,t-\tau)|\, dy \leq c\,[a(t-\tau)]^n\, b(t-\tau)\ ,$$

$$|w_2(x,t,\tau)| \leq C|\tau|_p^{n/2}$$

so that

$$|w(x,t,\tau)| \leq Cb(\tau)\max\left\{[a(\tau)]^n\ ,\ |\tau|_p^{n/2}\right\}\ ,\ |\tau|_p \geq p^k, \tag{2.76}$$

where a positive constant C depends on x,t and k.

If $|\tau|_p \leq p^\nu$ then due to the local constancy of z

$$w(x,t,\tau) = \int\limits_{\|y\|>p^\nu} \left\{\prod_{j=1}^n \chi_p\left(\frac{y_j^2}{4\tau a_j}\right)\right\}[z(x-y,t-\tau) - z(x,t)]\,dy$$

$$= w_{1,\nu}(x,t,\tau) + w_{2,\nu}(x,t,\tau),$$

where

$$w_{1,\nu}(x,t,\tau) = \int\limits_{\|y\|>p^\nu} \left\{\prod_{j=1}^n \chi_p\left(\frac{y_j^2}{4\tau a_j}\right)\right\} z(x-y,t-\tau)\,dy,$$

$$w_{2,\nu}(x,t,\tau) = z(x,t) \int\limits_{\|y\|>p^\nu} \prod_{j=1}^n \chi_p\left(\frac{y_j^2}{4\tau a_j}\right)\,dy.$$

Fix x,t and consider the function $\Phi(y,\tau) = z(x-y,t-\tau)$, $\|y\| > p^\nu$, $|\tau|_p \leq p^\nu$. Since Φ has a compact support on its domain of definition and is locally constant we come to the conclusion that Φ can be represented in the form

$$\Phi(y,\tau) = \sum_{k=1}^m c_k\zeta_k(y,\tau)\quad ,\quad 1 \leq m < \infty\ ,$$

where $c_k \in \mathbb{C}$, $\zeta_k(y,\tau)$ is a characteristic function of a set

$$V_k = \left\{(y,\tau) \in \mathbb{Q}_p^{n+1} :\ \|y - y^{(k)}\| \leq p^{\nu_k}\ ,\ |\tau - \tau^{(k)}|_p \leq p^{\nu_k}\right\}$$

where $\|y^{(k)}\| > p^\nu$, $|\tau^{(k)}|_p \leq p^\nu$, $\nu_k < \nu$, and the sets V_k do not intersect each other. Then we have

$$|w_{1,\nu}(x,t,\tau)| \leq \sum_{k=1}^m \left| c_k \int\limits_{\|y-y^{(k)}\|\leq p^{\nu_k}} \prod_{j=1}^n \chi_p\left(\frac{y_j^2}{4\tau a_j}\right) dy \right|$$

$$= \sum_{k=1}^m \left| c_k \prod_{j=1}^n \left\{ \chi_p\left(\frac{y_j^{(k)2}}{4\tau a_j}\right) \int\limits_{|z_j|_p\leq p^{\nu_k}} \chi_p\left(\frac{z_j^2}{4\tau a_j} + \frac{z_j y^{(k)}}{2\tau a_j}\right) dz_j \right\} \right|.$$

Using (1.38) we obtain that

$$w_{1,\nu}(x,t,\tau) = 0 \quad \text{if } |\tau|_p < \min_{j,k} p^{2\nu_k} |a_j|_p^{-1} .$$

It follows from (1.38) and (1.39) that

$$w_{2,\nu}(x,t,\tau) = 0 \quad \text{if } |\tau|_p < p^{2\nu} \min_j |a_j|_p^{-1} .$$

Hence $w(x,t,\tau) = 0$ when $|\tau|_p$ is small.This,together with the inequality (2.76), guarantees convergence of the external integral in (2.75).

Let us show that the right-hand sides of (2.71) and (2.75) are equal. Denote as before,

$$v(y,t) = F_{x\to y} z(x,t).$$

Then we obtain using (1.39) that

$$\begin{aligned} w_1(x,t,\tau) &= F^{-1}_{y\to x}\left\{ v(y,t-\tau) F_{z\to y}\left[\prod_{j=1}^n \chi_p\left(\frac{z_j^2}{4\tau a_j}\right)\right]\right\} \\ &= |\tau|_p^{n/2} \prod_{j=1}^n \left(\lambda_p\left(\frac{1}{\tau a_j}\right)|a_j|_p^{1/2}\right) \\ &\quad\times F^{-1}_{y\to x}\left\{\chi_p(-\tau(a_1x_1^2+\cdots+a_nx_n^2))v(y,t-\tau)\right\}, \end{aligned}$$

$$w_2(x,t,\tau) = |\tau|_p^{n/2} \prod_{j=1}^n \left(\lambda_p\left(\frac{1}{\tau a_j}\right)|a_j|_p^{1/2}\right) F^{-1}_{y\to x} v(y,t).$$

Note that for any $h \in \mathbb{Q}_p$, $h \neq 0$,

$$\lambda_p(h)\lambda_p(-h^{-1}) = 1 .$$

This identity is an easy consequence of the well-known properties [59] of the Legendre symbol.

Substitute the expression for $w = w_1 - w_2$ into (2.75). Then to reach the required formula (2.71) it remains only to interchange the inverse Fourier transform and integration with respect to τ. This is possible because $w(x,t,\tau) = 0$ when τ stays in a neighbourhood of zero,and $v(y,t)$ satisfies the condition (2.69). ■

2.10.2. The inhomogeneous equation. Let us consider the equation

(2.65) with $f \in \mathcal{D}(\mathbb{Q}_p^{n+1})$. Let $z(x,t)$ be the function

$$z(x,t) = \frac{1-p^{-\alpha}}{1-p^{\alpha-1}} \prod_{j=1}^{n} |a_j|_p^{-1/2} \int_{Q_p} \left\{ \prod_{k=1}^{n} \lambda_p(-\tau a_k) \right\} |\tau|_p^{\alpha-n/2-1}\, d\tau \times \int_{Q_p^n} f(x-y, t-\tau) \left\{ \prod_{l=1}^{n} \chi_p\left(\frac{y_l^2}{4\tau a_l}\right) \right\} dy, \quad (2.77)$$

if $\alpha \neq 1$, or

$$z(x,t) = \frac{1-p}{p\log p} \prod_{j=1}^{n} |a_j|_p^{-1/2} \int_{Q_p} \left\{ \prod_{k=1}^{n} \lambda_p(-\tau a_k) \right\} |\tau|_p^{-n/2} \log|\tau|_p\, d\tau \times \int_{Q_p^n} f(x-y, t-\tau) \left\{ \prod_{l=1}^{n} \chi_p\left(\frac{y_l^2}{4\tau a_l}\right) \right\} dy, \quad (2.78)$$

if $\alpha = 1$.

Lemma 2.5. *The integrals (2.77),(2.78) exist as iterated ones and define a function $z \in C_\alpha$ satisfying the equation (2.65).*

Proof. We confine ourselves to the case $\alpha \neq 1$. The case $\alpha = 1$ can be considered in a similar way.

Denote by $\omega(x,t,\tau)$ the internal integral from (2.77):

$$\omega(x,t,\tau) = \int_{Q_p^n} f(x-y, t-\tau) \left\{ \prod_{j=1}^{n} \chi_p\left(\frac{y_j^2}{4\tau a_j}\right) \right\} dy. \quad (2.79)$$

The integral (2.79) is absolutely convergent since $f \in \mathcal{D}(Q_p^{n+1})$.

Just as in the proof of Lemma 2.4, we can construct a finite covering of $\operatorname{supp} f$ by non-intersecting polydiscs obtaining the representation

$$f = \sum_{k=1}^{m} c_k \rho_k, \quad c_k \in \mathbb{C}\ ,\ 1 \le m < \infty,$$

where ρ_k is the indicator of the set

$$V_k = \left\{ (y,\tau) \in \mathbb{Q}_p^{n+1} : \|y - y^{(k)}\| \le p^{\nu_k},\ |\tau - \tau^{(k)}|_p \le p^{\nu_k} \right\} .$$

Calculating the integral (2.79) with the use of the identity (1.38) we easily obtain that

$$|\omega(x,t,\tau)| \le C|\tau|_p^{n/2}$$

for small $|\tau|_p$, which implies convergence of the iterated integral in (2.77). This calculation also shows that

$$|z(x,t)| \le C|t|_p^{\alpha-n/2-1} \quad \text{if } |t|_p > 1 \; ; \; x \in \mathbb{Q}_p^n; \tag{2.80}$$

$$\omega(x,t,\tau) = 0 \quad \text{if } \|x\| > \mu(1+|t|_p), \tag{2.81}$$

with some $\mu > 0$, which does not depend on τ. The property (2.81) enables us to calculate $v(y,t) = F_{x\to y}z(x,t)$ by interchanging the integration with respect to τ and the Fourier transform. Using (1.39) we obtain that

$$v(y,t) = \frac{1-p^{-\alpha}}{1-p^{\alpha-1}} \int\limits_{\mathbb{Q}_p} |\tau|_p^{\alpha-1} \chi_p(-\tau(a_1y_1^2+\cdots+a_ny_n^2))\widetilde{f}(y,t-\tau)\,d\tau, \tag{2.82}$$

where $\widetilde{f}(y,s) = F_{x\to y}f(x,s)$.

Since $\widetilde{f} \in \mathcal{D}(\mathbb{Q}_p^{n+1})$, it follows from (2.82) that $v \in \mathcal{B}_\alpha$ with $c(t) = \gamma|t|_p^{\alpha-1}$, $\gamma > 0$, for $|t|_p > 1$.The equality (2.81) implies that

$$z(x,t) = 0 \quad \text{if } \|x\| > \mu(1+|t|_p) \;,\; \mu > 0. \tag{2.83}$$

The properties (2.80) and (2.83) mean that $z \in \mathcal{C}_\alpha$.

The representation (2.82) can be written as

$$v(y,t) = \varphi_y^{(-\alpha)}(t)\chi_p(-t(a_1y_1^2+\cdots+a_ny_n^2)), \tag{2.84}$$

where $\varphi_y(t) = \chi_p(t(a_1y_1^2+\cdots+a_ny_n^2))\widetilde{f}(y,t)$, $\varphi^{(-\alpha)}$ is an order α (fractional) integral of the function φ, that is $\varphi^{(-\alpha)} = f_\alpha * \varphi$, and f_α is the Riesz kernel. Substituting (2.84) into (2.71) and using Proposition 2.2 we obtain that

$$\begin{aligned}(Az)(x,t) \\ = \frac{1-p^{\alpha}}{1-p^{-\alpha-1}} F^{-1}_{y\to x}\Big\{\chi_p(-t(a_1y_1^2+\cdots+a_ny_n^2)) \int\limits_{\mathbb{Q}_p} |\tau|_p^{-\alpha-1}\Big[\varphi_y^{(-\alpha)}(t-\tau) \\ - \varphi_y^{(-\alpha)}(t)\Big]\,d\tau\Big\} = F^{-1}_{y\to x}\Big\{\chi_p(-t(a_1y_1^2+\cdots+a_ny_n^2))(D^\alpha\varphi_y^{(-\alpha)})(t)\Big\} \\ = f(x,t). \quad \blacksquare\end{aligned}$$

2.10.3. The Cauchy problem. Of course, the formulas (2.77) and (2.78) can be interpreted as convolutions of a fundamental solution with the right-hand side of (2.65). However, in this case we can do more than that, and now we are ready to consider the problem (2.65)-(2.66) where $f \in \mathcal{D}(\mathbb{Q}_p^{n+1})$, $z_0 \in \mathcal{D}(\mathbb{Q}_p^n)$. We look for a solution $z \in \mathcal{B}_\alpha$ (if possible,$z \in \mathcal{C}_\alpha$) understanding the operator A in the sense of (2.71) or (2.75).

Theorem 2.9. *A solution $z \in \mathcal{C}_\alpha$ of the problem (2.65)-(2.66) is given by*

$$z = z_1 + z_2$$

where $z_1(x,t)$ is equal to the right-hand side of (2.77) (if $\alpha \neq 1$) or (2.78) (if $\alpha = 1$),

$$z_2(x,t) = |t|_p^{-n/2} \prod_{j=1}^{n} \left[\lambda_p(-ta_j)|a_j|_p^{-1/2}\right] \times \int_{\mathbb{Q}_p^n} \chi_p\left(\frac{y_1^2}{4ta_1} + \cdots + \frac{y_n^2}{4ta_n}\right) [z_0(x-y) - z_1(x-y,0)]\, dy.$$

Proof. According to Lemma 2.5, $z_1 \in \mathcal{C}_\alpha$, $Az_1 = f$. Since $z_1 \in \mathcal{C}_\alpha$, we have in particular $z_1(\cdot, 0) \in \mathcal{D}(\mathbb{Q}_p^n)$ so that the function $v_0(x) = z_0(x) - z_1(x,0)$ also belongs to $\mathcal{D}(\mathbb{Q}_p^n)$.

Let us investigate properties of z_2. Using the identity (1.39) we find the Fourier transform $w(y,t) = F_{x\to y} z_2(x,t)$:

$$w(y,t) = \chi_p(-t(a_1 y_1^2 + \cdots + a_n y_n^2))\widetilde{v}_0(y), \tag{2.85}$$

where $\widetilde{v}_0$ is the Fourier transform of v_0. It follows from (2.85) that $z_2 \in \mathcal{B}_\alpha$ with $c(t) = \text{const}$.

On the other hand we can write

$$v_0(x) = \sum_{k=1}^{m} c_k \rho_k(x) ,$$

where $c_k \in \mathbb{C}$, $\rho_k(x)$ is the indicator of the ball $V_k = \{y \in \mathbb{Q}_p^n : \|y - y^{(k)}\| \leq p^\nu\}$, and the balls V_k do not intersect each other. Now the identity (1.38) yields

$$z_2(x,t) = |t|_p^{-n/2} \prod_{j=1}^{n} [\lambda_p(-ta_j)|a_j|_p^{-1/2}]$$

$$\times \sum_{k=1}^{m} c_k \int\limits_{\|x-y-y^{(k)}\| \le p^\nu} \chi_p\left(\frac{y_1^2}{4ta_1} + \cdots + \frac{y_n^2}{4ta_n}\right) dy$$

$$= |t|_p^{-n/2} \prod_{j=1}^{n} [\lambda_p(-ta_j)|a_j|_p^{-1/2}] \sum_{k=1}^{m} c_k \prod_{l=1}^{n} \psi_{lk}(x,t),$$

where

$$\psi_{lk}(x,t) = \begin{cases} \lambda_p\left(\frac{1}{a_l t}\right) |a_l t|_p^{1/2} \omega\left(p^{-\nu}|x_l - y_l^{(k)}|_p\right), & \text{if } |t|_p < \frac{p^{2\nu}}{|a_l|_p} \\ \chi_p\left(\frac{(x_l - y_l^{(k)})^2}{4ta_l}\right) p^\nu \omega\left(p^\nu |\frac{x_l - y_l^{(k)}}{ta_l}|_p\right), & \text{if } |t|_p \ge \frac{p^{2\nu}}{|a_l|_p}. \end{cases}$$

Hence $z_2 \in \mathcal{C}_\alpha$ with $a(t) = \mu(1 + |t|_p)$ for some $\mu > 0$, $b(t) = C(1 + |t|_p)^{-n/2}$, $C > 0$.

Substituting (2.85) into (2.71) we obtain that $Az_2 = 0$. It also follows from (2.85) that $z_2(x,0) = v_0(x)$. It remains to use Lemma 2.5 to complete the proof. ■

Now we shall prove the uniqueness result.

Theorem 2.10. *If* $z \in \mathcal{B}_\alpha$, $Az = 0$, $z(x,0) \equiv 0$, *then* $z(x,t) \equiv 0$.

Proof. It follows from (2.71) that

$$\int\limits_{\mathbb{Q}_p} |\tau|_p^{-\alpha-1} [\chi_p(-\tau(a_1 y_1^2 + \cdots + a_n y_n^2)) v(y, t-\tau) - v(y,t)]\, d\tau = 0 \quad (2.86)$$

for all $y \in \mathbb{Q}_p^n$, $\tau \in \mathbb{Q}_p$, where as before $v(y,t) = F_{x\to y} z(x,t)$. Fixing y consider the function $\psi_y(t) = \chi_p(t(a_1 y_1^2 + \cdots + a_n y_n^2)) v(y,t)$. The equality (2.86) can be written as

$$(D^\alpha \psi_y)(t) = 0 \ , \ t \in \mathbb{Q}_p.$$

By Theorem 2.1 this means that ψ_y does not depend on t. Since $\psi_y(0) = v(y,0) = 0$ we come to the conclusion that $\psi_y(t) \equiv 0$ whence $v(y,t) \equiv 0$ and $z(x,t) \equiv 0$. ■

2.11 On some models of p-adic quantum theory

The conventional description of the physical space-time is based on the use of the field $\mathbb{R}$ of real numbers. In most cases mathematical models based on $\mathbb{R}$ provide quite satisfactory descriptions of the physical reality. However one should not forget the fact that the result of a physical measurement is always a rational number, so that the use of $\mathbb{R}$, a completion of $\mathbb{Q}$, is not more than a mathematical idealization. On the other hand, by Ostrowski's theorem, the only reasonable alternative to $\mathbb{R}$ among completions of $\mathbb{Q}$ is $\mathbb{Q}_p$ for some p. Other related structures, which can be seen as alternatives to the structures based on $\mathbb{R}$, are adeles, non-archimedean division algebras etc., which employ $\mathbb{Q}_p$ as their main building block.

Thus it looks quite natural to try non-Archimedean models in various physical situations, where the conventional space-time geometry is known to fail, for example in the attempts to understand the matter at sub-Planck distances or time intervals. In order to do that, the first task is the development of non-Archimedean counterparts of quantum mechanics and quantum field theory.

There are two different approaches to the p-adic quantum mechanics: a wave function defined on $\mathbb{Q}_p$ or $\mathbb{Q}_p^n$ may take its values either in $\mathbb{Q}_p$ (or its extensions), or in $\mathbb{C}$.

This section is not, of course, a review of non-Archimedean quantum mechanics, which deserves at least a book, and fortunately a reader can be directed to three of them [154, 68, 69] containing many additional references, and also historical information. Our task here is much more modest: to indicate some areas of mathematical physics related to this chapter. Correspondingly, we shall deal only with complex-valued wave functions. The opposite case is expounded in [68, 69]; for some recent new approaches (which lead also to new results in non-Archimedean analysis) see [82, 86, 89].

The formalism of the p-adic quantum mechanics with complex-valued wave functions is based on the direct construction of the unitary propagator $U(t)$ on the Hilbert space $L_2(\mathbb{Q}_p)$ related to the classical dynamics via a representation of the Heisenberg-Weyl group. Both time and spatial variables are assumed to be p-adic. For the free one-dimensional particle one gets (with the appropriate choice of units)

$$(U(t)\psi)(x) = \int\limits_{\mathbb{Q}_p} K_t(x-y)\psi(y)\,dy, \quad \psi \in L_2(\mathbb{Q}_p),$$

where

$$K_t(\xi) = \lambda_p(-t)|t|_p^{-1/2}\chi_p(t^{-1}\xi^2).$$

Note that the Fourier transform

$$\widetilde{K_t}(z) = \int\limits_{Q_p} \chi_p(z\xi)K_t(\xi)\,d\xi$$

has the simple form $\widetilde{K_t}(z) = \chi_p(-tz^2)$.

The standard Hamiltonian approach fails in the p-adic situation since $\mathbb{Q}_p$ is totally disconnected, which excludes any possibility to define infinitesimal generators of space or time translations. Nevertheless there is a natural desire to find some equation for wave functions. The first attempt in this direction was made in [153], where the equation $D_t^1\psi - D_x^2\psi = 0$ was considered with this purpose. Unfortunately, the solution of the Cauchy problem for that equation is not unique, even in the space $\mathcal{D}(\mathbb{Q}_p^2)$, which is in fact too narrow to include the wave function, whose support with respect to t is not compact.

Returning to the Schrödinger-type equation (2.65) we see that the free wave function $\psi(x,t) = (U(t)\psi_0)(x)$ satisfies the equation $A\psi = 0$ with $n = 1$, $a_1 = 1/4$, and the initial condition $\psi(x,0) = \psi_0(x)$, $x \in \mathbb{Q}_p$. Together with Theorem 2.10, this shows that (2.65) is indeed a p-adic counterpart of the Schrödinger equation for a free particle. The problem of constructing Schrödinger-type equations for more complicated dynamics like the p-adic harmonic oscillator [154] remains open.

Among various approaches to the p-adic quantum field theory there is an analogue [152] of the Euclidean approach [44], which is based on the construction of a Gaussian measure μ on the space $\mathcal{D}_r'(\mathbb{Q}_p^n)$ of real Bruhat-Schwartz distributions, with the characteristic functional

$$\int \exp[i\Phi(\varphi)]\,d\mu(\Phi) = \exp\left[-\frac{1}{2}G(\varphi,\varphi)\right], \quad \varphi \in \mathcal{D}_r(\mathbb{Q}_p^n),\ \Phi \in \mathcal{D}_r'(\mathbb{Q}_p^n),$$

where

$$G(\varphi,\varphi) = \iint\limits_{\mathbb{Q}_p^n \times \mathbb{Q}_p^n} G(x-y)\varphi(x)\varphi(y)\,dx\,dy,$$

$G(x)$ is the Green function of some pseudo-differential operator, for example the one considered in Sect. 2.8. This application shows the importance of the positivity property of the Green function, which is crucial for the construction of a Gaussian measure.

2.12 Comments

The Riesz kernels over non-Archimedean local fields were studied in [39, 128, 143, 157] as the main examples of homogeneous distributions. The most detailed treatment (for $K = \mathbb{Q}_p$) was given by Vladimirov [150, 154], who also initiated the systematic study of the fractional differential operator D^α. In particular, he proved Theorem 2.1. The connection with zeroes of the Riemann zeta (Theorem 2.2) was found by Haran [52].

The results of Sections 2.5–2.9 were obtained by the author [79, 80, 87]; here we have corrected a minor error contained in the proof of Theorem 2.4 given in [79], and also some misprints in the formulas in [79, 87]. Lemma 2 appeared in [144]; our proof is taken from [79]. The first result on the asymptotics of the Green function (for $n = 2$, $\alpha = 1$, and the quadratic form $h(\xi_1, \xi_2) = \xi_1^2 + \xi_2^2$) is due to Bikulov [12]. In particular, if $p \equiv 3 \pmod 4$, this form is anisotropic, and Bikulov's theorem (proved by a different method) is a special case of Theorem 2.5. The asymptotics of the Green functions for some non-elliptic pseudo-differential operators over $\mathbb{Q}_p$ was found by Jang [62].

The idea of using the spaces of functions and distributions defined outside the singularities of a symbol (as we did in the hyperbolic case) was employed in [70] for arbitrary pseudo-differential operators. However for the hyperbolic equation we managed (Theorem 2.8, see [87]) to extend the fundamental solution onto the singular set.

The equation of the Schrödinger type (Sect. 2.10) was introduced and studied by the author in [77].

Chapter 3

Spectral Theory

3.1 Fractional differentiation operator

Let us consider the operator D^α, $\alpha > 0$, as an unbounded operator on $L_2(K)$. As before, K is a local field, and we shall maintain the notation of Chapter 2. *We shall assume that the characteristic of the residue field of K is different from 2.*

The domain of the operator D^α is the set of those $u \in L_2(K)$, for which $\|\xi\|^\alpha \widetilde{u}(\xi) \in L_2(K)$. Since D^α is unitarily equivalent to the operator of multiplication by $\|\xi\|^\alpha$, it is self-adjoint, its spectrum consists of the eigenvalues $\lambda_N = q^{\alpha N}$ ($N \in \mathbb{Z}$) of infinite multiplicity, and their accumulation point $\lambda = 0$.

In this section we give an explicit construction of a complete orthonormal system of eigenfunctions of the operator D^α. This system of functions is indexed by several parameters and consists of the following subsystems:

$$\psi^l_{N,k,\varepsilon}(x) = q^{(N+1-l)/2}\delta(\|x\| - q^{l-N})\delta_K(x_0 - k)\chi(\varepsilon\beta^{l-2N}x^2), \tag{3.1}$$

where $N \in \mathbb{Z}$, $l \in \mathbb{Z}$, $l \geq 2$, $\varepsilon = \varepsilon_0 + \varepsilon_1\beta + \cdots + \varepsilon_{l-2}\beta^{l-2}$, $\varepsilon_j \in S$, $\varepsilon_0 \notin P$, $k \in S$, $k \notin P$;

$$\psi^1_{N,k,0}(x) = q^{(N-1)/2}\omega(q^{N-1}\|x\|)\chi(-k\beta^{-N}x), \tag{3.2}$$

where $N \in \mathbb{Z}$, $k \in S$, $k \notin P$.

As before, q is the cardinality of the residue field O/P, $S \subset O$ is the full set of representatives of classes from O/P, χ is a rank zero additive character; if $s,t \in \mathbb{R}$, $s \geq 0$, $a \in O$, then

$$\delta(t) = \begin{cases} 1, & t = 0, \\ 0, & t \neq 0; \end{cases} \quad \omega(s) = \begin{cases} 1, & 0 \leq s \leq 1, \\ 0, & s > 1; \end{cases} \quad \delta_K(a) = \begin{cases} 1, & a \in P, \\ 0, & a \notin P. \end{cases}$$

In (3.1) x_0 is the coefficient from the canonical representation (1.7) of an element x with $||x|| = q^{l-N}$.

We shall call the functions (3.1) and (3.2) the *Vladimirov functions.*

Lemma 3.1. *The Fourier transforms of the Vladimirov functions are as follows:*

$$\widetilde{\psi^l_{N,k,\varepsilon}}(\xi) = q^{-(N-1)/2}\rho\delta(||\xi|| - q^N)\delta_K(\xi_0 + 2\varepsilon_0 k)\chi\left(-\frac{1}{4\varepsilon}\beta^{2N-l}\xi^2\right), \tag{3.3}$$

$l \geq 2$;

$$\widetilde{\psi^1_{N,k,0}}(\xi) = q^{-(N-1)/2}\delta(||\xi|| - q^N)\delta_K(\xi_0 - k), \tag{3.4}$$

where ξ_0 is the coefficient from the canonical representation (1.7) of an element ξ with $||\xi|| = q^N$,

$$\rho = \begin{cases} 1, & \text{if } l \text{ is even}, \\ q^{-1/2}\theta^{(2)}(\widehat{\varepsilon})\, G_K, & \text{if } l \text{ is odd}, \end{cases}$$

$\theta^{(2)}$ is the quadratic multiplicative character on the finite field O/P, $\widehat{\varepsilon}$ is the class of an element $\varepsilon \in O$ in O/P; the number G_K was defined in Sect. 1.3.

Proof. Integrals, which emerge in the course of computing $\widetilde{\psi^l_{N,k,\varepsilon}}$ ($l \geq 2$), after simple transformations of variables are reduced to the Gaussian integrals I_5 and I_6 from (1.40). The proof of (3.3) consists of a numerous use of Lemma 1.4 and Lemma 1.5. The equality (3.4) is a consequence of (1.27). ■

Theorem 3.1. *The system of Vladimirov functions (3.1)–(3.2) is a complete orthonormal system of eigenfunctions of the operator D^α in $L_2(K)$.*

Proof. It is clear from (3.3)–(3.4) that

$$||\xi||^\alpha \widetilde{\psi^l_{N,k,\varepsilon}}(\xi) = q^{\alpha N}\widetilde{\psi^l_{N,k,\varepsilon}}(\xi),$$

so that $\psi^l_{N,k,\varepsilon}$ is an eigenfunction of D^α with the eigenvalue λ_N.

It follows from (1.22) that the L_2-norms of all the functions $\psi^l_{N,k,\varepsilon}$ are equal to 1. The orthogonality of the eigenfunctions with different N or different k is the consequence of (3.3), (3.4), while the orthogonality of two eigenfunctions with the same N and different values of l is seen from (3.1), (3.2).

The proof of the orthogonality of the eigenfunctions $\psi^l_{N,k,\varepsilon'}$ and $\psi^l_{N,k,\varepsilon''}$, $l \geq 2$, $\varepsilon' \neq \varepsilon''$, is reduced to the calculation of several integrals of the form of I_5 or I_6 (see (1.40)). One should take into account that $\|\varepsilon' - \varepsilon''\| \geq q^{2-l}$.

In order to prove the completeness, it is sufficient to show that for each N the functions $\widetilde{\psi^l_{N,k,\varepsilon}}$ form a complete system in $L_2(S_N)$, $S_N = \{\xi \in K : \|\xi\| = q^N\}$. It follows from the Parseval identity that the functions $u_a(\xi) = \chi(\xi a)$, where either $a = 0$, or $\|a\| \geq q^{1-N}$, form a total system in $L_2(S_N)$. Therefore it is sufficient to prove that each of these functions can be expanded into a series convergent in $L_2(S_N)$ by the system of functions $\widetilde{\psi^l_{N,k,\varepsilon}}$, which is clearly orthonormal in $L_2(S_N)$.

It is obvious that

$$1 \equiv q^{(N-1)/2} \sum_{k \in S, k \notin P} \widetilde{\psi^1_{N,k,0}}(\xi), \quad \xi \in S_N.$$

If $\|a\| = q^{1-N}$, $a = \beta^{N-1}(a_0 + a_1\beta + \cdots)$, $\xi = \beta^{-N}(\xi_0 + \xi_1\beta + \cdots)$, then we obtain for $\xi \in S_N$ that

$$\begin{aligned}\chi(\xi a) = \chi(\xi a) \sum_{k \in S, k \notin P} \delta_K(\xi_0 - k) &= \sum_{k \in S, k \notin P} \delta_K(\xi_0 - k)\chi(\xi_0 a_0 \beta^{-1}) \\ &= q^{(N-1)/2} \sum_{k \in S, k \notin P} \chi(k a_0 \beta^{-1}) \widetilde{\psi^1_{N,k,0}}(\xi).\end{aligned}$$

Suppose that $\|a\| = q^{n-N}$, $n \geq 2$. Denote by $(\cdot,\cdot)_N$ the inner product in $L_2(S_N)$. We have

$$\left(u_a, \widetilde{\psi^l_{N,k,\varepsilon}}\right)_N = \int_K \overline{\chi(-a\xi)\psi^l_{N,k,\varepsilon}(\xi)}\, d\xi = \psi^l_{N,k,\varepsilon}(a),$$

which means that

$$\left|\left(u_a, \widetilde{\psi^l_{N,k,\varepsilon}}\right)_N\right|^2 = \begin{cases} q^{N+1-n}\delta_K(a_0 - k), & \text{if } l = n; \\ 0, & \text{if } l \neq n. \end{cases}$$

The quantity of the functions $\widetilde{\psi^l_{N,k,\varepsilon}}$ with fixed N, $k = a_0$, $l = n \geq 2$, is equal to $(q-1)q^{n-2}$. Therefore

$$\sum_{\varepsilon} \left| \left(u_a, \widetilde{\psi^l_{N,k,\varepsilon}} \right)_N \right|^2 = (q-1)q^{(N-1)} \quad (k = a_0).$$

On the other hand, by virtue of (1.21),

$$(u_a, u_a)_N = \int\limits_{\|\xi\|=q^N} d\xi = (q-1)q^{N-1},$$

which implies our assertion. ■

3.2 An operator of the Schrödinger type

3.2.1. General properties. Let $v(x)$ be a locally bounded measurable real-valued function on K. Define an operator H' on $L_2(K)$ with the domain $\mathfrak{D}(H') = \mathcal{D}(K)$ setting

$$(H'\varphi)(x) = (D^\alpha \varphi)(x) + v(x)\varphi(x).$$

It is clear that the operator H' is densely defined and symmetric. Let H be its closure. The domain of the adjoint operator H^* consists of those $f \in L_2(K)$, for which $H^*f = D^\alpha f + v(x)f \in L_2(K)$. Here D^α is understood as a pseudo-differential operator (in the distribution sense). In fact, if $f \in \mathcal{D}(K)$, then $D^\alpha f \in L_2^{\text{loc}}(K)$.

Theorem 3.2. *The operator H is self-adjoint.*

Proof. Denote

$$a_1(\xi) = \begin{cases} \|\xi\|^\alpha, & \text{if } \|\xi\| \leq 1; \\ 0, & \text{if } \|\xi\| > 1; \end{cases} \qquad a_2(\xi) = \begin{cases} 0, & \text{if } \|\xi\| \leq 1; \\ \|\xi\|^\alpha, & \text{if } \|\xi\| > 1. \end{cases}$$

We have $Hf = F^{-1}M_{a_1}Ff + F^{-1}M_{a_2}Ff + vf$, where, as before, M_{a_j} is the operator of multiplication by the function a_j. The operator $F^{-1}M_{a_1}F$ is bounded and does not influence self-adjointness. Next, $F^{-1}M_{a_2}F = \widetilde{a_2} * f$.

Let us show that the support of the distribution $\widetilde{a_2}$ is contained in the ring of integers O. Indeed, let

$$a_{2,N}(\xi) = \begin{cases} 0, & \text{if } \|\xi\| \leq 1 \text{ or } \|\xi\| \geq q^N; \\ \|\xi\|^\alpha, & \text{if } 1 < \|\xi\| < q^N, \end{cases} \qquad N \geq 2.$$

Then $a_{2,N} \in \mathcal{D}(K)$ and $a_{2,N} \to a_2$ in $\mathcal{D}'(K)$. Thus it suffices to prove that $\operatorname{supp} \widetilde{a_{2,N}} \subset O$ for any N. We have

$$\widetilde{a_{2,N}}(x) = \int\limits_{1<\|\xi\|<q^N} \chi(x\xi)\|\xi\|^\alpha \, d\xi.$$

If $\|x\| > 1$, there exists such $\xi_0 \in O$ that $\chi(x\xi_0) \neq 1$ (since χ is a rank zero character). After the change of variables $\eta = \xi + \xi_0$ we find that

$$\widetilde{a_{2,N}}(x) = \chi(x\xi_0) \int\limits_{1<\|\eta\|<q^N} \chi(x\eta)\|\eta\|^\alpha \, d\eta = \chi(x\xi_0)\widetilde{a_{2,N}}(x),$$

whence $\widetilde{a_{2,N}}(x) = 0$.

This property of $\widetilde{a_2}$ means that for any ball $B_{x_0} = \{x \in K : \|x - x_0\| \leq 1\}$ the subspace $\mathfrak{H}(x_0) = \{f \in L_2(K) : f(x) = 0 \text{ for } x \notin B_{x_0}\}$ is invariant with respect to the operator $H_2 = F^{-1}M_{a_2}F + M_v$ defined just in the same way as the operator H above. Moreover, H_2 commutes with the orthoprojector onto the subspace $\mathfrak{H}(x_0)$, so that $\mathfrak{H}(x_0)$ reduces the operator H_2 (see [1]).

The field K can be covered by a sequence of non-intersecting balls $B_{x_0^{(n)}}$ (recall that due to the ultra-metric property two balls of the same radius either coincide or do not intersect). This leads to the orthogonal decomposition

$$L_2(K) = \bigoplus_n \mathfrak{H}\left(x_0^{(n)}\right),$$

where H_2 is reduced by each subspace $\mathfrak{H}\left(x_0^{(n)}\right)$. The convolution with $\widetilde{a_2}$ is a self-adjoint operator on each of them, while the restriction of M_v is bounded. Thus H_2 is self-adjoint as an orthogonal sum of self-adjoint operators. ■

The conditions, under which the spectrum of the operator H is discrete, are given in the following theorem.

Theorem 3.3. *Suppose that v is non-negative, measurable, locally bounded, and $v(x) \to \infty$ as $\|x\| \to \infty$. Then the operator H has a compact resolvent, so that its spectrum is discrete.*

The proof is similar to the one for the conventional Schrödinger operators (see e.g. [124]); the main tools are the analogues of the Riesz and Rellich compactness criteria for subsets of $L_2(K)$. In fact those criteria (well-known for $\mathbb{R}^n$) carry over virtually unchanged to our situation. For the details and generalizations see [154].

3.2.2. The case of a radial potential. Let us consider the case of a "radial" function $v(x) = V(\|x\|)$, where V is a real-valued function defined on the sequence $\{q^l, l \in \mathbb{Z} \cup \{-\infty\}\}$. In this case we shall give a deeper analysis of spectral properties of the operator H.

As before, we assume that V is locally bounded, that is the sequence $\{V(q^l)\}$ is bounded as $l \to -\infty$.

Denote by $\mathfrak{H}_1$ and $\mathfrak{H}_2$ the closed subspaces of $L_2(K)$ spanned by the Vladimirov functions $\psi^l_{N,k,\varepsilon}$ with $l = 1$ and $l \geq 2$ respectively. We have $L_2(K) = \mathfrak{H}_1 \oplus \mathfrak{H}_2$.

Lemma 3.2. *The subspaces $\mathfrak{H}_1$ and $\mathfrak{H}_2$ reduce the operator H.*

Proof. Let P_2 be the orthoprojector onto $\mathfrak{H}_2$ in $L_2(K)$. It is sufficient to show [1] that for any $f \in \mathfrak{D}(H)$ we have $P_2 f \in \mathfrak{D}(H)$, $P_2 H f = H P_2 f$.

If $f = \sum\limits_{l=1}^{\infty} \sum\limits_{N,k,\varepsilon} c_{l,N,k,\varepsilon} \psi^l_{N,k,\varepsilon}$, then $P_2 f = \sum\limits_{l=2}^{\infty} \sum\limits_{N,k,\varepsilon} c_{l,N,k,\varepsilon} \psi^l_{N,k,\varepsilon}$. For any $\varphi \in \mathcal{D}(K)$

$$< D^\alpha f, \overline{\varphi} >= \int_K \|\xi\|^\alpha \widetilde{f}(\xi) \overline{\widetilde{\varphi}(\xi)}\, d\xi = (f, D^\alpha \varphi),$$

where $(\cdot,\cdot)$ is the inner product in $L_2(K)$ (recall that $D^\alpha : \mathcal{D}(K) \to L_2(K)$). Since the series for f converges in $L_2(K)$ and $\psi^l_{N,k,\varepsilon} \in \mathcal{D}(K)$, we get that

$$\begin{aligned} < D^\alpha f, \overline{\varphi} >= &\sum_{l=1}^{\infty} \sum_{N,k,\varepsilon} c_{l,N,k,\varepsilon} \left(\psi^l_{N,k,\varepsilon}, D^\alpha \varphi\right) \\ &= \sum_{l=1}^{\infty} \sum_{N,k,\varepsilon} c_{l,N,k,\varepsilon} q^{\alpha N} \left\langle \psi^l_{N,k,\varepsilon}, \overline{\varphi} \right\rangle, \end{aligned}$$

so that

$$D^\alpha f = \sum_{l=1}^{\infty} \sum_{N,k,\varepsilon} c_{l,N,k,\varepsilon} q^{\alpha N} \psi^l_{N,k,\varepsilon},$$

where the series converges in $\mathcal{D}'(K)$. Since $V\varphi \in L_2(K)$, we may write a similar expansion for $Vf \in L_2^{\mathrm{loc}}(K) \subset \mathcal{D}'(K)$, which results in a representation of the function $h = (D^\alpha + V) f \in L_2(K)$ in the form of a series convergent in $\mathcal{D}'(K)$:

$$h(x) = \sum_{l=1}^{\infty} \sum_{N,k,\varepsilon} c_{l,N,k,\varepsilon} [q^{\alpha N} + V(\|x\|)] \psi^l_{N,k,\varepsilon}(x). \tag{3.5}$$

A similar representation (with $l \geq 2$) is obtained for the function $(D^\alpha + V)P_2 f$. Note that for $l \geq 2$

$$V(\|x\|)\psi^l_{N,k,\varepsilon}(x) = V(q^{l-N})\psi^l_{N,k,\varepsilon}(x), \quad x \in K.$$

Let us show that $V\psi^1_{N,k,0} \in \mathfrak{H}_1$. Indeed, let

$$V^{(N)}(\|x\|) = \begin{cases} V(\|x\|), & \text{if } \|x\| < q^{-N+2}, \\ 0, & \text{if } \|x\| \geq q^{-N+2}. \end{cases}$$

Then $V^{(N)}\psi^1_{N,k,0} = V\psi^1_{N,k,0}$. The operator of multiplication by $V^{(N)}$ is bounded and self-adjoint in $L_2(K)$. The subspace $\mathfrak{H}_2$ is invariant with respect to the multiplication by $V^{(N)}$ (as well as by an arbitrary bounded function depending only on $\|x\|$). Therefore $\mathfrak{H}_1$ is also invariant with respect to the multiplication by $V^{(N)}$, whence $V\psi^1_{N,k,0} \in \mathfrak{H}_1$.

Now it follows from (3.5) that

$$(h, \psi^l_{N,k,\varepsilon}) = c_{l,N,k,\varepsilon}[q^{\alpha N} + V(q^{l-N})], \quad l \geq 2,$$

so that

$$P_2 h = \sum_{l=2}^{\infty} \sum_{N,k,\varepsilon} c_{l,N,k,\varepsilon}[q^{\alpha N} + V(\|x\|)]\psi^l_{N,k,\varepsilon}. \tag{3.6}$$

Since the right-hand side of (3.6) coincides with the expression for $(D^\alpha + V)P_2 f$, we see that $P_2 f \in \mathfrak{D}(H)$, $P_2 H f = H P_2 f$. ■

Let H_j be the part of the operator H on the subspace $\mathfrak{H}_j$, $j = 1, 2$. It is clear that the operator H_2 is self-adjoint and possesses a complete (in $\mathfrak{H}_2$) systems of eigenvectors $\psi^l_{N,k,\varepsilon}$, $l \geq 2$, corresponding to the eigenvalues $q^{\alpha N} + V(q^{l-N})$.

Let us consider the operator H_1. Fixing $\nu \in \mathbb{Z}$ denote

$$V_\nu(\|x\|) = \begin{cases} V(\|x\|), & \|x\| \leq q^\nu; \\ 0, & \|x\| > q^\nu; \end{cases}$$

$$W_\nu(\|x\|) = V(\|x\|) - V_\nu(\|x\|).$$

Denote by $\mathfrak{H}_{1,-}$ and $\mathfrak{H}_{1,+}$ the closed subspaces in $\mathfrak{H}_1$ spanned by the functions $\psi^1_{N,k,0}$ with $N \leq -\nu$ and $N \geq -\nu + 1$ respectively. If $N \geq 1 - \nu$, then $\psi^1_{N,k,0}(x) = 0$ for $\|x\| > q^\nu$, that is $W_\nu \psi^1_{N,k,0} = 0$ if $N \geq 1 - \nu$. This means that

$$(H_1 - V_\nu) f = D^\alpha f, \quad f \in \mathfrak{H}_{1,+}. \tag{3.7}$$

Lemma 3.3. *The subspaces $\mathfrak{H}_{1,-}$ and $\mathfrak{H}_{1,+}$ reduce the operator $H_1 - V_\nu$.*

The *proof* is similar to that of Lemma 3.2. ■

Let $H_{1,-}$ and $H_{1,+}$ be the parts of the operator $H_1 - V_\nu$ in $\mathfrak{H}_{1,-}$ and $\mathfrak{H}_{1,+}$ respectively. The operator $H_{1,+}$ is self-adjoint and possesses a complete system of eigenvectors $\psi^1_{N,k,0}$, $N \geq 1-\nu$, corresponding to the eigenvalues $q^{\alpha N}$ of the multiplicity $q-1$.

If $f \in \mathfrak{H}_{1,-}$, then $H_{1,-}f = D^\alpha f + W_\nu f$. The restriction of D^α to the subspace $\mathfrak{H}_{1,-}$ has an eigenbasis $\left\{\psi^1_{N,k,0}\right\}_{N\leq -\nu}$ with the eigenvalues $q^{\alpha N}$ of the multiplicity $q-1$; therefore it is an operator of trace class. Of course the operator of multiplication by the function W_ν is self-adjoint.

The above considerations can be summarized in the following lemma.

Lemma 3.4. *The decomposition $L_2(K) = \mathfrak{H}_{1,-} \oplus \mathfrak{H}_{1,+} \oplus \mathfrak{H}_2$ corresponds to the following decomposition of the operator H:*

$$H = (V_\nu + (W_\nu + A) \oplus H_{1,+}) \oplus H_2,$$

where V_ν is the operator on $\mathfrak{H}_1 = \mathfrak{H}_{1,-} \oplus \mathfrak{H}_{1,+}$ of multiplication by the function $V_\nu(\|x\|)$, W_ν is the operator on $\mathfrak{H}_{1,-}$ of multiplication by the function $W_\nu(\|x\|)$, A is a non-negative trace class operator on $\mathfrak{H}_{1,-}$, $H_{1,-}$ and H_2 are the diagonal operators described above.

Since the spectrum and the eigenfunctions of the operator H_2 are known, in order to study spectral properties of the operator H it suffices to investigate the operator

$$H_1 = V_\nu + (W_\nu + A) \oplus H_{1,+}. \tag{3.8}$$

Lemma 3.5. *The essential spectrum of the operator W_ν on $\mathfrak{H}_{1,-}$ coincides with the set of finite accumulation points of the sequence $\{V(q^l)\}_{l>\nu}$.*

Proof. Observe first of all that the operator of multiplication by any real-valued bounded function of $\|x\|$, which equals a constant for $\|x\| \leq q^\nu$, acts as a bounded operator on $\mathfrak{H}_{1,-}$ (since obviously it is defined and bounded on $\mathfrak{H}_{1,+}$ and $\mathfrak{H}_2$). This means that any point $\omega \in \mathbb{R}$, which does not belong to the closure of the range of the function $V(\|x\|)$, $\|x\| \geq q^\nu$, is regular for W_ν: the resolvent $(W_\nu - \omega I)^{-1}$ is the operator on $\mathfrak{H}_{1,-}$ of multiplication by the function $\dfrac{1}{W_\nu(\|x\|) - \omega}$.

Suppose that ω is one of the elements of the sequence $\{V(q^l)\}_{l>\nu}$, that is $\omega = V(q^{l_0}) = W_\nu(q^{l_0})$, $l_0 > \nu$. Then ω is an eigenvalue of the operator

W_ν. Indeed, since χ is a rank zero character,

$$\psi^1_{N,k,0}(x) = \begin{cases} 0, & \text{if } \|x\| \geq q^{-N+2}, \\ q^{(N-1)/2}, & \text{if } \|x\| \leq q^{-N}. \end{cases}$$

For a fixed N the functions $\psi^1_{N,k,0}(x)$ $(k \in S,\ k \notin P)$ are linearly independent on the set $\{x \in K : \|x\| = q^{-N+1}\}$, since their Gram determinant (computed with the use of (1.27)) is equal to

$$q^{1-q} \begin{vmatrix} q-1 & -1 & \dots & -1 \\ -1 & q-1 & \dots & -1 \\ \dots & \dots & \dots & \dots \\ -1 & -1 & \dots & q-1 \end{vmatrix} = 1$$

(see [98]; note that the number of rows of the determinant equals $q-1$). Therefore every function of the form $\sum\limits_{k\in S, k\notin P} c_k \psi^1_{1-l_0,k,0}$, where $c_k \in \mathbb{C}$, $\sum\limits_{k\in S,k\notin P} |c_k| \neq 0$, $\sum\limits_{k\in S,k\notin P} c_k = 0$, does not vanish identically, its support is contained in the set $\{x \in K : \|x\| = q^{l_0}\}$, so that it is an eigenfunction for the operator W_ν with the eigenvalue ω. We can make the same conclusion regarding the value $\omega = W_\nu(q^\nu) = 0$. It is clear that W_ν has no other eigenvalues.

If a number ω is repeated in the sequence $\{V(q^l)\}_{l>\nu}$ infinite number of times, then ω is an eigenvalue of the infinite multiplicity. Clearly all other finite limit points of this sequence also belong to the essential spectrum of the operator W_ν. In order to complete the proof, it remains to show that those eigenvalues, which appear in the sequence $\{V(q^l)\}_{l>\nu}$ a finite number of times (including possibly $\omega = 0$) have finite multiplicities. In its turn, that would be proved if we verify that for each fixed $l_0 > \nu$ the subspace

$$M(l_0) = \{\varphi \in \mathfrak{H}_{1,-} : \varphi(x) = 0 \text{ for } \|x\| \geq q^{l_0}\}$$

is finite-dimensional.

Let $\varphi \in M$, $\varphi = \sum\limits_{N=-\infty}^{-\nu} \sum\limits_{k\in S,k\notin P} c_{Nk}\psi^1_{N,k,0}$. If $\|x\| = q^l$, $l \geq l_0$, then $\varphi(x) = 0$, whence

$$\sum_{N=-\infty}^{-l} \sum_{k\in S,k\notin P} q^{(N-1)/2} c_{Nk} + \sum_{k\in S,k\notin P} c_{-l+1,k}\psi^1_{1-l,k,0}(x) = 0. \qquad (3.9)$$

Let us integrate the identity (3.9) over the set $\{x \in K : \|x\| = q^l\}$ using (1.27). Denoting $\sigma_N = \sum_{k\in S, k\notin P} c_{Nk}$, we get

$$q^l(1-q^{-1}) \sum_{N=-\infty}^{-l} q^{(N-1)/2}\sigma_N - q^{(l-2)/2}\sigma_{-l+1} = 0,$$

so that

$$\sigma_{-l+1} = q^{(l-1)/2}(q-1) \sum_{N=-\infty}^{-l} q^{N/2}\sigma_N. \tag{3.10}$$

Substitute in (3.10) $l+1$ for l, multiply the resulting equality by $q^{-1/2}$ and subtract it from (3.10). Then we find that

$$\sigma_{-l+1} - q^{-1/2}\sigma_{-l} = q^{(l-1)/2}(q-1)q^{-l/2}\sigma_{-l},$$

whence $\sigma_{-l+1} = q^{1/2}\sigma_{-l}$, so that for all $l \geq l_0$

$$\sigma_{-l} = \sigma_{-l_0} q^{-(l-l_0)/2}. \tag{3.11}$$

Suppose that all the coefficients $c_{-l_0,k}$ ($k \in S, k \notin P$) are given. It follows from (3.11), (3.9), and the above linear independence property, that one can recover from them in a unique way all the coefficients $c_{-l,k}$, $l > l_0$. This means that $\dim M < \infty$ ■.

Lemma 3.5, together with the representation (3.8), enabled us to obtain a detailed information about the spectrum of the operator H_1. As we know, the spectrum of H_1 (and H) is discrete if $V(\|x\|) \to \infty$, $\|x\| \to \infty$. The following result covers also the case when $V(\|x\|) \to -\infty$.

Theorem 3.4. *If the sequence $\{V(q^l)\}_{l>\nu}$ has no finite accumulation points, then the spectrum of H_1 is discrete, and the operator H has a complete system of eigenfunctions.*

The *proof* follows from (3.8), Lemma 3.5, and the discretness of the spectrum of $H_{1,+}$. Indeed, W_ν is in this case an unbounded self-adjoint operator with a discrete spectrum, A its compact perturbation, and V_ν a relatively compact perturbation of $(W_\nu + A) \oplus H_{1,+}$. ■

Using (3.8) we can also obtain information about the asymptotic behavior of eigenvalues of the operator H_1. Let $N(\lambda)$ be the distribution function of its eigenvalues, that is the quantity of those eigenvalues (counted with their multiplicities), which are smaller than $\lambda > 0$.

Theorem 3.5. *Suppose that $V(q^l) \sim Cq^{\gamma l}$, $l \to \infty$, where $C > 0$, $\gamma > 0$, and there exists such a natural number l_0 that*

$$V(q^{l_1}) \neq V(q^{l_2}) \quad \text{if } l_1, l_2 \geq l_0,\ l_1 \neq l_2.$$

Then

$$N(\lambda) = (q-1)\left(\frac{1}{\alpha} + \frac{1}{\gamma}\right) \log_q \lambda + O(1), \quad \lambda \to \infty. \tag{3.12}$$

Proof. Let us use the formula (3.8) with $\nu > l_0$. First we shall show that the multiplicity of each eigenvalue $\omega = W_\nu(q^l)$ $(l > \nu)$ of the operator W_ν equals $q-1$, or in other words, such is the dimension of the subspace $Z_l \subset \mathfrak{H}_{1,-}$ consisting of functions with supports in the set $\{x \in K : \|x\| = q^l\}$. If $f \in Z_l$, then

$$f(x) = \begin{cases} \sum\limits_{k \in S, k \notin P} c_k \psi^1_{1-l,k,0}(x) + c_0, & \|x\| = q^l, \\ 0, & \|x\| \neq q^l, \end{cases}$$

where $c_k, c_0 \in \mathbb{C}$, since all the functions $\psi^1_{N,k,0}$ with $N \neq 1-l$ are constant for $\|x\| = q^l$. If $\|x\| = q^l$, then $\sum\limits_{k \in S, k \notin P} \psi^1_{1-l,k,0}(x) \equiv \text{const} \neq 0$ as a consequence of (1.3). It means that $\dim Z_l \leq q-1$.

On the other hand, the set Z_l contains all the functions from $\mathfrak{H}_{1,-}$ of the form $\sum\limits_{k \in S, k \notin P} c_k \psi^1_{1-l,k,0}$, where $\sum\limits_{k \in S, k \notin P} c_k = 0$. Next, consider the function $\varphi(x) = \delta(\|x\| - q^l)$. It is orthogonal to $\mathfrak{H}_2$ since the support of each function $\psi^j_{N,k,\varepsilon}$, $j \geq 2$, either does not intersect, or coincides with $\operatorname{supp} \varphi$, and

$$\int_K \psi^j_{N,k,\varepsilon}(x)\, dx = \widetilde{\psi^j_{N,k,\varepsilon}}(0) = 0$$

by Lemma 3.1. Thus $\varphi \in \mathfrak{H}_1$. If $N \geq 1 - \nu$, then

$$\operatorname{supp} \psi^1_{N,k,0} \subset \{x \in K : \|x\| \leq q^{1-N}\} \subset \{x \in K : \|x\| \leq q^\nu\}.$$

Since $l > \nu$, we have $\varphi \perp \mathfrak{H}_{1,+}$, so that $\varphi \in \mathfrak{H}_{1,-}$, which means that $\varphi \in Z_l$. Computing the inner product, we find that

$$\left(\varphi, \sum_{k \in S, k \notin P} c_k \psi^1_{1-l,k,0}\right) = \sum_{k \in S, k \notin P} c_k q^{-l/2} \int\limits_{\|x\|=q^l} \overline{\chi(k\beta^{l-1}x)}\, dx$$
$$= -q^{\frac{l}{2}-1} \sum_{k \in S, k \notin P} c_k = 0$$

if $\sum_{k\in S, k\notin P} c_k = 0$. Hence $\dim Z_l = q-1$.

Let $N_W(\lambda)$ be the distribution function of eigenvalues of the operator W_ν; denote also

$$n(\lambda) = \operatorname{card}\{N \in \mathbb{Z} : \ V(q^N) < \lambda, N > \nu\}.$$

It follows from the above arguments that

$$N_W(\lambda) = (q-1)n(\lambda) + O(1), \quad \lambda \to \infty,$$

where the bounded remainder term corresponds to the zero eigenvalue. By our assumption,

$$n(\lambda) = \frac{1}{\gamma}\log_q \lambda + O(1), \quad \lambda \to \infty,$$

whence

$$N_W(\lambda) = \frac{q-1}{\gamma}\log_q \lambda + O(1), \quad \lambda \to \infty. \tag{3.13}$$

Let $\{\lambda_n\}$, $\{\mu_n\}$ be the sequences of all the eigenvalues (counted with their multiplicities) of the operators W_ν and $W_\nu + A$ arranged in the ascending order. It follows from the minimax principle [124] that

$$\lambda_n \le \mu_n \le \lambda_n + \|A\|,$$

so that

$$N_W(\lambda - \|A\|) \le N_\mu(\lambda) \le N_W(\lambda),$$

where $N_\mu(\lambda) = \sum_{\mu_n<\lambda} 1$. Now (3.13) yields

$$N_\mu(\lambda) = \frac{q-1}{\gamma}\log_q \lambda + O(1), \quad \lambda \to \infty.$$

The eigenvalue distribution function for an orthogonal sum of two self-adjoint operators equals the sum of corresponding distribution functions. Using again the minimax principle we come to (3.12). ■

3.2.3. Essential spectrum. Let us consider some classes of potentials $V(\|x\|)$, for which the essential spectrum $\sigma_{\text{ess}}(H_1)$ is not empty and can be described explicitly.

Theorem 3.6. *If $V(q^l) \to 0$ as $l \to \infty$, then $\sigma_{\text{ess}}(H_1) = \{0\}$.*

Proof. Let $H_{1,0}$ be the restriction of D^α to the subspace $\mathfrak{H}_1$, V, V_ν be the operators on $\mathfrak{H}_1$ of multiplication by $V(\|x\|)$, $V_\nu(\|x\|)$ respectively. It is sufficient to prove [124] that $V(H_{1,0}+I)^{-1}$ is a compact operator on $\mathfrak{H}_1$. Since $V_\nu \to V$ as $\nu \to \infty$, in the uniform operator topology, it suffices to prove the compactness of $V_\nu(H_{1,0}+I)^{-1}$ for any ν. In fact we shall show that the latter operator belongs to the Hilbert-Schmidt class.

The operator $V_\nu(H_{1,0}+I)^{-1}$ has the following matrix elements with respect to the basis $\left\{\psi^1_{N,k,0}\right\}$:

$$z_{N,k;M,l} = (q^{\alpha M}+1)^{-1} w_{N,k;M,l}, \quad M, N \in \mathbb{Z};\ k, l \in S, k, l \notin P,$$

where

$$w_{N,k;M,l} = \int\limits_K V_\nu(\|x\|)\psi^1_{N,k,0}(x)\overline{\psi^1_{M,l,0}(x)}\,dx.$$

It is clear that the integrand vanishes outside the set $\{x \in K : \|x\| \le q^{\min(1-N,1-M,\nu)}\}$.

If $N \ge 1-\nu$, $M \ge N$, then

$$|z_{N,k;M,l}| \le C(q^{\alpha M}+1)^{-1} q^{\frac{N+M}{2}-1} \int\limits_{\|x\|\le q^{1-M}} dx = C(q^{\alpha M}+1)^{-1} q^{\frac{N-M}{2}},$$

$C > 0$, whence

$$\begin{aligned}
&\sum_{N=1-\nu}^{\infty}\sum_{M=N}^{\infty}\sum_{k,l\in S;k,l\notin P} |z_{N,k;M,l}|^2 \\
&\qquad \le C^2(q-1)^2 \sum_{N=1-\nu}^{\infty}\sum_{M=N}^{\infty} (q^{\alpha M}+1)^{-2} q^{N-M} \\
&\qquad = C^2(q-1)^2 \sum_{M=1-\nu}^{\infty} (q^{\alpha M}+1)^{-2} q^{-M} \sum_{N=1-\nu}^{M} q^N \\
&\qquad \le C_1 \sum_{M=1-\nu}^{\infty} (q^{\alpha M}+1)^{-2} < \infty.
\end{aligned}$$

If $N \ge 1-\nu$, $M < N$, then

$$|z_{N,k;M,l}| \le C(q^{\alpha M}+1)^{-1} q^{\frac{N+M}{2}-1} \int\limits_{\|x\|\le q^{1-N}} dx = C(q^{\alpha M}+1)^{-1} q^{\frac{M-N}{2}},$$

so that as before,

$$\sum_{N=1-\nu}^{\infty} \sum_{M=-\infty}^{N-1} \sum_{k,l \in S; k,l \notin P} |z_{N,k;M,l}|^2 < \infty.$$

Similarly,

$$\sum_{N=-\infty}^{-\nu} \sum_{M=N}^{\infty} \sum_{k,l} |z_{N,k;M,l}|^2 < \infty.$$

Finally, if $M < N \leq -\nu$, then

$$|z_{N,k;M,l}| \leq Cq^{\frac{N+M}{2}-1} \int_{\|x\| \leq q^\nu} dx \leq C_1 q^{\frac{N+M}{2}},$$

$$\sum_{N=-\infty}^{-\nu} \sum_{M=-\infty}^{N-1} \sum_{k,l} |z_{N,k;M,l}|^2 \leq C_2 \sum_{N=-\infty}^{-\nu} q^N \sum_{M=-\infty}^{N-1} q^M$$

$$\leq C_3 \sum_{N=-\infty}^{-\nu} q^{2N} < \infty. \quad \blacksquare$$

The next result deals with the situation, in which the essential spectrum of the operator H_1 may have a much more complicated structure.

Theorem 3.7. *If $V(q^l) \to 0$ as $l \to -\infty$, then the essential spectrum of the operator H_1 coincides with the set of finite limit points of the sequence $\{V(q^l)\}$, $l \to \infty$. If in addition*

$$\sum_{l=-\infty}^{0} |V(q^l)| < \infty, \tag{3.14}$$

then the spectrum of H_1 (and that of H) is purely singular.

Proof. Let us use again the formula (3.8) with an arbitrary fixed number ν. Let $V_{\nu,\mu}$ ($\mu \in \mathbb{Z}$) be the operator on $\mathfrak{H}_1$ of multiplication by the function

$$V_{\nu,\mu}(\|x\|) = \begin{cases} V_\nu(\|x\|), & \text{if } \|x\| > q^\mu, \\ 0, & \text{if } \|x\| \leq q^\mu. \end{cases}$$

As in the proof of Theorem 3.6, we show that $V_{\nu,\mu}$ is a Hilbert-Schmidt operator. Under our assumptions, $V_{\nu,\mu} \to V_\nu$ in the uniform operator topology, as $\mu \to -\infty$. Therefore the operator V_ν is compact, and the first assertion of the theorem is a consequence of the formula (3.8), Lemma 3.5, and the Weyl theorem on the invariance of an essential spectrum.

Let us assume (3.14). Then V_ν is a trace class operator. In order to prove this, it is sufficient to verify the inequality

$$\sum_{N=-\infty}^{\infty} \sum_{k,l\in S;k,l\notin P} \int_K |V_\nu(\|x\|)| \cdot |\psi^1_{N,k,0}(x)|^2\, dx < \infty. \tag{3.15}$$

Denote by $a_{N,k}$ a term of the series (3.15). We have

$$a_{N,k} \le q^{N-1} \int_{\|x\|\le \min(q^{1-N},q^\nu)} |V_\nu(\|x\|)|\, dx.$$

If $N \le -\nu$ then $a_{N,k} \le Cq^N$, so that

$$\sum_{N=-\infty}^{-\nu} \sum_k a_{N,k} < \infty.$$

If $N \ge 1-\nu$, then

$$\int_{\|x\|\le q^{1-N}} |V_\nu(\|x\|)|\, dx = \sum_{l=-\infty}^{1-N} \int_{\|x\|=q^l} |V_\nu(\|x\|)|\, dx$$
$$= (1-q^{-1}) \sum_{l=-\infty}^{1-N} q^l |V(q^l)|,$$

whence

$$\sum_{N=1-\nu}^{\infty} \sum_k a_{N,k} \le C \sum_{N=1-\nu}^{\infty} \sum_{l=-\infty}^{1-N} q^{l+N} |V(q^l)|$$
$$= C \sum_{l=-\infty}^{\nu} q^l |V(q^l)| \sum_{N=1-\nu}^{1-l} q^N \le C_1 \sum_{l=-\infty}^{\nu} |V(q^l)| < \infty,$$

and we have proved (3.15).

Since V_ν and A are of trace class, and $H_{1,+}$ has only a discrete spectrum, it follows [123] from (3.8) that the absolutely continuous parts of the spectra of the operators H_1 and W_ν coincide. The absolutely continuous spectrum of the operator W_ν (on $\mathfrak{H}_{1,-}$) coincides with the absolutely continuous spectrum of the operator of multiplication by $W_\nu(||x||)$ on $L_2(K)$ (since the restrictions of the latter to $\mathfrak{H}_{1,+}$ and $\mathfrak{H}_2$ have no absolutely continuous spectra).

On the other hand, the spectrum of the latter multiplication operator is purely singular. Indeed, its spectral projection $E(\Delta)$ ($\Delta \subset \mathbb{R}$ is an arbitrary Borel set) is the operator of multiplication by the function $I_\Delta(W_\nu(||x||))$, where I_Δ is the indicator of the set Δ (see [121]). For any function $f \in L_2(K)$

$$||E(\Delta)f||^2_{L_2(K)} = \int_{\Phi(\Delta)} |f(x)|^2\, dx, \tag{3.16}$$

where $\Phi(\Delta) = \{x \in K : W_\nu(||x||) \in \Delta\}$. In particular, let Δ consist of the origin and the sequence $\{V(q^l)\}_{l>\nu}$. Then $\Phi(\Delta) = K$. If f is an absolutely continuous vector, then $E(\Delta)f = 0$ (since the Lebesgue measure of the set Δ equals zero), and by virtue of (3.16) we find that $f = 0$. ■

Example. Let $V(||x||) = \sin a||x||$, $a \in \mathbb{R}$. We have

$$\sum_{l=-\infty}^{0} |V(q^l)| \leq |a| \sum_{l=-\infty}^{0} q^l < \infty,$$

so that the essential spectrum of the operator H_1 is purely singular and coincides with the set Λ of all the limit points of the sequence $\{\sin aq^l,\ l \to \infty\}$. It is clear that $\sin aq^l = \sin\left(2\pi\left\{\frac{a}{2\pi}q^l\right\}\right)$, where $\{r\}$ is the fractional part of a number $r \in \mathbb{R}$. If the function $\theta(l) = \frac{a}{2\pi}q^l$ is uniformly distributed mod 1 (see [93]), then the set $\left\{\frac{a}{2\pi}q^l\right\}$ is dense in $[0,1]$, so that $\Lambda = [-1,1]$. By the metric theorems of the uniform distribution theory [93] $\Lambda = [-1,1]$ for almost all $a \in \mathbb{R}$. Some specific values of a can be found in [92, 117, 94].

3.3 Spectral properties of D^α on open subsets of a local field

3.3.1. Structure of open subsets. Let G be an open subset of a local field K. In order to define the operator D^α as a densely defined operator on $L_2(G)$, we consider first the restriction of D^α from $\mathcal{D}(K)$ to the set $\mathcal{D}(G) \subset \mathcal{D}(K)$ of test functions with supports in G. Now we are interested only in the values of $(D^\alpha\varphi)(x)$ for $x \in G$, $\varphi \in \mathcal{D}(G)$. It will be shown below that $\mathcal{D}(G)$ is dense in $L_2(G)$. Thus D^α is a non-negative symmetric operator on $L_2(G)$; denote by A_G its closure.

In this section we shall study relations between the geometry of a set G and the spectral properties of the operator A_G. We start from some results regarding the structure of open subsets of a local field.

Any open subset $G \subset K$ can be represented as a union

$$G = \bigcup_{k=1}^{N} V_k, \quad N \leq \infty, \tag{3.17}$$

of non-intersecting balls $V_k = \{x \in K : \|x - x_k\| \leq q^{r_k}\}$, $r_k \in \mathbb{Z}$. A set G of the form (3.17) is compact if and only if $N < \infty$.

The *type of a compact open set* G is defined as the smallest non-negative residue of the number N modulo $(q-1)$. It is known [136, 137, 135] that the type of G does not depend on the representation of G in the form (3.17) and determines G uniquely up to a locally linear diffeomorphism. Therefore the type is a basic geometric characteristic of compact open sets.

Suppose now that G is not compact, so that $N = \infty$ in the representation (3.17).

Proposition 3.1. *The set G is closed if and only if the sequence $\{x_k\}$ has no finite limit points.*

Proof. Let G be closed. Suppose that there exists a subsequence $\{x_{k_n}\}$, which converges to an element $\xi \in K$, as $n \to \infty$. Then $\xi \in G$, that is $\xi \in V_m$ for some m:

$$\|x_m - \xi\| \leq q^{r_m}.$$

Choose m in such a way that $k_n > m$, $\|x_{k_n} - \xi\| \leq q^{r_m}$. Then

$$\|x_{k_n} - x_m\| \leq \max\{\|x_{k_n} - \xi\|, \|\xi - x_m\|\} \leq q^{r_m},$$

so that $x_{k_n} \in V_m$, which is impossible since $V_m \cap V_{k_n} = \varnothing$.

Conversely, suppose that the sequence $\{x_k\}$ has no finite limit points, but the set G is not closed. The set G contains a sequence $\{y_n\}$ converging to an element $\eta \notin G$. Choose a sequence $\{k_n\}$ of natural numbers in such a way that all the numbers k_n are different, and $y_{k_n} \in V_{k_n}$. Then $r_{k_n} \to -\infty$ as $n \to \infty$. Indeed, if not, then there exists a subsequence $\{r'_j\} \subset \{r_{k_n}\}$, such that $r'_j \geq R > -\infty$. For the corresponding subsequences $\{y'_j\} \subset \{y_{k_n}\}$, $\{x'_j\} \subset \{x_{k_n}\}$ we obtain

$$\|y'_{j_1} - y'_{j_2}\| = \|(y'_{j_1} - x'_{j_1}) + (x'_{j_1} - x'_{j_2}) + (x'_{j_2} - y'_{j_2})\| = \|x'_{j_1} - x'_{j_2}\|$$
$$\geq \max\left\{q^{r'_{j1}}, q^{r'_{j2}}\right\} \geq q^R$$

(we have used that the balls are disjoint), which contradicts the convergence of the sequence $\{y_n\}$.

Hence $r_{k_n} \to -\infty$. Now

$$\|x_{k_n} - \eta\| \le \max\{\|x_{k_n} - y_{k_n}\|, \|y_{k_n} - \eta\|\} \le \max\{q^{r_{k_n}}, \|y_{k_n} - \eta\|\} \to 0$$

as $n \to \infty$, and we have come to a contradiction. ■

Let us call an open set G *regular* if

$$\inf_{i \ne j} \|x_i - x_j\| > 0.$$

Corollary 3.1. *If an open set G is regular, then it is closed.*

3.3.2. Model examples. Let us start the investigation of spectral properties of the operator A_G from two model examples: the case when G is a ball $B_r = \{x \in K : \|x\| \le q^r\}$, and the case of a sphere $S_r = \{x \in K : \|x\| = q^r\}$, $r \in \mathbb{Z}$.

Let $G = B_r$. It can be shown by an easy calculation that the function

$$\psi_0(x) = \begin{cases} q^{-r/2}, & x \in B_r, \\ 0, & x \notin B_r, \end{cases}$$

is an eigenfunction of A_{B_r} with the eigenvalue

$$\lambda_0 = \frac{q-1}{q^{\alpha+1}-1} q^{\alpha(1-r)}, \tag{3.18}$$

and $\|\psi_0\|_{L_2(B_r)} = 1$.

The Vladimirov functions $\psi^l_{N,j,\varepsilon}$ with $l \ge 2$, $l - N \le r$, are eigenfunctions of D^α in $L_2(K)$, and their supports are contained in B_r. Therefore they are eigenfunctions of A_{B_r}. Other Vladimirov functions $\psi^l_{N,j,\varepsilon}$, $l \ge 2$, vanish on B_r. Similarly, the Vladimirov functions $\psi^1_{N,j,0}$ with $1 - N \le r$, are eigenfunctions of A_{B_r}, while those with $1 - N > r$ are constant on B_r. Hence if a function from $L_2(B_r)$ is orthogonal to ψ_0 and those Vladimirov functions, which are eigenfunctions of A_{B_r}, then its extension onto K by zero is orthogonal in $L_2(K)$ to all the Vladimirov functions. This means that A_{B_r} has a complete orthonormal system of eigenfunctions, consisting of

$$\begin{gathered}\psi_0, \quad \text{with the eigenvalue } \lambda_0 \text{ (multiplicity 1)},\\ \psi^1_{1-r,j,0}, j \in S, j \notin P, \quad \text{with the eigenvalue}\\ \lambda_1 = q^{\alpha(1-r)} \text{(multiplicity } q-1),\\ \psi^l_{\nu-r,j,\varepsilon}\ (2 \le l \le \nu, j \in S, j \notin P, \varepsilon = \varepsilon_0 + \varepsilon_1\beta + \cdots + \varepsilon_{l-2}\beta^{l-2}) \text{ and}\\ \psi^1_{\nu-r,j,0}(j \in S, j \notin P), \quad \text{with the eigenvalue } \lambda_\nu = q^{\alpha(\nu-r)}\\ \text{of the total multiplicity } q^{\nu-1}(q-1), \nu = 2,3,\dots .\end{gathered}$$

In the case $G = S_r$ we proceed in a similar way. The function

$$\psi_0'(x) = \begin{cases}(1-q^{-1})^{-1/2}q^{-r/2}, & x \in S_r,\\ 0, & x \notin S_r,\end{cases}$$

is a normalized eigenfunction of A_{S_r} with the eigenvalue

$$\lambda_0' = \frac{q^\alpha + q - 2}{q^{\alpha+1} - 1} q^{\alpha(1-r)}. \tag{3.19}$$

The Vladimirov functions $\psi^l_{l-r,j,\varepsilon}$, $l \ge 2$, with supports in S_r, are eigenfunctions of A_{S_r} with the eigenvalue $\lambda_l' = q^{\alpha(l-r)}$. Other Vladimirov functions $\psi^l_{N,j,\varepsilon}$, $l \ge 2$ (with $l - N \ne r$) vanish on S_r. Considering the functions $\psi^1_{N,j,0}$, we see that their non-zero restrictions to S_r are proportional to the functions

$$1,\ v_j(x) = \chi(-j\beta^{r-1}x), \quad x \in S_r,\ j \in S, j \notin P. \tag{3.20}$$

The functions (3.20) are linearly dependent on S_r since

$$\sum_{j \in S, j \notin P} \chi(-j\beta^{-N}x) \equiv -1$$

by virtue of (1.3). On the other hand, the functions $v_j(x)$ are linearly independent (see the proof of Lemma 3.5). Thus the dimension of the subspace Ψ in $L_2(S_r)$ spanned by the functions (3.20) is equal to $q-1$.

Let us enumerate the elements $j \in S, j \notin P$ in an arbitrary way, so that the functions (3.20) are written as

$$1,\ \chi(-j_\nu\beta^{r-1}x), \quad \nu = 1, \dots, q-1.$$

We can choose the basis of the subspace Ψ consisting of the functions 1,

$$\chi(-j_\nu\beta^{r-1}x) - \chi(-j_{\nu+1}\beta^{r-1}x), \quad \nu = 1, \dots, q-2. \tag{3.21}$$

The latter differences vanish for $\|x\| < q^r$. Since they are proportional to the restrictions to S_r of the functions $\psi^1_{1-r,j_\nu,0} - \psi^1_{1-r,j_\nu+1,0}$, which vanish for $\|x\| > q^r$, we find that the functions (3.21) are eigenfunctions of A_{S_r} with the eigenvalue $\lambda'_1 = q^{\alpha(1-r)}$.

Thus we have found that the spectrum of A_{S_r} consists of the eigenvalue (3.19) of the multiplicity 1, the eigenvalue $\lambda'_1 = q^{\alpha(1-r)}$ with the multiplicity $q-2$, and the eigenvalues $\lambda'_l = q^{\alpha(l-r)}$, whose multiplicities can be easily computed and equal $(q-1)^2 q^{l-2}$, $l = 2, 3, \dots$.

Another construction of eigenfunctions of the operator A_{S_1} will be given in Sect. 3.5.

Note that in both cases (B_r and S_r) the eigenvalues have finite multiplicities and no finite accumulation points. Of course, both operators are self-adjoint.

3.3.3. Invariant subspaces. Since the operator D^α is non-local, the representation (3.17) of a set G as a disjoint union of balls does not lead to an orthogonal decomposition of the operator A_G. Nevertheless the above eigenbasis for A_{B_r} will be our principal tool in the general case too.

It follows from the translation invariance of the operator D^α that in order to obtain an eigenbasis in $L_2(V_k)$ (for a fixed k) it is sufficient to substitute $x - x_k$ for x, and r_k for r in the functions ψ_0, $\psi^1_{1-r,j,0}$, $\psi^l_{\nu-r,j,\varepsilon}$. The resulting functions will be denoted $\psi^k_0(x)$, $\psi^{l,k}_{\nu-r_k,j,\varepsilon}(x)$. Note that all these functions belong to $\mathcal{D}(K)$, and their supports are contained in V_k (the function ψ^k_0 is assumed extended onto K by zero). This implies the density of $\mathcal{D}(G)$ in $L_2(G)$ for any open subset $G \subset K$, since the decomposition (3.17) implies the orthogonal decomposition

$$L_2(G) = \bigoplus_{k=1}^{N} L_2(V_k), \quad N \le \infty.$$

Denote by $\mathfrak{H}_1$ the subspace in $L_2(G)$ spanned by all the functions $\psi^{l,k}_{\nu-r_k,j,\varepsilon}$, $l \ge 1$, $1 \le k \le N$. As we saw, these functions are eigenfunctions of D^α over K; thus they are also eigenfunctions over G. Therefore $\mathfrak{H}_1$ is an invariant subspace for the operator A_G, which induces on $\mathfrak{H}_1$ a self-adjoint operator $A_{G,1}$. The spectrum of $A_{G,1}$ consists of the eigenvalues

$$\lambda_{\nu k} = q^{\alpha(\nu - r_k)}, \quad \nu = 1, 2, \dots; k = 1, \dots, N,$$

and possibly also their accumulation points. The multiplicity of an eigenvalue $\lambda_{\nu k}$ corresponding to a fixed index k equals $q^{\nu-1}(q-1)$. The total multiplicity should be computed taking into account that the numbers r_k can coincide for certain k; thus it may be infinite if G is not compact.

Let $\mathfrak{H}_2 \subset L_2(G)$ be the subspace spanned by the functions ψ_0^k ($k = 1, \dots, N$) extended by zero:

$$\psi_0^k(x) = \begin{cases} q^{-r_k/2}, & x \in V_k, \\ o, & x \in G \setminus V_k. \end{cases}$$

As we have seen,

$$\left(D^\alpha \psi_0^k\right)(x) = \frac{q-1}{q^{\alpha+1}-1} q^{\alpha(1-r_k)} \psi_0^k(x), \quad x \in V_k. \tag{3.22}$$

If $x \in V_j$ $(j \neq k)$, then $\psi_0^k(x) = 0$,

$$\left(D^\alpha \psi_0^k\right)(x) = \frac{1-q^\alpha}{1-q^{-\alpha-1}} q^{-r_k/2} \int\limits_{\|y-x_k\| \leq q^{r_k/2}} \|x-y\|^{-\alpha-1}\, dy,$$

and since the balls V_k do not intersect,

$$\|x-y\| = \|(x-x_j)+(x_j-x_k)+(x_k-y)\| = \|x_j-x_k\|,$$

whence

$$\left(D^\alpha \psi_0^k\right)(x) = \frac{1-q^\alpha}{1-q^{-\alpha-1}} q^{r_k/2} \|x_j-x_k\|^{-\alpha-1}, \quad x \in V_j. \tag{3.23}$$

Lemma 3.6. *The subspaces $\mathfrak{H}_1$ and $\mathfrak{H}_2$ reduce the operator A_G.*

Proof. It is sufficient to verify that the orthogonal projection to $\mathfrak{H}_1$ does not send a function $f \in \mathfrak{D}(A_G)$ out of the domain $\mathfrak{D}(A_G)$ of the operator A_G.

It is obvious that D^α acts continuously from $\mathcal{D}(G)$ to $L_2(G)$; extending it to distributions we obtain a continuous action $D^\alpha : L_2(G) \to \mathcal{D}'(G)$. It follows that $A_G f = D^\alpha f$ for all $f \in \mathfrak{D}(A_G)$, where D^α is understood in the distribution sense.

On the other hand, let $f \in \mathfrak{D}(A_G)$. Expanding with respect to our basis we find that $f = f_1 + f_2$, where

$$f_1 = \sum_{k=1}^N \sum_{\nu,l,j,\varepsilon} c_{\nu l j \varepsilon k} \psi_{\nu-r_k,j,\varepsilon}^{l,k}, \quad f_2 = \sum_{k=1}^N d_k \psi_0^k,$$

and the series converge in $L_2(G)$. Applying D^α in the distribution sense we obtain the expansions, which are convergent in $\mathcal{D}'(G)$:

$$D^\alpha f = D^\alpha f_1 + D^\alpha f_2 \tag{3.24}$$

$$D^\alpha f_1 = \sum_{k,\nu,l,j,\varepsilon} c_{\nu l j \varepsilon k} q^{\alpha(\nu - r_k)} \psi^{l,k}_{\nu - r_k, j, \varepsilon},$$

$$D^\alpha f_2 = \sum_{k=1}^{N} d_k D^\alpha \psi_0^k.$$

By the above assumption, $D^\alpha f \in L_2(G)$, and it has an expansion convergent in $L_2(G)$:

$$D^\alpha f = \sum_{k,\nu,l,j,\varepsilon} \tilde{c}_{\nu l j \varepsilon k} \psi^{l,k}_{\nu - r_k, j, \varepsilon} + \sum_k \tilde{d}_k \psi_0^k. \tag{3.25}$$

It is clear from (3.24) that

$$\left\langle D^\alpha f, \overline{\psi^{l,k}_{\nu - r_k, j, \varepsilon}} \right\rangle = c_{\nu l j \varepsilon k} q^{\alpha(\nu - r_k)},$$

because

$$\left\langle D^\alpha \psi_0^\varkappa, \overline{\psi^{l,k}_{\nu - r_k, j, \varepsilon}} \right\rangle = q^{\alpha(\nu - r_k)} \left(\psi_0^\varkappa, \psi^{l,k}_{\nu - r_k, j, \varepsilon} \right) = 0.$$

Now it follows from (3.25) that

$$\tilde{c}_{\nu l j \varepsilon k} = q^{\alpha(\nu - r_k)} c_{\nu l j \varepsilon k},$$

so that the series for $D^\alpha f_1$ in (3.24) is a part of the expansion (3.25), it converges in $L_2(G)$, whence (since elements of the basis belong to $\mathcal{D}(G)$) $f_1 \in \mathfrak{D}(A_G)$. ■

Lemma 3.6 reduces the investigation of spectral properties of the operator A_G to the study of its part $A_{G,2}$ on the subspace $\mathfrak{H}_2$. Using (3.22) and (3.23), we can compute the matrix elements of the operator $A_{G,2}$ with respect to the basis $\{\psi_0^k\}$:

$$a_{kk} = \left(A_{G,2} \psi_0^k, \psi_0^k \right) = \frac{q-1}{q^{\alpha+1} - 1} q^{\alpha(1 - r_k)}; \tag{3.26}$$

$$a_{kj} = \left(A_{G,2} \psi_0^k, \psi_0^j \right) = \frac{1 - q^\alpha}{1 - q^{-\alpha-1}} q^{(r_k + r_j)/2} \|x_j - x_k\|^{-\alpha-1}, \quad j \neq k. \tag{3.27}$$

3.3.4. Self-adjointness. Let us find some classes of open sets G, for which the operator A_G is self-adjoint.

Theorem 3.8. *If an open set G is regular, then the operator A_G is self-adjoint.*

Proof. By Lemma 3.6, it is sufficient to prove that $A_{G,2}$ is self-adjoint. Since $\dim \mathfrak{H}_2 = N$, we need to consider only the case when $N = \infty$. Let $\mathfrak{S}$ be an operator on $\mathfrak{H}_2$ defined on finite linear combinations of basis functions by the diagonal matrix with the diagonal elements a_{kk}, $\mathfrak{T}$ be an operator on $\mathfrak{H}_2$ defined on the same set by the matrix $\left(a_{kj}^0\right)$, where

$$a_{kj}^0 = \begin{cases} 0, & k = j, \\ -a_{kj}, & k \neq j, \end{cases} \geq 0.$$

Evidently, $\mathfrak{S} - \mathfrak{T} \subset A_{G,2}$, and $\mathfrak{S}$ is essentially self-adjoint. We shall prove that $\mathfrak{T}$ is bounded, which will imply the self-adjointness of $A_{G,2}$.

Let us use the Schur test [51]: the operator $\mathfrak{T}$ corresponding to a symmetric matrix $\left(a_{kj}^0\right)$ with non-negative elements is bounded if there exists such a sequence $\{h_j\}$ of positive numbers that

$$\sum_{j=1}^{\infty} a_{kj}^0 h_j \leq \mu h_k, \quad k = 1, 2, \ldots; \mu > 0. \tag{3.28}$$

Moreover, $\|\mathfrak{T}\| \leq \mu$.

Set $h_j = q^{r_j/2}$. By our assumption, there exists such $m \in \mathbb{Z}$ that $\|x_j - x_k\| \geq q^m$ for any $j \neq k$. If $x \in G \setminus V_k$, then $x \in V_j$ for some $j \neq k$, and since the balls are disjoint,

$$\|x - x_k\| = \|(x - x_j) + (x_j - x_k)\| = \|x_j - x_k\| \geq q^m. \tag{3.29}$$

We have (using (3.23) and (3.29))

$$\begin{aligned} \sum_{j=1}^{\infty} a_{kj}^0 h_j &= \sum_{j \neq k} \left(-D^\alpha \psi_0^k, \psi_0^j\right) q^{r_j/2} = - \int_{G \setminus V_k} \left(D^\alpha \psi_0^k\right)(x)\, dx \\ &= \frac{q^\alpha - 1}{1 - q^{-\alpha-1}} q^{r_k/2} \int_{G \setminus V_k} \|x - x_k\|^{-\alpha-1}\, dx \\ &\leq \frac{q^\alpha - 1}{1 - q^{-\alpha-1}} q^{r_k/2} \int_{\|x - x_k\| \geq q^m} \|x - x_k\|^{-\alpha-1}\, dx = \mu h_k, \end{aligned}$$

where

$$\mu = \frac{q^\alpha - 1}{1 - q^{-\alpha-1}} \int_{\|x - x_k\| \geq q^m} \|x - x_k\|^{-\alpha-1}\, dx,$$

so that we have proved (3.28). ■

The next theorem (as well as Proposition 3.4 below) shows that A_G can be self-adjoint even for a non-closed open set G, when the centres of the balls V_k have finite accumulation points but converge to them not too rapidly.

Theorem 3.9. *If there exist such numbers $a > 0$, $\gamma \in \left(0, \frac{2\alpha+1}{2(\alpha+1)}\right)$, that*

$$\|x_i - x_k\| \geq a q^{\gamma r_k}, \quad i \neq k; \tag{3.30}$$

$$\sum_{i=1}^{\infty} q^{\eta r_i} < \infty, \quad \eta = \min\{1, 2\alpha + 1 - 2\gamma(\alpha + 1)\}, \tag{3.31}$$

then the operator A_G is self-adjoint.

Proof. Let us keep the notations from the proof of Theorem 3.8. It is sufficient to show that the operator $\mathfrak{T}(\mathfrak{S} + I)^{-1}$ can be extended to a compact operator on $\mathfrak{H}_2$ (see Corollary 2 to Theorem XIII.14 in [124]). In fact, it follows from (3.30) and (3.31) that $\mathfrak{T}(\mathfrak{S}+I)^{-1}$ is a Hilbert-Schmidt operator. Indeed, by (3.26) and (3.27) its matrix elements b_{kj} are equal to zero for $j = k$, while for $j \neq k$

$$b_{kj} = C_1 \left[q^{\alpha(1-r_k)} + C_2\right]^{-1} q^{(r_k+r_j)/2} \|x_j - x_k\|^{-\alpha-1},$$

where C_1 and C_2 are positive constants.

It follows from (3.31) that $r_i \to -\infty$ as $i \to \infty$. Since $\eta \leq 1$, the inequality (3.31) implies the inequality $\sum_{i=1}^{\infty} q^{r_i} < \infty$. Taking into account (3.30) we find that

$$\sum_{k,j=1}^{\infty} b_{kj}^2 \leq C_3 \sum_{j=1}^{\infty} q^{r_j} \sum_{k=1}^{\infty} q^{\eta r_k} < \infty, \quad C_3 > 0,$$

which was required. ■.

Example. Let $\|x_i\| = q^{M_i}$, where the sequence $\{M_i\} \in \mathbb{Z}$ tends strictly monotonically to $-\infty$, and for large i

$$r_i < M_i < \infty, \quad \frac{M_i}{r_i} \leq \gamma < \frac{2\alpha + 1}{2(\alpha + 1)}, \quad \sum_{i=1}^{\infty} q^{\eta r_i} < \infty.$$

Then the conditions of Theorem 3.9 are fulfilled – if $i \neq k$, then

$$\|x_i - x_k\| = \begin{cases} q^{M_i}, & i < k, \\ q^{M_k}, & i > k, \end{cases} \geq q^{M_k} \geq a q^{\gamma r_k}$$

for some $k > 0$.

3.3.5. Spectral properties. The arguments used in the proofs of Theorems 3.8 and 3.9 can be used in order to investigate the structure of the spectrum of the operator A_G. Assume first that the set G is non-compact.

Proposition 3.2. *Let the set G be regular:*

$$\inf_{j \neq k} \|x_j - x_k\| = q^m. \quad m \in \mathbb{Z}.$$

If $r_k \to -\infty$ as $k \to \infty$, then the spectrum of the operator A_G is discrete. If $\{\mu_n\}$ is a sequence of all the eigenvalues of the operator $A_{G,2}$ arranged in the ascending order (including the multiplicities), while the sequence $\{r_k\}$ is arranged in the descending order, then

$$\left| \mu_n - \frac{q-1}{q^{\alpha+1}-1} q^{\alpha(1-r_n)} \right| \leq \frac{q^{\alpha-1}(q-1)}{1-q^{-\alpha-1}} q^{-\alpha m}, \quad n = 1, 2, \ldots . \tag{3.32}$$

Proof. It is quite clear that the operators $A_{G,1}$ and $\mathfrak{S}$ have discrete spectra. Since $\mathfrak{T}$ is bounded, the spectrum of $A_{G,2}$ is discrete too. The inequality (3.32) is a consequence of the minimax principle and the inequality

$$\begin{aligned} \|\mathfrak{T}\| &\leq \frac{q^\alpha - 1}{1-q^{-\alpha-1}} \int_{\|x\| \geq q^m} \|x\|^{-\alpha-1}\, dx = \frac{q^\alpha - 1}{1-q^{-\alpha-1}} \sum_{l=m}^{\infty} \int_{\|x\|=q^l} dx \\ &= \frac{(q^\alpha-1)(q-1)}{q(1-q^{-\alpha-1})} \sum_{l=m}^{\infty} q^{-\alpha l} = \frac{q^{\alpha-1}(q-1)}{1-q^{-\alpha-1}} q^{-\alpha m}, \end{aligned}$$

and the proof is complete. ■.

Note that the spectrum of A_G is discrete also under the conditions of Theorem 3.9 (the proof is actually contained in the proof of that theorem).

The condition $r_k \to -\infty$ is necessary for the spectrum of A_G to be discrete, when G is non-compact (otherwise the set of eigenvalues of the operator $A_{G,1}$ has finite limit points). On the other hand, the operator $A_{G,1}$ always possesses a system of eigenfunctions, which is complete in $\mathfrak{H}_1$.

Proposition 3.3. *If* $\sum_{k=1}^{\infty} q^{-\alpha r_k} < \infty$, *then the operator* $A_{G,2}$ *is compact, so that the operator* A_G *has a complete system of eigenfunctions.*

Proof. The operator $\mathfrak{S}$ is obviously compact. Since the balls are disjoint, we have

$$\|x_j - x_k\| > q^{\max\{r_k, r_j\}} \geq q^{(r_k+r_j)/2}, \quad j \neq k,$$

whence $a_{kj} \leq Cq^{-\alpha(r_k+r_j)/2}$, so that $\mathfrak{T}$ is a Hilbert-Schmidt operator. ■

Proposition 3.4. *If*

$$\sum_{j,k=1}^{\infty} q^{(r_k+r_j)/2} \|x_j - x_k\|^{-\alpha-1} < \infty, \tag{3.33}$$

then the operator A_G *is self-adjoint, and its spectrum is purely singular*

Proof. It follows from (3.27) and (3.33) that $\mathfrak{T}$ is a trace class operator. Now it is evident that $A_{G,2}$ is self-adjoint. The singularity of the spectrum is a consequence of the Rosenblum-Kato theorem (see [66]). ■

3.3.6. Compact open sets. Let us consider the case of a compact open set G. Now in the representation (3.17) we have $N < \infty$, and $\dim \mathfrak{H}_2 = N$. The spectrum of A_G is discrete.

The first natural issue for this case is the relation between geometric characteristics of the set G and the spectral properties of the operator A_G. It is sufficient to consider the eigenvalues of the operator $A_{G,1}$.

Since a ball of the radius q^n in K can be presented as a union of q non-intersecting balls of the radius q^{n-1}, we may assume that

$$r_1 = \ldots = r_N = r$$

in the representation of a compact open set G in the form (3.17). Now the eigenvalues of the operator $A_{G,1}$ take the form $q^{\alpha(\nu-r)}$, $\nu = 1, 2, \ldots$; the multiplicity of each eigenvalue equals $N(q-1)q^{\nu-1}$.

If $q^{\alpha(\nu-r)} < \lambda$, then $\nu < \alpha^{-1} \log_q \lambda + r$ $(\lambda > 0)$. Therefore the distribution function $N_1(\lambda)$ of the eigenvalues of the operator $A_{G,1}$ is equal to

$$N_1(\lambda) = \sum_{\nu=1}^{[\alpha^{-1}\log_q \lambda]+r} N(q-1)q^{\nu-1} = (\operatorname{Vol} G) q^{[\alpha^{-1}\log_q \lambda]} + \text{const},$$

where $\operatorname{Vol} G = Nq^r = \int_G dx$ is the "volume" of the set G. Thus $\operatorname{Vol} G$ can be reconstructed from the asymptotic behaviour of the eigenvalues of the operator A_G.

Let N_0 be the type of a set G. We shall show that the type can also be reconstructed from the spectral characteristics of the operator A_G. Let $\varkappa$ be the multiplicity of an arbitrary (large enough) eigenvalue of the operator A_G. Then N_0 is the minimal non-negative residue of the number $\varkappa/(q-1)$ modulo $(q-1)$.

Indeed, $\varkappa = N(q-1)q^k$, where k is some natural number. We have

$$\frac{\varkappa}{q-1} - N_0 = Nq^k - N_0 = (N - N_0) + N(q^k - 1)$$
$$= (N - N_0) + N(q-1)(q^{k-1} + q^{k-2} + \cdots + 1) \equiv 0 \pmod{(q-1)},$$

and it remains to recall the definition of N_0.

Let $\lambda_{\min}$ be the smallest eigenvalue of the operator A_G.

Theorem 3.10. *The smallest eigenvalue of the operator A_G on an open compact set G is simple; the corresponding eigenfunction is strictly positive. The inequalities*

$$(N-1)\frac{q^\alpha - 1}{1 - q^{-\alpha-1}} q^{r-M(\alpha+1)} \le \frac{q-1}{q^{\alpha+1} - 1} q^{\alpha(1-r)} - \lambda_{\min}$$
$$\le (N-1)\frac{q^\alpha - 1}{1 - q^{-\alpha-1}} q^{r-m(\alpha+1)} \qquad (3.34)$$

hold, where

$$q^m = \min_{i \neq j} \|x_i - x_j\|, \quad q^M = \max \|x_i - x_j\|.$$

Proof. The operator A_2 is represented with respect to the basis $\{\psi_0^k\}_{k=1}^N$ by the matrix

$$\frac{q-1}{q^{\alpha+1} - 1} q^{\alpha(1-r)} I - q^r \frac{q^\alpha - 1}{1 - q^{-\alpha-1}} C,$$

where I is the unit matrix, $C = (c_{ij})$,

$$c_{ij} = \begin{cases} 0, & i = j, \\ \|x_i - x_j\|^{-\alpha-1}, & i \neq j. \end{cases}$$

Let $\widetilde{\lambda}_{\min}$ be the smallest eigenvalue of the operator A_2. Then

$$\widetilde{\lambda}_{\min} = \frac{q-1}{q^{\alpha+1} - 1} q^{\alpha(1-r)} - q^r \frac{q^\alpha - 1}{1 - q^{-\alpha-1}} \mu,$$

where μ is the largest eigenvalue of the matrix C. Since $\operatorname{tr} C = 0$, we have $\mu > 0$, whence

$$\widetilde{\lambda}_{\min} \leq \frac{q-1}{q^{\alpha+1}-1} q^{\alpha(1-r)} < q^{(k-r)\alpha}, \quad k \geq 1.$$

This means that $\lambda_{\min} = \widetilde{\lambda}_{\min}$.

Elements of the matrix C are non-negative; since all the elements of the matrix $I + C$ are strictly positive, C is indecomposable ([57], Lemma 8.4.1). By the Perron-Frobenius theorem (see [57]) the number μ coincides with the spectral radius of the matrix C; μ is a simple eigenvalue, and the corresponding eigenvector has positive coordinates. This implies our assertions regarding $\lambda_{\min}$.

In order to prove (3.34) we shall find upper and lower bounds for μ. Since the spectral radius depends monotonically on a matrix (see Corollary 8.1.19 in [57]), it suffices to find the spectral radius of the matrix

$$\begin{pmatrix} 0 & a & a \\ a & 0 & a \\ \cdots & \cdots & \cdots \\ a & a & 0 \end{pmatrix}, \quad a > 0.$$

Its eigenvalues can be found with the aid of the formula for $\det(B - \lambda I)$ given in [98] – they equal $-a$ (with the multiplicity $N-1$) and $(N-1)a$. Therefore

$$(N-1)q^{-M(\alpha+1)} \leq \mu \leq (N-1)q^{-m(\alpha+1)},$$

which implies (3.34). ■

Note that for the sphere $G = S_R$ we can write the representation (3.17) with $N = q-1$, $x_j = \xi_j \beta^{-R}$ (where ξ_j, $j = 1, \ldots q-1$, are all different elements of S, the full system of representatives of the classes from O/P, $\xi_j \notin P$), $r = R-1$. In this case $m = M = R$, the inequalities (3.34) turn into equalities, and we obtain again the formula (3.19) for $\lambda_{\min}$.

The estimates (3.34) are based on the representation of the set G as a union of balls of the same radius. While such a representation of a compact open set is always possible, its use can lead to a loss of a part of the information about the impact of the geometry of G upon the quantity $\lambda_{\min}$ (for example, if for G there is a representation (3.17) with $r_1 \gg r_k$ for $k \neq 1$). In this connection we shall give some estimates of $\lambda_{\min}$ of a somewhat different nature.

Suppose that we have a representation (3.17) with $N < \infty$ and with numbers r_k, which may in general be different. It will be assumed without loss of generality that

$$r_1 \geq r_2 \geq \ldots \geq r_N.$$

As before, we shall use the notation $q^M = \max \|x_i - x_j\|$.

Proposition 3.5. $\lambda_{\min}$ *satisfies the inequalities*

$$\frac{q-1}{q^{\alpha+1}-1} q^{\alpha(1-M)} \leq \lambda_{\min} \leq \frac{q-1}{q^{\alpha+1}-1} q^{\alpha(1-r_1)}. \tag{3.35}$$

If $r_1 = r_2$, *then*

$$\lambda_{\min} \leq \frac{q-1}{q^{\alpha+1}-1} q^{\alpha(1-r_1)} - \frac{q^\alpha - 1}{1-q^{-\alpha-1}} q^{r_1} \|x_1 - x_2\|^{-\alpha-1}. \tag{3.36}$$

Proof. Let G', G'' be compact open subsets of K. Consider the operators $A_{G'}$ and $A_{G''}$ on the Hilbert spaces $L_2(G')$ and $L_2(G'')$ respectively. Let $\{\lambda'_n\}$ and $\{\lambda''_n\}$ be the sequences of their eigenvalues (including the multiplicities) arranged in the ascending order. Then the inclusion $G' \subset G''$ implies the inequality $\lambda'_n \geq \lambda''_n$.

Indeed, extending functions from G' and G'' by zero we may assume that $L_2(G') \subset L_2(G'')$, $\mathcal{D}(G') \subset \mathcal{D}(G'')$. Since D^α is essentially self-adjoint on $\mathcal{D}(G')$ and $\mathcal{D}(G'')$, it remains to use standard variational arguments (see Sect. XIII.15 in [124]).

Now the inequalities (3.35) follow from the formula (3.18) for $\lambda_{\min}$ for the case of a ball.

Let us prove (3.36). Let $\mathfrak{P}$ be the orthoprojection in $L_2(G)$ onto the subspace X spanned by the functions ψ_0^1, ψ_0^2. Let μ be the smallest eigenvalue of the operator A_X on X defined as the restriction of the operator $\mathfrak{P} A_G$ onto X. By the Rayleigh-Ritz scheme (see Theorem XIII.3 in [124]), $\lambda_{\min} \leq \mu$.

With respect to the basis $\{\psi_0^1, \psi_0^2\}$ in X

$$A_X = \begin{pmatrix} (A_G\psi_0^1, \psi_0^1) & (A_G\psi_0^1, \psi_0^2) \\ (A_G\psi_0^2, \psi_0^1) & (A_G\psi_0^2, \psi_0^2) \end{pmatrix}.$$

The formulas (3.22), (3.23) give that

$$(A_G\psi_0^1, \psi_0^1) = (A_G\psi_0^2, \psi_0^2) = \frac{q-1}{q^{\alpha+1}-1} q^{\alpha(1-r_1)};$$

$$(A_G\psi_0^1, \psi_0^2) = (A_G\psi_0^2, \psi_0^1) = \frac{1-q^\alpha}{1-q^{-\alpha-1}} q^{r_1} \|x_1 - x_2\|^{-\alpha-1},$$

so that

$$A_X = \frac{q-1}{q^{\alpha+1}-1} q^{\alpha(1-r_1)} I - \frac{q^\alpha - 1}{1-q^{-\alpha-1}} q^{r_1} \|x_1 - x_2\|^{-\alpha-1} \begin{pmatrix} 0 & 1 \\ 1 & 0 \end{pmatrix},$$

and μ is equal to the right-hand side of (3.36). ■

3.4 An analogue of the Hamiltonian of point interactions

3.4.1. Point interactions. Point interaction Hamiltonians in quantum mechanics (see [2]) are the operator realizations in $L_2(\mathbb{R}^n)$ of differential expressions of the form $-\Delta + \delta_Y$, where δ_Y is the delta-function concentrated on the subset $Y \subset \mathbb{R}^n$. Such realizations are usually taken to be self-adjoint extensions of the operator H_0 obtained by restricting the self-adjoint operator $-\Delta$ (with the Sobolev space $H^2(\mathbb{R}^n)$ as its domain) to the set of functions vanishing on a neighbourhood of the set Y.

More general operators, for example $(-\Delta)^k + \delta_Y$, have also been considered. In the latter case, if Y is a finite set consisting of m points, then the operator H_0 is essentially self-adjoint if and only if $n \geq 4k$. If $n < 4k$, then the deficiency index of the operator H_0 depends on k, and for $n = 3$, for example, equals (m, m) when $k = 1$, and $(10m, 10m)$ when $k = 2$.

Below we consider similar constructions for the operator D^α. In order to describe self-adjoint extensions, we use the techniques proposed in [73]. In contrast to the real case, here the deficiency index does not depend on α.

3.4.2. Deficiency index. Let $\Omega \subset K$ be an open set, such that the operator A_Ω is self-adjoint. Let us write a representation (3.17) for the set Ω:

$$\Omega = \bigcup_{\varkappa=1}^{N} V_\varkappa, \quad N \leq \infty,$$

$$V_\varkappa = \{x \in K : \ \|x - x_\varkappa\| \leq q^{r_\varkappa}\}.$$

We shall consider an open subset

$$G = \Omega \setminus \{x_1, \dots, x_m\},$$

where $m \leq N$ is a fixed number, $1 \leq m < \infty$.

Theorem 3.11. *If $\alpha \leq 1/2$, then the operator A_G is self-adjoint, so that $A_G = A_\Omega$. If $\alpha > 1/2$, then the deficiency index of the operator A_G is equal to (m, m).*

It is convenient to single out the basic technical part of the proof in the form of a lemma.

Lemma 3.7. *Let $\delta_k \in \mathcal{D}'(\Omega)$ be the delta-function concentrated at the point x_k. If $\alpha > 1/2$, then the equation*

$$D^\alpha u + u = \delta_k, \quad u \in L_2(\Omega), \tag{3.37}$$

where $D^\alpha : L_2(\Omega) \to \mathcal{D}'(\Omega)$ is understood in the distribution sense, has a unique solution. If $\alpha \leq 1/2$, then the equation

$$D^\alpha u + u = \sum_{k=1}^{m} \rho_k \delta_k, \tag{3.38}$$

where $\rho_k \in \mathbb{C}$, $\sum\limits_{k=1}^{m} |\rho_k| \neq 0$, has no generalized solutions belonging to $L_2(\Omega)$.

Proof. Let $\varphi \in \mathcal{D}(\Omega)$. Expand φ in a series with respect to the basis in $L_2(\Omega)$ described in Sect. 3.3.3 (notations used there in connection with the set G are applied now to Ω):

$$\varphi(x) = \sum_{\varkappa=1}^{N} \sum_{\nu=1}^{\infty} \sum_{l=1}^{\nu} \sum_{j\in S, j\notin P} \sum_{\varepsilon} \left(\varphi, \psi^{l,\varkappa}_{\nu-r_\varkappa, j,\varepsilon}\right) \psi^{l,\varkappa}_{\nu-r_\varkappa,j,\varepsilon}(x) + \sum_{\varkappa=1}^{N} \left(\varphi, \psi_0^\varkappa\right) \psi_0^\varkappa(x). \tag{3.39}$$

Since the support $\operatorname{supp}\varphi$ is compact, it is covered by a finite number of balls $V_\varkappa$, so that the series with respect to $\varkappa$ in (3.39) are finite sums even when $N = \infty$. Let us choose a natural number n in such a way that $n\alpha > 1/2$. Obviously,

$$\left(\varphi, \psi^{l,\varkappa}_{\nu-r_\varkappa,j,\varepsilon}\right) = q^{-n\alpha(\nu-r_\varkappa)} \left(D^{n\alpha}\varphi, \psi^{l,\varkappa}_{\nu-r_\varkappa,j,\varepsilon}\right),$$

where $D^{n\alpha}\varphi \in L_2(\Omega)$, so that

$$\sum_{\varkappa,\nu,l,j\varepsilon} \left|\left(D^{n\alpha}\varphi, \psi^{l,\varkappa}_{\nu-r_\varkappa,j,\varepsilon}\right)\right|^2 < \infty,$$

It is clear from the formulas for the basis functions that

$$\left|\psi^{l,\varkappa}_{\nu-r_\varkappa,j,\varepsilon}(x)\right| \leq C q^{(\nu-l)/2}.$$

Now it is easy to see that the series in (3.39) is uniformly convergent: the absolute value of the general term of the series with respect to ν does not exceed

$$\sum_{l=1}^{\nu} \sum_{j\in S, j\notin P} \sum_{\varepsilon} q^{(\nu-l)/2-n\alpha\nu} \left|\left(D^{n\alpha}\varphi, \psi^{l,\varkappa}_{\nu-r_\varkappa,j,\varepsilon}\right)\right|$$

$$\leq C \left\{\sum_{l,j,\varepsilon} \left|\left(D^{n\alpha}\varphi, \psi^{l,\varkappa}_{\nu-r_\varkappa,j,\varepsilon}\right)\right|^2\right\}^{1/2} \left\{\sum_{l=1}^{\nu} \sum_{j\in S, j\notin P} \sum_{\varepsilon} q^{\nu-l-2n\alpha\nu}\right\}^{1/2},$$

where (since the number of terms in the sum over ε equals $(q-1)q^{l-2}$)

$$\sum_{l=1}^{\nu}\sum_{j\in S, j\notin P}\sum_{\varepsilon} q^{\nu-l-2n\alpha\nu} = (1-q^{-1})^2 \nu q^{-\nu(2n\alpha-1)}.$$

In particular, the equality (3.39) holds for $x = x_k$. Note that

$$\psi^{l,\varkappa}_{\nu-r_\varkappa,j,\varepsilon}(x_k) = \begin{cases} 0, & k=\varkappa, l>1, \\ q^{(\nu-r_k-1)/2}, & k=\varkappa, l=1, \\ 0, & k\neq\varkappa, \end{cases}$$

$$\psi_0^{\varkappa}(x_k) = \begin{cases} q^{-r_k/2}, & k=\varkappa, \\ 0, & k\neq\varkappa. \end{cases}$$

Therefore

$$\varphi(x_k) = \sum_{\nu=1}^{\infty}\sum_{j\in S,j\notin P}\left(\varphi, \psi^{1,k}_{\nu-r_k,j,0}\right) q^{(\nu-r_k-1)/2} + \left(\varphi,\psi_0^k\right) q^{-r_k/2}. \tag{3.40}$$

The equality (3.40) means that

$$\delta_k = \sum_{\nu=1}^{\infty}\sum_{j\in S,j\notin P} q^{(\nu-r_k-1)/2}\overline{\psi^{1,k}_{\nu-r_k,j,0}} + q^{-r_k/2}\psi_0^k, \tag{3.41}$$

where the series converges in $\mathcal{D}'(\Omega)$.

Suppose that a function $u \in L_2(\Omega)$ is represented as a convergent series in $L_2(\Omega)$:

$$u = \sum_{\varkappa=1}^{N}\sum_{\nu=1}^{\infty}\sum_{l=1}^{\nu}\sum_{j\in S,j\notin P}\sum_{\varepsilon} c_{\varkappa\nu lj\varepsilon}\overline{\psi^{l,\varkappa}_{\nu-r_\varkappa,j,\varepsilon}} + \sum_{\varkappa=1}^{N} d_\varkappa \psi_0^\varkappa. \tag{3.42}$$

Applying the operator D^α termwise, we get a series converging in $\mathcal{D}'(\Omega)$. Note that

$$D^\alpha\overline{\psi^{l,\varkappa}_{\nu-r_\varkappa,j,\varepsilon}} = q^{(\nu-r_\varkappa)\alpha}\overline{\psi^{l,\varkappa}_{\nu-r_\varkappa,j,\varepsilon}}, \quad D^\alpha\psi_0^\varkappa = \sum a_{\varkappa j}\psi_0^j$$

(see (3.26), (3.27)). If $\varphi \in \mathcal{D}(\Omega)$, then

$$\left\langle \sum_{\varkappa=1}^{N} d_\varkappa D^\alpha \psi_0^\varkappa, \varphi \right\rangle = \sum_{\varkappa=1}^{N} d_\varkappa \sum_{\mu=1}^{N} a_{\varkappa\mu}\langle \psi_0^\mu, \varphi\rangle$$

$$= \sum_{\mu=1}^{N}\left(\sum_{\varkappa=1}^{N} d_\varkappa a_{\varkappa\mu}\right)\langle \psi_0^\mu, \varphi\rangle,$$

where the order of summation can be changed because the series with respect to μ contains only a finite number of nonzero terms (due to the compactness of $\operatorname{supp}\varphi$), while the series with respect to $\varkappa$ is absolutely convergent: $\{d_k\} \in l_2$ and $\{a_{\varkappa\mu}\} \in l_2$ for each fixed μ. Thus

$$D^\alpha\left(\sum_{\varkappa=1}^N d_\varkappa \psi_0^\varkappa\right) = \sum_{\mu=1}^N \left(\sum_{\varkappa=1}^N d_\varkappa a_{\varkappa\mu}\right) \psi_0^\mu .$$

For the distribution $D^\alpha u + u$ we get the expansion

$$D^\alpha u + u = \sum_{\varkappa,\nu,l,j,\varepsilon} \left[q^{(\nu-r_\varkappa)\alpha} + 1\right] c_{\varkappa\nu lj\varepsilon} \overline{\psi^{l,\varkappa}_{\nu-r_\varkappa,j,\varepsilon}} + \sum_{\mu=1}^N \left(\sum_{\varkappa=1}^N d_\varkappa a_{\varkappa\mu} + d_\mu\right) \psi_0^\mu, \tag{3.43}$$

which converges in $\mathcal{D}'(\Omega)$.

Let us consider the case when $\alpha > 1/2$. In order to construct a solution of the equation (3.37), we take a function (3.42), where $c_{\varkappa\nu lj\varepsilon} = 0$ for $\varkappa \neq k$, and also for $\varkappa = k$, $l > 1$,

$$c_{k\nu 1jo} = q^{(\nu-r_k-1)/2}\left[q^{(\nu-r_k)\alpha} + 1\right]^{-1}, \tag{3.44}$$

and the coefficients $d_\varkappa$ are determined from the system of equations

$$\sum_{\varkappa=1}^N d_\varkappa a_{\varkappa\mu} + d_\mu = \begin{cases} q^{-r_k/2}, & \mu = k, \\ 0, & \mu \neq k, \end{cases} \tag{3.45}$$

$\mu = 1, \dots, N$.

The system (3.45) is solvable in l_2 because the operator $A_{\Omega,2}$ is self-adjoint and non-negative. Next,

$$\sum_{\nu=1}^\infty \sum_{j\in S, j\notin P} |c_{k\nu 1j0}|^2 \leq C \sum_{\nu=1}^\infty q^{-(2\alpha-1)\nu} < \infty,$$

so that $u \in L_2(\Omega)$. Comparing (3.41), (3.43), (3.44), (3.45), we see that u is a solution of (3.37). The uniqueness follows from the fact that A_Ω is self-adjoint and non-negative.

Suppose now that $\alpha < 1/2$. Assume that a function $u \in L_2(\Omega)$ of the form (3.42) satisfies the equation (3.38). Arguing as above, we find that

$$c_{\varkappa\nu lj\varepsilon} = \begin{cases} 0, & \varkappa > m, \\ \rho_\varkappa q^{(\nu-r_\varkappa-1)/2}\left[q^{(\nu-r_\varkappa)\alpha} + 1\right]^{-1}, & \varkappa \leq m, l = 1, \\ 0, & \varkappa \leq m, l > 1, \end{cases}$$

and at the same time $\sum |c_{\varkappa\nu lj\varepsilon}|^2 < \infty$. However for each $\varkappa$

$$|c_{\varkappa\nu 1j0}|^2 \geq hq^{\nu(1-2\alpha)}|\rho_\varkappa|^2, \quad h > 0,$$

whence

$$\sum_{\nu,l,j,\varepsilon} |c_{\varkappa\nu lj\varepsilon}|^2 = \infty$$

if $\rho_\varkappa \neq 0$, and we have arrived at a contradiction. ■

Proof of Theorem 3.11. Let $\mathfrak{N}$ be the orthogonal complement in $L_2(G)$ of the range of the operator $A_G + I$. It suffices to verify that

$$\dim \mathfrak{N} = \begin{cases} 0, & \alpha \leq 1/2, \\ m, & \alpha > 1/2. \end{cases}$$

If $u \in \mathfrak{N}$, then for any $\varphi \in \mathcal{D}(G)$

$$(u, D^\alpha \varphi + \varphi) = 0.$$

This means that the distribution $D^\alpha u + u \in \mathcal{D}'(\Omega)$ is concentrated on the finite set $\{x_1, \dots, x_m\}$.

Let

$$\varphi_k(x) = \begin{cases} 1, & x \in V_k, \\ 0, & x \notin V_k, \end{cases} \qquad k = 1, \dots, m.$$

Then $D^\alpha u + u = \sum_{k=1}^{m} v_k$ on Ω, where $v_k = \varphi_k(D^\alpha u + u)$. It is clear that $v_1, \dots, v_m \in \mathcal{D}'(K)$ are concentrated at the points $\{x_1, \dots, x_m\}$ respectively. By Theorem 1.9

$$v_k = \rho_k \delta_k, \quad \rho_k \in \mathbb{C} \ (k = 1, \dots, m).$$

Thus the function $u \in \mathfrak{N}$ satisfies the equation (3.38). If $\alpha \leq 1/2$, then $u = 0$ by Lemma 3.7, and this is what was required. Let $\alpha > 1/2$. Denote by u_k the solution of the equation (3.37). Obviously, $u_k \in \mathfrak{N}$ ($k = 1, \dots, m$), and the functions u_k are linearly independent. If $w = \sum_{k=1}^{m} \rho_k u_k$, then$D^\alpha(u - w) + (u - w) = 0$ on Ω, $u - w \in L_2(\Omega)$, so that $u = w$ because A_Ω is self-adjoint and non-negative. Hence $\{u_1, \dots, u_m\}$ is a basis in $\mathfrak{N}$. ■

The self-adjointness of the operator A_G for $\alpha < 1/2$ can be deduced also from the next proposition, which contains a simple method of approximating a function from $\mathcal{D}(\Omega)$ by functions from $\mathcal{D}(G)$ (see Chapter X, Problem 9 in [122] for a similar result for the Laplace operator on $\mathbb{R}^n \setminus \{0\}$).

Denote

$$\eta(r) = \begin{cases} 0, & r \le 1, \\ 1, & r > 1, \end{cases}$$

$$\varphi_n(x) = \prod_{j=1}^{m} \eta(q^n \|x - x_j\|), \quad x \in \Omega, \ n = 1, 2, \ldots .$$

Proposition 3.6. *If $\alpha < 1/2$, then for any function $f \in \mathcal{D}(\Omega)$*

$$\varphi_n f \to f, \quad D^\alpha(\varphi_n f) \to D^\alpha f, \quad n \to \infty,$$

in the space $L_2(\Omega)$.

Proof. The first limit property is verified in a quite simple way: if n is large enough, then

$$\|\varphi_n f - f\|_{L_2(\Omega)} = \sum_{k=1}^{m} \int_{V_k} |\varphi_n(x) f(x) - f(x)|^2 \, dx$$
$$= \sum_{k=1}^{m} \int_{\|x - x_k\| \le q^{-n}} |f(x)|^2 \, dx \longrightarrow 0, \quad n \to \infty.$$

Let us consider the expression for $D^\alpha(\varphi_n f)$. We have

$$(D^\alpha(\varphi_n f))(x) = \varphi_n(x)(D^\alpha f)(x) + \frac{1 - q^\alpha}{1 - q^{-\alpha-1}} \int_K f(x - y) \frac{\varphi_n(x - y) - \varphi_n(x)}{\|y\|^{\alpha+1}} \, dy. \quad (3.46)$$

As before, the first summand in the right-hand side of (3.46) tends to $D^\alpha f$ in $L_2(\Omega)$ as $n \to \infty$. It follows from the identity

$$a_1 \ldots a_m - b_1 \ldots b_m = (a_1 - b_1) a_2 \ldots a_m + b_1(a_2 - b_2) a_3 \ldots a_m + b_1 b_2 (a_3 - b_3) \ldots a_m + b_1 b_2 b_3 \ldots (a_m - b_m)$$

that the square of the $L_2(\Omega)$-norm of the second summand in the right-hand side of (3.46) does not exceed

$$C\sum_{j=1}^{m}\int_{\Omega}\left[\int_{K}\frac{|\eta(q^n\|x-x_j-y\|)-\eta(q^n\|x-x_j\|)|}{\|y\|^{\alpha+1}}\,dy\right]^2 dx$$

$$\leq Cm\int_{K}\left[\int_{K}\frac{|\eta(q^n\|x-y\|)-\eta(q^n\|x\|)|}{\|y\|^{\alpha+1}}\,dy\right]^2 dx$$

$$= Cmq^{-n(1-2\alpha)}\int_{K}\left[\int_{K}\frac{|\eta(\|\xi-\zeta\|)-\eta(\|\xi\|)|}{\|\zeta\|^{\alpha+1}}\,d\zeta\right]^2 d\xi,$$

where we have still to show that the iterated integral on the right is finite.

Denote

$$I(\xi)=\int_{K}\frac{|\eta(\|\xi-\zeta\|)-\eta(\|\xi\|)|}{\|\zeta\|^{\alpha+1}}\,d\zeta.$$

It is clear that $\|I(\xi)\| = \text{const}$ if $\|\xi\| \leq 1$. Let $\|\xi\| > 1$. Then the integrand is non-zero only for $\|\zeta\| > 1$, so that

$$I(\xi)=\int_{\|\zeta\|>1}|\eta(\|\xi-\zeta\|)-1|\cdot\|\zeta\|^{-\alpha-1}\,d\zeta$$

$$=\int_{\|\zeta\|=\|\xi\|}|\eta(\|\xi-\zeta\|)-1|\cdot\|\zeta\|^{-\alpha-1}\,d\zeta$$

$$=\|\xi\|^{-\alpha}\int_{\|\tau\|=1}|\eta(\|\xi\|\cdot\|1-\tau\|)-1|\,d\tau.$$

Let us consider the canonical representation of an element τ,

$$\tau=a_0+a_1\beta+a_2\beta^2+\cdots;\quad a_0,a_1,\ldots\in S,\ a_0\notin P.$$

It is convenient to assume that $0, 1 \in S$. If $a_0 \neq 1$, then $\|1-\tau\| = 1$, and the integrand equals zero. If $a_0 = 1$, $a_1 \neq 0$, then $\|1-\tau\| = q^{-1}$,

$$|\eta(\|\xi\|\cdot\|1-\tau\|)-1|=\begin{cases}1, & \text{if } \|\xi\|\leq q,\\ 0, & \text{if } \|\xi\|>q.\end{cases}$$

Further, if $a_0 = 1$, $a_1 = \ldots = a_{l-1} = 0$, $a_l \neq 0$, then $\|1-\tau\| = q^{-l}$, that is

$$|\eta(\|\xi\|\cdot\|1-\tau\|)-1|=\begin{cases}1, & \text{if } \|\xi\|\leq q^l,\\ 0, & \text{if } \|\xi\|>q^l.\end{cases}$$

It follows from the invariance of the Haar measure that the measure of this set of elements τ equals $(1-q^{-1})q^{-l}$.

Therefore if $\|\xi\| = q^k$, $k > 0$, then

$$I(\xi) = \|\xi\|^{-\alpha}(1-q^{-1})\sum_{l=k}^{\infty} q^{-l} = \|\xi\|^{-\alpha-1}.$$

Now we find from the above estimates that the $L_2(\Omega)$-norm of the second summand in the right-hand side of (3.46) tends to zero as $n \to \infty$. ■

3.4.3. Self-adjoint extensions. Let $\alpha > 1/2$. For this case we shall describe all self-adjoint extensions of the operator A_G. Let, as before, $u_k \in L_2(\Omega)$ be the solution of the equation (3.37), $k = 1, 2, \dots, m$. Denote by $\mathfrak{H}_0$ the subspace of $L_2(\Omega)$ spanned by the functions $u_1, \dots, u_m$. It was shown in the course of proving Theorem 3.11 that $\mathfrak{H}_0 = \ker(A_G^* + I)$. For the domain of the operator A_G^* (or $A_G^* + I$, which is the same) we obtain the equality

$$\mathfrak{D}(A_G^*) = \mathfrak{D}(A_\Omega) \dotplus \mathfrak{H}_0.$$

If $u \in \mathfrak{D}(A_G^*)$, then $u = f + \tau$, $f \in \mathfrak{D}(A_\Omega)$, $\tau \in \mathfrak{H}_0$,

$$A_G^* u = (A_G^* + I)u - u = (A_\Omega + I)f - u.$$

Let $\mathfrak{P}$ be the orthoprojection on $\mathfrak{H}_0$ in the space $L_2(\Omega)$.

Proposition 3.7. *The domain of every self-adjoint extension of the operator A_G is a part of $\mathfrak{D}(A_G^*)$ specified by the condition*

$$(k - I_0)\mathfrak{P}(A_\Omega + I)f + i(k + I_0)\tau = 0, \quad f + \tau \in \mathfrak{D}(A_G^*), \tag{3.47}$$

where k is an arbitrary unitary operator on $\mathfrak{H}_0$, I_0 and I are the unit operators $\mathfrak{H}_0$ and $L_2(\Omega)$ respectively.

The *proof* is a consequence of general results from the extension theory of symmetric operators (see [45, 46, 96]) applied to the positive definite symmetric operator $A_G + I$. ■

As in [73], it is easy to describe some other classes of extensions – maximal dissipative extensions, solvable extensions etc.

Example. Let $\Omega = K$, $m = 1$, $x_1 = 0$. For this case the condition (3.47) can be written in a simpler form. The subspace $\mathfrak{H}_0$ is now one-dimensional; if $u_0 = \|u_1\|_{L_2(\Omega)}^{-1} u_1$, then the orthoprojection $\mathfrak{P}$ has the form

$$\mathfrak{P}u = (u, u_0)u_0, \quad u \in L_2(K).$$

Thus for any $f \in \mathcal{D}(K)$

$$\mathfrak{P}(D^\alpha + I)f = f(0)u_0, \tag{3.48}$$

where we write D^α instead of A_Ω.

By arguing as in the proof of Lemma 3.7 it is not hard to see that the expansion of any function $f \in \mathfrak{D}(D^\alpha)$ with respect to the eigenbasis is uniformly convergent. Hence f is continuous, and its value at a point makes sense. Moreover, the same arguments prove the continuity of the imbedding of $\mathfrak{D}(D^\alpha)$ (with the graph norm) into the space of bounded continuous functions. Since D^α coincides with the closure of its restriction to $\mathcal{D}(K)$, the equality (3.48) holds for any $f \in \mathfrak{D}(D^\alpha)$.

An element $\tau \in \mathfrak{H}_0$ has a unique representation in the form $\tau = au_0$, $a \in \mathbb{C}$. Denoting

$$c = i(k-1)^{-1}(k+1), \quad k \neq 1,$$

we come to the following description of all self-adjoint extensions. One of them (corresponding to $k = 1$) is the operator D^α. The others are parametrized by the number $c \in \mathbb{R}$. For a fixed c the domain of the corresponding extension consists of the functions $u = f + au_0$ ($f \in \mathfrak{D}(D^\alpha)$, $a \in \mathbb{C}$), for which $f(0) + ca = 0$.

Note that the function u_0 can be computed explicitly by applying the Fourier transform to (3.37) and using integration formulas of Chapter 1.

3.5 Multiplicative fractional differentiation

Let us consider the operator D^α on the sphere S_1, that is on the group of units U of the field K. As we saw, the spectrum of the operator A_U consists of the eigenvalues

$$\lambda_0 = \frac{q^{\alpha-1} - 2q^{-1} + 1}{1 - q^{-\alpha-1}} \quad \text{(multiplicity 1)},$$

$$\lambda_1 = q^\alpha \quad \text{(multiplicity } q-2\text{)},$$

$$\lambda_k = q^{k\alpha} \quad \text{(multiplicity } (q-1)^2 q^{k-2}\text{)}, \; k = 2, 3, \ldots .$$

One can notice that the multiplicity of λ_k coincides with the number of (multiplicative) characters of the group U with the ramification degree k (see Sect. 1.3).

In this section we show that the coincidence is not accidental – the multiplicative characters are actually the eigenfunctions of the operator A_U, it admits a hyper-singular integral representation based on the multiplicative structure.

Theorem 3.12. (i) *If θ is a character of U with the ramification degree k, then $A_U\theta = \lambda_k\theta$.*

(ii) *If $\varphi \in \mathcal{D}(U)$, then*

$$(A_U\varphi)(x) = \frac{1-q^\alpha}{1-q^{-\alpha-1}} \int_U \|1-y\|^{-\alpha-1}[\varphi(xy^{-1}) - \varphi(x)]\,dy + \lambda_0\varphi(x), \quad x \in U. \tag{3.49}$$

Proof. (i) It is sufficient to consider the case $k \geq 1$, since the case $k = 0$ has been dealt with above. According to Theorem 1.14, the Fourier transform

$$(F\theta)(\xi) = \int_U \chi(x\xi)\theta(x)\,dx = 0$$

if $\|\xi\| \neq q^k$. Now it follows from the definition of D^α as a pseudo-differential operator that

$$A_U\theta = q^{\alpha k}F^{-1}(F\theta) = \lambda_k\theta.$$

(ii) In order to prove (3.49) denote the expression in the right-hand side by $M_\alpha\varphi$. It suffices to show that

$$M_\alpha\theta = \lambda_k\theta. \tag{3.50}$$

for each character θ with the ramification degree k.

The equality (3.50) is obvious for $k = 0$. If $k \geq 1$, then

$$(M_\alpha\theta)(x) = \theta(x)\left\{\frac{1-q^\alpha}{1-q^{-\alpha-1}} \int_U \|1-y\|^{-\alpha-1}[\overline{\theta(y)} - 1]\,dy + \lambda_0\right\}, \quad x \in U. \tag{3.51}$$

Consider the function

$$f(z) = \int_U \|1-y\|^{z-1}[\overline{\theta(y)} - 1]\,dy.$$

It is an entire function due to the local constancy of a character.

As in the proof of Proposition 3.6, we can consider the canonical representation of an element $y \in U$,

$$y = a_0 + a_1\beta + a_2\beta^2 + \cdots; \ a_0, a_1, \ldots \in S, \ a_0 \notin P,$$

assuming that $0, 1 \in S$. It follows from the invariance of the Haar measure that the set of such $y \in U$ that $a_0 \neq 1$ has the measure $1 - \frac{2}{q}$. On this set $\|1 - y\| = 1$. Further, for any $l \geq 1$, the set of such elements y that $a_0 = 1$, $a_j = 0$ for $0 < j < l$, $a_l \neq 0$ (on this set $\|1 - y\| = q^{-l}$) has the measure $(1 - q^{-1})q^{-l}$. Therefore if $z > 0$, then

$$\int_U \|1 - y\|^{z-1}\, dy = 1 - \frac{2}{q} + (1 - q^{-1}) \sum_{l=1}^{\infty} q^{-l(z-1)} \cdot q^{-l} = \frac{1 - q^{-1}}{1 - q^{-z}} - q^{-1}.$$

In order to calculate the integral containing $\overline{\theta(y)}$ we use the decomposition

$$U = \bigcup_{j=0}^{\infty} (V_j \setminus V_{j+1}),$$

where $V_0 = U$, $V_j = \{y \in U : \|1 - y\| \leq q^{-j}\}$, $j \geq 1$. Using the identities (1.30) we find that

$$\int_U \|1 - y\|^{z-1} \overline{\theta(y)}\, dy = -q^{-(z-1)(k-1)-k} + (1 - q^{-1}) \sum_{j=k}^{\infty} q^{-jz}$$
$$= q^{-kz} \frac{1 - q^{z-1}}{1 - q^{-z}}$$

if $z > 0$. Thus

$$f(z) = q^{-kz} \frac{1 - q^{z-1}}{1 - q^{-z}} - \frac{1 - q^{-1}}{1 - q^{-z}} + q^{-1}, \quad z > 0.$$

Due to the analyticity of f this equality remains valid for $z = -\alpha$. Substituting it into (3.51) we come to (3.50). ■

3.6 Comments

The Vladimirov functions were introduced for $K = \mathbb{Q}_p$ by Vladimirov (see [154]), and for a general local field K by the author [81]. Our assumption that the characteristic of the residue field is different from 2 excludes the case $K = \mathbb{Q}_2$, which was also considered in [154].

Theorem 3.2 is contained essentially in the paper [60] by Ismagilov, Theorem 3.3 is a special case of results by Vladimirov [151, 154]. Other results of Sect. 3.2 are taken from [75]. A detailed analysis of the spectrum and eigenfunctions of the Schrödinger type operator $D^\alpha + V(|x|_p)$ over $\mathbb{Q}_p$, with $V(r) \to \infty$ as $r \to \infty$, is given in [154]; the case of a "harmonic oscillator" ($\alpha = 2$, $V(r) = r^2$) was considered also by Blair [16].

The eigenvalues and eigenfunctions of D^α for a ball and a sphere were found by Vladimirov [151, 154]; the general theorems of Sect. 3.3 appeared for the first time in [78] (for $K = \mathbb{Q}_p$), as well as the results regarding a p-adic analogue of point interactions. Theorem 3.12 is taken from [85].

Chapter 4

Parabolic Equations

4.1 An analogue of the heat equation

4.1.1. Fundamental solution. Let us consider the Cauchy problem

$$\frac{\partial u(x,t)}{\partial t} + a(D^\alpha u)(x,t) = f(x,t), \quad x \in K, 0 < t \le T, \tag{4.1}$$

$$u(x,0) = \varphi(x), \quad x \in K, \tag{4.2}$$

where $a > 0$, $\alpha > 0$, K is a local field, the right-hand side f and the initial function φ are complex- or real-valued.

In this section we shall construct and investigate a *fundamental solution* of the Cauchy problem (4.1)-(4.2), that is such a function $Z(x,t)$, $x \in K$, $0 < t \le T$, that a solution $u(x,t)$ can be represented as

$$u(x,t) = \int_K Z(x-\xi,t)\varphi(\xi)\,d\xi + \int_0^t d\tau \int_K Z(x-\xi,t-\tau)f(\xi,\tau)\,d\tau, \tag{4.3}$$

provided φ and f satisfy some natural regularity conditions.

We set

$$Z(x,t) = \int_K \chi(-x\xi)e^{-at\|\xi\|^\alpha}\,d\xi. \tag{4.4}$$

As in the preceding chapters, we follow our notations regarding local fields; in particular, $\|\cdot\|$ is the normalized absolute value on K, χ is a rank zero additive character. As before, we shall denote various positive constants by the same letter C.

Using the integration formula (1.28) we find that

$$Z(x,t) = (1-q^{-1})\|x\|^{-1}\sum_{\nu=0}^{\infty} q^{-\nu}\exp\left(-atq^{-\alpha\nu}\|x\|^{-\alpha}\right) - \|x\|^{-1}\exp\left(-atq^{\alpha}\|x\|^{-\alpha}\right), \quad x\neq 0.$$

If we expand the exponential function in a power series, change the order of summation, and sum the geometric progression, we find that for $x\neq 0$

$$Z(x,t) = \sum_{m=1}^{\infty}\frac{(-1)^m}{m!}\cdot\frac{1-q^{\alpha m}}{1-q^{-\alpha m-1}}(at)^m\|x\|^{-\alpha m-1}. \tag{4.5}$$

It follows from (4.4) that $Z(x,t)$ is continuous with respect to $x\in K$. As we saw in Sect. 2.1, the function $\xi\mapsto\exp(-at\|\xi\|^{\alpha})$ is positive definite. Thus $Z(x,t)\geq 0$ for all x,t. This can be verified also by a direct calculation – according to (4.4),

$$\begin{aligned} Z(x,t) &= \sum_{k=-\infty}^{\infty}\exp(-atq^{k\alpha})\int_{\|\xi\|=q^k}\chi(-x\xi)\,d\xi \\ &= \sum_{k=-\infty}^{\infty}\exp(-atq^{k\alpha})\left[q^k\Omega_{-k}(x) - q^{k-1}\Omega_{-k+1}(x)\right] \\ &= \sum_{k=-\infty}^{\infty} q^k\left[\exp(-atq^{k\alpha}) - \exp(-atq^{k\alpha+\alpha})\right]\Omega_{-k}(x), \end{aligned}$$

where Ω_k is the indicator of the ball $\{x\in K:\ \|x\|\leq q^k\}$. It follows even that $Z(x,t)>0$ for all x,t.

Lemma 4.1.

$$Z(x,t)\leq Ct(t^{1/\alpha}+\|x\|)^{-\alpha-1},\quad x\in K,\ t>0. \tag{4.6}$$

Proof. By (4.4)

$$Z(x,t)\leq\int_K\exp(-at\|\xi\|^{\alpha})\,d\xi. \tag{4.7}$$

Let an integer k be such that $q^{k-1} \le t^{1/\alpha} \le q^k$. Then

$$Z(x,t) \le \int_K \exp(-aq^{\alpha(k-1)}\|\xi\|^\alpha)\,d\xi = \int_K \exp(-a\|\beta^{-(k-1)}\xi\|^\alpha)\,d\xi$$
$$= q^{-(k-1)} \int_K \exp(-a\|\eta\|^\alpha)\,d\eta = Ct^{-1/\alpha}.$$

On the other hand, it follows from (4.5) that if $\|x\| \ge t^{1/\alpha}$, then

$$Z(x,t) \le \|x\|^{-1} \sum_{m=1}^{\infty} \frac{C^m}{m!}(t\|x\|^{-\alpha})^m \le Ct\|x\|^{\alpha-1}.$$

The last two inequalities imply (4.6). Indeed, if $\|x\| \ge t^{1/\alpha}$, then

$$\|x\|^{-\alpha-1} \le \left(\frac{1}{2}\|x\| + \frac{1}{2}t^{1/\alpha}\right)^{-\alpha-1} = 2^{\alpha+1}\left(\|x\| + t^{1/\alpha}\right)^{-\alpha-1}.$$

If $\|x\| < t^{1/\alpha}$, then

$$\left(\|x\| + t^{1/\alpha}\right)^{-\alpha-1} \ge 2^{-\alpha-1}t^{-1-1/\alpha},$$

so that $t^{-1/\alpha} \le 2^{\alpha+1}t\left(\|x\| + t^{1/\alpha}\right)^{-\alpha-1}$. ■

The inequality (4.6) shows in particular that the function Z belongs (with respect to the variable x) to $L_2(K)$, so that we find from (4.4) and the Fourier inversion formula that

$$\int_K Z(x,t)\,dx = 1. \tag{4.8}$$

If φ is a locally constant function on K, and $|\varphi(x)| \le (1+\|x\|^\lambda)$, $\lambda < \alpha$, then

$$\int_K Z(x-\xi,t)\varphi(\xi)\,d\xi \longrightarrow \varphi(x), \quad t \to 0.$$

Indeed, by (4.8)

$$\int_K Z(x-\xi,t)\varphi(\xi)\,d\xi = \int_K Z(x-\xi,t)[\varphi(\xi) - \varphi(x)]\,d\xi + \varphi(x),$$

and it remains to use the inequality (4.6) and the local constancy of φ.

Let us consider the derivative $\dfrac{\partial Z}{\partial t}$. It is clear that one may differentiate in (4.4) under the integral sign:

$$\frac{\partial Z(x,t)}{\partial t} = -a \int\limits_K \|\xi\|^\alpha \chi(-x\xi) \exp(-at\|\xi\|^\alpha)\, d\xi. \tag{4.9}$$

If $x \neq 0$, then in accordance with (4.5)

$$Z(x+\xi, t) = Z(x,t), \quad \|\xi\| < \|x\|.$$

Therefore we can define (by (2.8)) the function

$$Z_\gamma(x,t) = (D_x^\gamma Z)(x,t), \quad 0 < \gamma \le \alpha.$$

In order to compute this function, denote

$$Z^{(m)}(x,t) = \int\limits_{\|\eta\| \le q^m} \chi(-x\eta) \exp(-at\|\eta\|^\alpha)\, d\eta.$$

$Z^{(m)}$ is bounded and locally constant with the exponent m in the variable x. We have

$$\begin{aligned}
&\left(D_x^\gamma Z^{(m)}\right)(x,t) \\
&\quad = \frac{1}{\Gamma_K(-\gamma)} \int\limits_{\|\xi\| > q^{-m}} \|\xi\|^{-\gamma-1} \left[Z^{(m)}(x-\xi,t) - Z^{(m)}(x,t)\right] d\xi \\
&= \int\limits_{\|\eta\| \le q^m} \exp(-at\|\eta\|^\alpha)\, d\eta \int\limits_{\|\xi\| > q^{-m}} \frac{\|\xi\|^{-\gamma-1}}{\Gamma_K(-\gamma)} [\chi(-(x-\xi)\eta) - \chi(-x\eta)]\, d\xi \\
&\quad = \int\limits_{\|\eta\| \le q^m} \exp(-at\|\eta\|^\alpha) \chi(-x\eta)\, d\eta \int\limits_K \frac{\|\xi\|^{-\gamma-1}}{\Gamma_K(-\gamma)} [\chi(\xi\eta) - 1]\, d\xi.
\end{aligned}$$

By Proposition 2.3 the inner integral equals $\|\eta\|^\gamma$, so that

$$\left(D_x^\gamma Z^{(m)}\right)(x,t) = \int\limits_{\|\eta\| \le q^m} \chi(-x\eta) \|\eta\|^\gamma \exp(-at\|\eta\|^\alpha)\, d\eta.$$

According to (1.28), $Z^{(m)}$ depends actually on $\|x\|$. Fixing $x \neq 0$ we find that

$$\left(D_x^\gamma Z^{(m)}\right)(x,t)$$
$$= \frac{1}{\Gamma_K(-\gamma)} \int\limits_{\|\xi\|\geq\|x\|} \|\xi\|^{-\gamma-1} \left[Z^{(m)}(x-\xi,t) - Z^{(m)}(x,t)\right] d\xi$$
$$\longrightarrow \left(D_x^\gamma Z\right)(x,t)$$

as $m \to \infty$, by the dominated convergence theorem. Now it follows that

$$\left(D_x^\gamma Z\right)(x,t) = \int\limits_K \chi(-x\eta)\|\eta\|^\gamma \exp(-at\|\eta\|^\alpha)\, d\eta, \quad x \neq 0. \tag{4.10}$$

The next lemma is proved in the same way as Lemma 4.1.

Lemma 4.2.

$$\left|\frac{\partial Z(x,t)}{\partial t}\right| \leq C(t^{1/\alpha} + \|x\|)^{-\alpha-1} \tag{4.11}$$

$$\left|\left(D_x^\gamma Z\right)(x,t)\right| \leq C(t^{1/\alpha} + \|x\|)^{-\gamma-1}. \tag{4.12}$$

It follows from (4.12) and the Fourier inversion formula that

$$\int\limits_K \left(D_x^\gamma Z\right)(x,t)\, dx = 0. \tag{4.13}$$

4.1.2. Heat potentials and Cauchy problem. Denote by $\mathfrak{M}_\gamma$ ($\gamma \geq 0$) the class of complex-valued locally constant functions $\psi(x)$ on K, such that

$$|\psi(x)| \leq C(1 + \|x\|^\gamma).$$

If a function ψ depends also on a parameter t, we shall say that $\psi \in \mathfrak{M}_\gamma$ *uniformly with respect to* t, if its constant C and its exponent of local constancy do not depend on t.

Let us study the "heat potential"

$$u(x,t,\tau) = \int\limits_\tau^t d\theta \int\limits_K Z(x-y, t-\theta) f(y,\theta)\, dy.$$

We shall assume that $f \in \mathfrak{M}_\lambda$ ($0 \leq \lambda < \alpha$) uniformly with respect to θ, and f is continuous in (y,θ).

Let us prove that $u \in \mathfrak{M}_\lambda$ uniformly with respect to t, τ. Writing $u(x,t,\tau)$ as

$$u(x,t,\tau) = \int_\tau^t d\theta \int_K Z(\xi, t-\theta) f(x-\xi, \theta)\, d\xi,$$

we see that $u(x,t,\tau)$ has the same exponent of local constancy N as f. Furthermore, by Lemma 4.1 we have

$$|u(x,t,\tau)| \leq C \int_\tau^t (t-\theta)\, d\theta \int_K [(t-\theta)^{1/\alpha} + \|x-y\|]^{-\alpha-1}(1+\|y\|^\lambda)\, dy. \tag{4.14}$$

Lemma 4.3. *If* $0 < b \leq B < \infty$, $0 \leq \lambda < \alpha$, *then*

$$\int_K (b + \|x-\xi\|)^{-\alpha-1}\|\xi\|^\lambda\, d\xi \leq Cb^{-\alpha}(1+\|x\|^\lambda), \quad x \in K, \tag{4.15}$$

where the constant C *does not depend on* b, x.

Proof. Denote by $I(b,x)$ the expression in the left-hand side of (4.15). Let an integer m be such that $q^{m-1} \leq b \leq q^m$. Then

$$I(b,x) \leq I(q^{m-1},x) = \int_K (\|\beta^{-m+1}\| + \|x-\xi\|)^{-\alpha-1}\|\xi\|^\lambda d\xi$$
$$= q^{(m-1)(\lambda-\alpha)} I(1, \beta^{m-1}x).$$

Let $\beta^{m-1}x = y$, $\|y\| = q^k$. We have

$$I(1,y) = I_1(y) + I_2(y) + I_3(y),$$

where

$$I_1(y) = \sum_{\nu=-\infty}^{k-1} \int_{\|\xi\|=q^\nu} (1+\|y-\xi\|)^{-\alpha-1}\|\xi\|^\lambda\, d\xi,$$

$$I_2(y) = \int_{\|\xi\|=q^k} (1+\|y-\xi\|)^{-\alpha-1}\|\xi\|^\lambda\, d\xi,$$

$$I_3(y) = \sum_{\nu=k+1}^{\infty} \int_{\|\xi\|=q^\nu} (1+\|y-\xi\|)^{-\alpha-1}\|\xi\|^\lambda\, d\xi.$$

Let us find a bound for each summand. Firstly,

$$I_1(y) = \sum_{\nu=-\infty}^{k-1} q^{\lambda\nu}(1+\|y\|)^{-\alpha-1} q^{\nu}(1-q^{-1})$$
$$= (1-q^{-1})(1+\|y\|)^{-\alpha-1} \sum_{\nu=-\infty}^{k-1} q^{(\lambda+1)\nu} \le C(1+\|y\|)^{-\alpha-1}\|y\|^{\lambda+1}.$$

Next,

$$I_2(y) = \|y\|^{\lambda} \int_{\|\xi\|=\|y\|} (1+\|y-\xi\|)^{-\alpha-1}\, d\xi$$
$$= \|y\|^{\lambda-\alpha} \int_{\|\eta\|=1} (\|y\|^{-1}+\|1-\eta\|)^{-\alpha-1}\, d\eta.$$

Represent an element η in the canonical form:

$$\eta = a_0 + a_1\beta + a_2\beta^2 + \cdots,$$

where $a_j \in S$, $a_0 \notin P$, and the series converges in K. As in the similar calculations performed in Chapters 2,3, it is convenient to assume that $0, 1 \in S$. If $a_0 \neq 1$, then $\|1-\eta\| = 1$. If $a_0 = 1$, $a_1 \neq 0$, then $\|1-\eta\| = q^{-1}$. If $a_0 = 1$, $a_1 = 0$, $a_2 \neq 0$, then $\|1-\eta\| = q^{-2}$ etc. The measure of the set of $\eta \in U$ with the above properties equals respectively $1-2q^{-1}$, $(q-1)q^{-2}$, $(q-1)q^{-3}$ etc. Therefore

$$I_2(y) = \|y\|^{\lambda-\alpha}\Big[(\|y\|^{-1}+1)^{-\alpha-1}(1-2q^{-1})$$
$$+\sum_{j=1}^{\infty}(\|y\|^{-1}+q^{-j})^{-\alpha-1}(q-1)q^{-j-1}\Big]$$
$$\le (1-q^{-1})\|y\|^{\lambda-\alpha}\sum_{j=0}^{\infty} q^{-j}(\|y\|^{-1}+q^{-j})^{-\alpha-1}$$
$$\le \|y\|^{\lambda-\alpha}\int_K (\|y\|^{-1}+\|\eta\|)^{-\alpha-1}\, d\eta = C\|y\|^{\lambda}.$$

Finally,

$$I_3(y) = \sum_{\nu=k+1}^{\infty} \int\limits_{\|\xi\|=q^\nu} (1+\|\xi\|)^{-\alpha-1}\|\xi\|^\lambda \, d\xi$$

$$\leq \int\limits_K (1+\|\xi\|)^{-\alpha-1}\|\xi\|^\lambda \, d\xi = C.$$

As a result we find that $I(1,y) \leq C(1+\|y\|^\lambda)$, whence

$$\begin{aligned} I(b,x) &\leq Cq^{(m-1)(\lambda-\alpha)}(1+\|\beta^{m-1}x\|^\lambda) \\ &= Cq^{(m-1)(\lambda-\alpha)} + Cq^{-(m-1)\alpha}\|x\|^\lambda \leq Cb^{-\alpha}(1+\|x\|^\lambda). \quad \blacksquare \end{aligned}$$

Now from (4.14) and (4.15) we get the required estimate of $u(x,t,\tau)$:

$$|u(x,t,\tau)| \leq C(1+\|x\|^\lambda), \tag{4.16}$$

where C does not depend on $t, \tau (\leq T)$, that is $u \in \mathfrak{M}_\lambda$ uniformly with respect to t, τ.

As in the classical theory of parabolic equations [97], it is easy to compute the derivative $\dfrac{\partial u(x,t,\tau)}{\partial t}$. Let

$$u_h(x,t,\tau) = \int\limits_\tau^{t-h} d\theta \int\limits_K Z(x-\xi, t-\theta) f(\xi,\theta) \, d\xi, \tag{4.17}$$

where h is a small positive number. We can differentiate u_h under the sign of integral, so that

$$\frac{\partial u_h}{\partial t} = \int\limits_\tau^{t-h} d\theta \int\limits_K \frac{\partial Z(x-\xi,t-\theta)}{\partial t} f(\xi,\theta)\, d\xi + \int\limits_K Z(x-\xi,h) f(\xi,t-h)\, d\xi$$

$$= \int\limits_\tau^{t-h} d\theta \int\limits_K \frac{\partial Z(x-\xi,t-\theta)}{\partial t} [f(\xi,\theta) - f(x,\theta)]\, d\xi$$

$$+ \int\limits_\tau^{t-h} f(x,\theta) d\theta \int\limits_K \frac{\partial Z(x-\xi,t-\theta)}{\partial t}\, d\xi + \int\limits_K Z(x-\xi,h)[f(\xi,t-h) - f(\xi,t)]\, d\xi$$

$$+ \int\limits_K Z(x-\xi,h) f(\xi,t)\, d\xi.$$

The first integral contains no singularity at $t = \theta$ due to (4.11) and the local constancy of f. The second integral is equal to zero by virtue of (4.8). The third integral can be written as the sum of the integrals over $\{\xi \in K : \|x-\xi\| \le q^N\}$ and the complement of this set; one of them is estimated on the basis of the uniform continuity of f, while the other contains no singularity. Finally, the fourth integral tends to $f(x,t)$ as $h \to 0$. As a result we obtain the formula

$$\frac{\partial u(x,t,\tau)}{\partial t} = f(x,t) + \int_\tau^t d\theta \int_K \frac{\partial Z(x-\xi,t-\theta)}{\partial t}[f(\xi,\theta) - f(x,\theta)]\, d\xi. \tag{4.18}$$

Let us consider the action of the operator D^γ, $0 < \gamma \le \alpha$, upon the heat potential $u(x,t,\tau)$. Note first of all (see (4.16)) that the function $D^\gamma u$ makes sense only for $\gamma > \lambda$. If $\gamma < \alpha$, then

$$(D^\gamma u)(x,t,\tau) = \int_\tau^t d\theta \int_K Z_\gamma(x-y,t-\theta) f(y,\theta)\, dy, \quad Z_\gamma = D_x^\gamma Z. \tag{4.19}$$

Indeed, let

$$Z_{\gamma,k}(x,t) = \frac{1}{\Gamma_K(-\gamma)} \int_{\|\xi\| \ge q^{-k}} \|\xi\|^{-\gamma-1}[Z(x-\xi,t) - Z(x,t)]\, d\xi,$$

and $u_h(x,t,\tau)$ be given by (4.17), $0 < h < t - \tau$. By the Fubini theorem

$$\frac{1}{\Gamma_K(-\gamma)} \int_{\|\xi\| \ge q^{-k}} \|\xi\|^{-\gamma-1}[u_h(x-\xi,t,\tau) - u_h(x,t,\tau)]\, d\xi$$
$$= \int_\tau^{t-h} d\theta \int_K Z_{\gamma,k}(x-y,t-\theta) f(y,\theta)\, dy \tag{4.20}$$

On the other hand,

$$Z_{\gamma,k}(x,t) = \int_K \chi(-x\eta) P_k(\eta) \exp(-at\|\eta\|^\alpha)\, d\eta,$$

where

$$P_k(\eta) = \frac{1}{\Gamma_K(-\gamma)} \int\limits_{\|\xi\| \ge q^{-k}} [\chi(\xi\eta) - 1] \|\xi\|^{-\gamma-1} \, d\xi$$

$$= \frac{\|\eta\|^\gamma}{\Gamma_K(-\gamma)} \int\limits_{\|\xi\| \ge q^{-k}\|\eta\|} [\chi(\zeta) - 1] \|\zeta\|^{-\gamma-1} \, d\zeta.$$

It is clear that

$$|P_k(\eta)| \le \frac{\|\eta\|^\gamma}{|\Gamma_K(-\gamma)|} \int\limits_K |\chi(\zeta) - 1| \cdot \|\zeta\|^{-\gamma-1} \, d\zeta = C\|\eta\|^\gamma,$$

whence

$$|Z_{\gamma,k}(x,t)| \le C$$

with a constant, which does not depend on x, $t \ge h + \tau$, k. Writing the right-hand side of (4.20) as

$$\int\limits_\tau^{t-h} d\theta \int\limits_{\|x-y\| \ge q^{-l}} Z_{\gamma,k}(x-y, t-\theta) f(y,\theta) \, dy$$

$$+ \int\limits_\tau^{t-h} d\theta \int\limits_{\|x-y\| < q^{-l}} Z_{\gamma,k}(x-y, t-\theta) f(y,\theta) \, dy$$

(l is a fixed natural number), it is easy to pass to the limit in (4.20) for $k \to \infty$, since for $\|x\| \ge q^{-l}$, $k > l$,

$$Z_{\gamma,k}(x,t) = Z_\gamma(x,t).$$

We find that

$$(D^\gamma u_h)(x,t,\tau) = \int\limits_\tau^{t-h} d\theta \int\limits_K Z_\gamma(x-y, t-\theta) f(y,\theta) \, dy. \tag{4.21}$$

It follows from Lemma 4.3 that $u_h \in \mathfrak{M}_\lambda$, uniformly with respect to h. Together with Lemma 4.2 and the dominated convergence theorem that makes it possible to pass to the limit in (4.21) for $h \to 0$, which gives the required equality (4.19).

Suppose now that $\gamma = \alpha$. The equality (4.21) is valid in this case too. Using (4.13) we may rewrite it as

$$(D^\alpha u_h)(x,t,\tau) = \int\limits_\tau^{t-h} d\theta \int\limits_K Z_\alpha(x-y,t-\theta)[f(y,\theta) - f(x,\theta)]\,dy. \quad (4.22)$$

Since f is locally constant, we may as before pass to the limit in (4.22) for $h \to 0$, so that

$$(D^\alpha u)(x,t,\tau) = \int\limits_\tau^{t} d\theta \int\limits_K Z_\alpha(x-y,t-\theta)[f(y,\theta) - f(x,\theta)]\,dy. \quad (4.23)$$

Now we are ready to construct a solution of the Cauchy problem (4.1), (4.2). We assume that $\varphi \in \mathfrak{M}_\lambda$, $f \in \mathfrak{M}_\lambda$ uniformly with respect to t, and f is continuous in (x,t). Here $0 \le \lambda < \alpha$. This sort of restriction, which is absent in the Euclidean case, is inevitable here because of the nonlocal character of the operator D^α.

We shall look for a solution to the problem (4.1), (4.2) in the class of functions $u(x,t)$, which are continuous on $K \times [0,T]$, continuously differentiable in t, and as a function of $x \in K$ belong to the class $\mathfrak{M}_\lambda$ uniformly with respect to t.

Theorem 4.1. *The Cauchy problem (4.1), (4.2) has a solution, which can be represented in the form (4.3).*

The *proof* is a direct consequence of (4.18), (4.23), (4.9), and (4.10). ■

4.2 Probabilistic interpretation

It follows from (4.4) that

$$Z(x,t_1+t_2) = \int\limits_K Z(x-y,t_1)Z(y,t_2)\,dy, \quad t_1,t_2 > 0,\ x \in K. \quad (4.24)$$

Theorem 4.2. *The fundamental solution $Z(x,t)$ is a transition density of a time- and space-homogeneous non-exploding right-continuous strict Markov process without second kind discontinuities.*

Proof. The identity (4.24) is equivalent to the semigroup property for the family of operators

$$(\Theta(t)f)(x) = \int_K Z(x-y,t)f(y)\,dy.$$

We know that $Z(x,t) > 0$, and that $\Theta(t)$ preserves the function $f(x) \equiv 1$ (see (4.8)). Thus $\Theta(t)$ is a Markov semigroup. The required properties of the corresponding Markov process follow from the inequality (4.6) and general theorems of the theory of Markov processes [26]. ■

The process $\xi_\alpha(t)$ constructed in Theorem 4.2 is the non-Archimedean counterpart of the symmetric stable process on $\mathbb{R}$. Below when speaking about $\xi_\alpha(t)$, we shall assume for simplicity that $a = 1$.

It is often useful to consider $\xi_\alpha(t)$ as a stochastic process with independent increments (see [56, 142, 49] for the basics on stochastic processes on locally compact groups). It follows from Proposition 2.3 that the Lévy measure of the process $\xi_\alpha(t)$ is

$$\pi_\alpha(dx) = \frac{q^\alpha - 1}{1 - q^{-\alpha-1}} \|x\|^{-\alpha-1}\,dx.$$

Now, considering D^α as a self-adjoint operator on $L_2(K)$ (as in Chapter 3) we can obtain an analogue of the Feynman-Kac formula.

Theorem 4.3. *Suppose that $V: K \to \mathbb{R}$ is continuous and bounded from below. Then for any $f \in L_2(K)$*

$$\left(e^{-t(D^\alpha+V)}f\right)(x) = \int_{D(0,t;K)} f(\omega(t)) \exp\left(-\int_0^t V(\omega(s))\,ds\right) \mu_x(d\omega), \tag{4.25}$$

where μ_x is the measure on the Skorokhod space of paths $[0,t] \to K$, right continuous with left limits, corresponding to the Markov process ξ_α starting at a point $x \in K$.

Proof. Let us consider first the case where V is compactly supported.

By the product formula for semigroups of operators

$$\left(e^{-t(D^\alpha+V)}f\right)(x) = \lim_{n\to\infty}\left[\left(e^{-(t/n)D^\alpha}e^{-(t/n)V}\right)^n f\right](x)$$

$$= \int_K \dots \int_K Z\left(x - x_n, \frac{t}{n}\right) \dots Z\left(x - x_n, \frac{t}{n}\right) Z\left(x_2 - x_1, \frac{t}{n}\right) f(x_1)$$

$$\times \exp\left(-\sum_{j=0}^{n-1}\frac{t}{n}V(x_j)\right) dx_1 \dots dx_n$$

$$= \lim_{n\to\infty} \int_{D(0,t;K)} \exp\left(-\frac{t}{n}\sum_{j=0}^{n-1} V\left(\omega\left(\frac{jt}{n}\right)\right)\right) f(\omega(t))\mu_x(d\omega).$$

Since the paths in $D(0,t;K)$ are right continuous with left limits, and V is uniformly continuous, the composition $V \circ \omega$ is Riemann integrable, and

$$\frac{t}{n}\sum_{j=0}^{n-1} V\left(\omega\left(\frac{jt}{n}\right)\right) \longrightarrow \int_0^t V(\omega(s))\,ds, \quad n \to \infty.$$

Next, since $V(x) \geq V_0$,

$$\int_{D(0,t;K)} \left|\exp\left(-\frac{t}{n}\sum_{j=0}^{n-1} V\left(\omega\left(\frac{jt}{n}\right)\right)\right) f(\omega(t))\right| \mu_x(d\omega)$$

$$\leq e^{-tV_0} \int_{D(0,t;K)} |f(\omega(t))|\,\mu_x(d\omega) = e^{-tV_0}\left(e^{-tD^\alpha}|f|\right)(x) < \infty$$

for almost all x. By the dominated convergence theorem we come to (4.25).

For general V we can consider the cutoffs

$$V_k(x) = \begin{cases} V(x), & \text{if } \|x\| \leq q^k, \\ 0, & \text{otherwise,} \end{cases}$$

apply (4.25) with V_k substituted instead of V (note that V_k is continuous!), and use the monotone convergence theorem for integrals, and the strong resolvent convergence of $D^\alpha + V_k$ (see Theorem VIII.20 of [121]) as $k \to \infty$. ■

4.3 Stabilization

Let us consider the Cauchy problem (4.1)-(4.2) with $f(x,t) \equiv 0$ and $\varphi \in \mathfrak{M}_0$. Above we considered solutions of (4.1)-(4.2) on a finite "time" interval $[0,T]$. In this section we shall find necessary and sufficient conditions for the solution

$$u(x,t) = \int\limits_K Z(x-\xi,t)\varphi(\xi)\,d\xi, \quad x \in K, 0 \le t < \infty, \tag{4.26}$$

to stabilize as $t \to \infty$.

It is clear that $u(x,t)$ belongs to $\mathfrak{M}_0$ uniformly with respect to $t \in [0,\infty)$. In particular, the solution is bounded on $K \times [0,\infty)$.

First we shall find another representation of $u(x,t)$. Denote

$$M_x^{(\nu)}\varphi = q^{-\nu} \int\limits_{\|y\| \le q^\nu} \varphi(x+y)\,dy, \quad \nu = 0, \pm 1, \pm 2, \dots . \tag{4.27}$$

Lemma 4.4.

$$u(x,t) = \sum_{\nu=-\infty}^{\infty} \left(M_x^{(\nu)}\varphi\right)[c_\nu(t) - c_{\nu-1}(t)], \quad x \in K, 0 < t < \infty, \tag{4.28}$$

where

$$C_\nu(t) = \exp(-atq^{-\alpha\nu}).$$

Proof. Let us rewrite (4.26) using (4.5) as

$$u(x,t) = \sum_{\nu=-\infty}^{\infty} q_\nu(t)h_\nu(x),$$

where

$$g_\nu(t) = \sum_{n=1}^{\infty} \frac{(-1)^n}{n!} \cdot \frac{1-q^{\alpha n}}{1-q^{-\alpha n-1}}(at)^n q^{-\nu(\alpha n+1)},$$

$$h_\nu(x) = \int\limits_{\|y\|=q^\nu} \varphi(x+y)\,dy.$$

We apply the Abel transform

$$\sum_{\nu=-N}^{M} g_\nu(t)h_\nu(x) = \sum_{\nu=-N}^{M-1} H_\nu^{(N)}(x)[g_\nu(t) - g_{\nu+1}(t)] + H_M^{(N)}(x)g_M(t), \tag{4.29}$$

where M, N are natural numbers, $H_\nu^{(N)}(x) = \sum_{n=-N}^{\nu} h_n(x)$.

Let

$$H_\nu(x) = \sum_{n=-\infty}^{\nu} h_n(x) = \int_{\|y\|\le q^\nu} \varphi(x+y)\,dy.$$

We find that

$$|H_\nu(x)| \le Cq^\nu, \quad |H_\nu^{(N)}(x) - H_\nu(x)| \le C \int_{\|y\|<q^{-N}} dy \le Cq^{-N}$$

(C does not depend on ν). Next,

$$g_\nu(t) - g_{\nu+1}(t) = \sum_{n=1}^{\infty} \frac{(-1)^n}{n!}(1-q^{\alpha n})(at)^n q^{-\nu(\alpha n+1)}$$
$$= q^{-\nu}[c_\nu(t) - c_{\nu-1}(t)].$$

If $\nu < 0$, then

$$|g_\nu(t)| \le Cq^{-\nu}\exp(-atq^{-\alpha\nu}),$$
$$|g_\nu(t) - g_{\nu+1}| \le Cq^{-\nu}\exp(-atq^{-\alpha\nu}), \quad |h_\nu(x)| \le Cq^\nu.$$

It follows from these estimates that

$$\left|\sum_{\nu=-N}^{-1} H_\nu^{(N)}(x)[g_\nu(t) - g_{\nu+1}(t)] - \sum_{\nu=-\infty}^{-1} H_\nu(x)[g_\nu(t) - g_{\nu+1}(t)]\right|$$
$$\le \sum_{\nu=-N}^{-1} \left|H_\nu^{(N)}(x) - H_\nu^{(N)}(x)H_\nu(x)\right| \cdot |g_\nu(t) - g_{\nu+1}(t)|$$
$$+ \sum_{\nu=-\infty}^{-1} |H_\nu(x)| \cdot |g_\nu(t) - g_{\nu+1}(t)| \longrightarrow 0$$

as $N \to \infty$. Passing to the limit $N \to \infty$ in (4.29) we obtain the equality

$$\sum_{\nu=-\infty}^{M} g_\nu(t)h_\nu(x) = \sum_{\nu=-\infty}^{M-1} H_\nu(x)q^{-\nu}[c_\nu(t) - c_{\nu-1}(t)] + H_M(x)g_M(t). \tag{4.30}$$

For $\nu > 0$ we have

$$|g_\nu(t)| \le Cq^{-\nu(\alpha+1)}, \quad |g_\nu(t) - g_{\nu+1}(t)| \le Cq^{-\nu(\alpha+1)}.$$

Note that $q^{-\nu}H_\nu(x) = M_x^{(\nu)}\varphi$. Passing to the limit $M \to \infty$ in (4.30), we come to (4.28). ∎

Now we formulate the stabilization conditions. As before, we assume that $\varphi \in \mathfrak{M}_0$.

Theorem 4.4. *In order that $u(x,t) \to l$ (=const) as $t \to \infty$ for all $x \in K$, it is necessary and sufficient that $M_x^{(n)}\varphi \to l$ as $n \to \infty$, for every $x \in K$.*

Proof. The sufficiency follows from the representation (4.28). Indeed, the series over negative values of ν on the right in (4.28) converges uniformly with respect to $t \in [t_0, \infty)$, $t_0 > 0$, and hence its sum tends to zero as $t \to \infty$. Let

$$u_1(x,t) = \sum_{\nu=0}^{\infty} \left(M_x^{(\nu)}\varphi\right)[c_\nu(t) - c_{\nu-1}(t)].$$

Obviously,

$$\sum_{\nu=0}^{N} \left(M_x^{(\nu)}\varphi\right)[c_\nu(t) - c_{\nu-1}(t)] = \sum_{\nu=0}^{N-1}\left[M_x^{(\nu)}\varphi - M_x^{(\nu+1)}\varphi\right] c_\nu(t)$$
$$+ \left(M_x^{(0)}\varphi\right) c_{-1}(t) + \left(M_x^{(N)}\varphi\right) c_N(t). \quad (4.31)$$

The series

$$\sum_{\nu=0}^{\infty}\left[M_x^{(\nu)}\varphi - M_x^{(\nu+1)}\varphi\right] c_\nu(t)$$

converges uniformly with respect to t, by the Abel convergence test, since $c_\nu(t) \uparrow 1$ as $\nu \to \infty$, $|c_\nu(t)| \le 1$. Taking the limit in (4.31) as $N \to \infty$, we see that $u_1(x,t) = l + u_2(x,t)$, where $u_2(x,t) \to 0$ as $t \to 0$.

Let us prove the necessity. Suppose that $u(x,t) \to l$ as $t \to \infty$, for any $x \in K$. In particular, for any $\tau > 0$ we have

$$u(x, \tau q^{\alpha n}) \longrightarrow l, \quad n \to \infty.$$

According to (4.28),

$$\begin{aligned} u(x,\tau q^{\alpha n}) &= \sum_{\nu=-\infty}^{\infty} \left(M_x^{(\nu)}\varphi\right)\left[\exp(-a\tau q^{-\alpha(\nu-n)}) - \exp(-a\tau q^{-\alpha(\nu-1-n)})\right] \\ &= \sum_{m=-\infty}^{\infty} \left(M_x^{(n-m)}\varphi\right)\left[\exp(-a\tau q^{\alpha m}) - \exp(-a\tau q^{\alpha(m+1)})\right] \\ &= (v * w_\tau)_n, \end{aligned}$$

where the sequences $v = \{v_m\}_{-\infty}^{\infty}$ and $w_\tau = \{w_{\tau,m}\}_{-\infty}^{\infty}$ are given by

$$v_m = M_x^{(m)}\varphi, \quad w_{\tau,m} = \exp(-a\tau q^{\alpha m}) - \exp(-a\tau q^{\alpha(m+1)}).$$

It is easy to verify that $v \in l_\infty(\mathbb{Z})$, $w_\tau \in l_1(\mathbb{Z})$, and $w_{\tau,m} > 0$,

$$\sum_{\nu=-\infty}^{\infty} w_{\tau,m} = \lim_{m\to-\infty} \exp(-a\tau q^{\alpha m}) - \lim_{m\to\infty} \exp(-a\tau q^{\alpha m}) = 1.$$

Let us consider the Fourier transform (in the sense of harmonic analysis on the group $\mathbb{Z}$):

$$\widehat{w_\tau}(\mu) = \sum_{\nu=-\infty}^{\infty} e^{-im\mu} w_{\tau,m}, \quad 0 \le \mu < 2\pi.$$

We shall show that for any $\mu \in [0, 2\pi)$ there exists such $\tau > 0$ that $\widehat{w_\tau}(\mu) \neq 0$. Suppose the contrary, i.e. that for some $\mu_0 \in [0, 2\pi)$

$$\sum_{m=-\infty}^{\infty} e^{-im\mu_0} \left[\exp(-a\tau q^{\alpha m}) - \exp(-a\tau q^{\alpha(m+1)})\right] = 0 \tag{4.32}$$

for all $\tau > 0$. It is clear that $\mu_0 \neq 0$.

Differentiating (4.32) termwise with respect to τ we find that

$$\sum_{m=-\infty}^{\infty} e^{-im\mu_0} \left[aq^{\alpha(m+1)} \exp(-a\tau q^{\alpha(m+1)}) - aq^{\alpha m} \exp(-a\tau q^{\alpha m})\right] = 0. \tag{4.33}$$

This series is absolutely convergent, uniformly with respect to $\tau \in [\tau_1, \tau_2]$, $0 < \tau_1 < \tau_2 < \infty$, so that the differentiation was legitimate. Denote the right-hand side of (4.33) by $F(\tau)$. It is important that in (4.33) (in contrast to (4.32)) one can expand the square brackets and present the series as the difference of two convergent series. After an obvious change of variables we arrive at the representation

$$F(\tau) = \sum_{m=-\infty}^{\infty} f_m e^{\lambda_m \tau},$$

where $f_m = e^{-im\mu_0}(e^{i\mu_0} - 1)q^{\alpha m}$, $\lambda_m = -aq^{\alpha m}$.

The function $F(\tau)$ extends analytically into the half-plane $\tau + i\theta$, $\tau > 0$, $\theta \in \mathbb{R}$. For each fixed $\tau > 0$ the series

$$F(\tau + i\theta) = \sum_{m=-\infty}^{\infty} f_m e^{\lambda_m \tau} \cdot e^{i\lambda_m \theta}$$

converges uniformly with respect to θ. Therefore $F(\tau + i\theta)$ is Bohr almost periodic in θ. On the other hand, by (4.33) and the uniqueness theorem for analytic functions, $F(\tau + i\theta) \equiv 0$. Now it follows from the Parseval identity for almost periodic functions that $f_m = 0$ for all $m \in \mathbb{Z}$, whence $\mu_0 = 0$, and we have got a contradiction.

Thus we have proved that

$$\bigcap_{\tau>0} \{\mu \in [0, 2\pi) : \ \widehat{w_\tau}(\mu) = 0\} = \varnothing. \tag{4.34}$$

Consider the sequence

$$\delta \in l_1(\mathbb{Z}), \quad \delta_n = \begin{cases} 1, & n = 0, \\ 0, & n \neq 0. \end{cases}$$

By Wiener's Tauberian theorem for the group $\mathbb{Z}$ (see [55]), it follows from (4.34) that

$$M_x^{(n)}\varphi = (v * \delta)_n \longrightarrow l \sum_{n=-\infty}^{\infty} \delta_n = l,$$

as $n \to \infty$. The theorem is proved. ■

Theorem 4.4 can be compared with similar properties of classical parabolic equations [29].

4.4 General uniqueness theorem

Let us consider a more general parabolic equation of the form

$$\frac{\partial u(x,t)}{\partial t} + a_0(x,t)(D^\alpha u)(x,t) + \sum_{k=1}^{n} a_k(x,t)(D^{\alpha_k} u)(x,t)$$
$$+ b(x,t)u(x,t) = f(x,t), \quad x \in K, \ t \in (0,T], \tag{4.35}$$

$0 < \alpha_1 < \alpha_2 < \ldots < \alpha_n < \alpha$.

In this section we assume that the coefficients $a_k(x,t)$, $k = 0, 1, \ldots, n$, are non-negative bounded continuous functions, $b(x,t)$ is a continuous bounded function. Let $0 \leq \gamma < \alpha_1$ (if $a_1(x,t) = \ldots = a_n(x,t) \equiv 0$, then we shall assume that $0 \leq \gamma < \alpha$).

Theorem 4.5. *If $u(x,t)$ is a solution of the equation (4.5) with $f(x,t) \equiv 0$, such that $u \in \mathfrak{M}_\gamma$ uniformly with respect to t, and $u(x,0) \equiv 0$, then $u(x,t) \equiv 0$.*

Proof. We may assume that $b(x,t) \geq 0$ (otherwise we would be able to consider the equation for $u(x,t)e^{\lambda t}$ with an appropriate λ). Let us prove that $u(x,t) \geq 0$. Suppose the contrary, that is

$$u(x',t') = \varkappa < 0$$

for some $x' \in K$, $t' \in (0,T]$.

Denote

$$\omega_r(s) = \begin{cases} 1, & 0 \leq s \leq r, \\ 0, & s > r. \end{cases}$$

Fixing a natural number l consider the function

$$\delta_l(x) = q^l \omega_{q^{-l}}(\|x\|), \quad x \in K.$$

As we know, $\delta_l \in D(K)$, $\int\limits_K \delta_l(x)\,dx = 1$. Let us choose $\eta > 0$ in such a way that $\gamma < \eta < \alpha_1$, or if $a_1(x,t) = \ldots = a_n(x,t) \equiv 0$, then $\gamma < \eta < \alpha$. Denote

$$\psi(x) = \int\limits_K \delta_l(x-\xi)\|\xi\|^\eta \, d\xi.$$

The function ψ is locally constant, and $\psi(x) = \|x\|^\eta$, if $\|x\| > q^{-l}$. Further,

$$D^\alpha \psi = (f_{-\alpha} * \delta_l * f_{\eta+1})\Gamma_K(\eta+1) = \Gamma_K(\gamma+1) f_{\eta-\alpha+1} * \delta_l \tag{4.36}$$

(see Chapter 2); recall that $f_s \in \mathcal{D}'(K)$ ($s \neq 0,1$) is a distribution, which corresponds (if $s < 0$, after the regularization), to the function $\dfrac{\|x\|^{s-1}}{\Gamma_K(s)}$; also $f_0 = \delta$. Similarly,

$$D^{\alpha_k}\psi = \Gamma_K(\eta+1) f_{\eta-\alpha_k+1} * \delta_l. \tag{4.37}$$

It follows from (4.36) and (4.37) that $(D^\alpha\psi)(x), D^{\alpha_k}\psi(x) \to 0$ as $\|x\| \to \infty$, and that $D^\alpha\psi$, $D^{\alpha_k}\psi$ are locally constant functions on K.

Let

$$M = \sup_{x \in K, 0 \leq t \leq T} \left\{ a(x,t)|(D^\alpha\psi)(x)| + \sum_{k=1}^n a_k(x,t)|D^{\alpha_k}\psi(x)| \right\}.$$

Choose $\rho > 0$ in such a way that $\varkappa + T\rho < 0$. Then choose a number $\sigma > 0$ so small that

$$\varkappa + T\rho + \sigma\psi(x') < 0, \tag{4.38}$$

$$\rho - \sigma M > 0. \tag{4.39}$$

Consider the function

$$v(x,t) = u(x,t) + \rho t + \sigma \psi(x).$$

It is seen from (4.38) that $v(x', t') < 0$, so that

$$\inf_{x \in K, 0 \le t \le T} v(x,t) < 0.$$

Since $u(x,0) = 0$, we have $v(x,0) = \sigma\psi(x) \geq 0$. As $\|x\| \to \infty$, $u(x,t) = o(\psi(x))$ because $u \in \mathfrak{M}_\gamma$ and $\gamma < \eta$. Thus $v(x,t) > 0$ if $\|x\|$ is large enough. This implies the existence of such $x_0 \in K$, $t_0 \in (0,T]$, that

$$\inf v(x,t) = v(x_0, t_0) < 0.$$

On the other hand,

$$\frac{\partial v(x,t)}{\partial t} + a_0(x,t)(D^\alpha v)(x,t) + \sum_{k=1}^{n} a_k(x,t)(D^{\alpha_k} v)(x,t) + b(x,t)v(x,t)$$

$$= \rho + \sigma \left[a_0(x,t)(D^\alpha \psi)(x) + \sum_{k=1}^{n} a_k(x,t)(D^{\alpha_k}\psi)(x) \right] + b(x,t)[\rho t + \sigma\psi(x)]$$

$$\geq \rho - \sigma M > 0$$

according to (4.39). However it follows from the hypersingular representation (2.8) for the operator D^α that at the point (x_0, t_0) of the global minimum we have $(D^\alpha v)(x_0,t_0) \leq 0$, $(D^{\alpha_k} v)(x_0, t_0) \leq 0$. It is clear that also $\dfrac{\partial v(x_0,t_0)}{\partial t} \leq 0$. We have come to the contradiction.

Thus we have proved that $u(x,t) \geq 0$. Considering $-u(x,t)$ instead of $u(x,t)$, we find similarly that $u(x,t) \leq 0$. Therefore $u(x,t) \equiv 0$. ■

4.5 Fundamental solutions of parabolic equations

4.5.1. Parametrix. In this section we construct a fundamental solution of the Cauchy problem for the general parabolic equation (4.35). We shall assume that $\alpha > 1$, $0 < \alpha_1 < \ldots < \alpha_n < \alpha$, the coefficients $a_0(x,t)$, $a_1(x,t), \ldots, a_n(x,t)$, $b(x,t)$ belong (with respect to $x \in K$) to the class $\mathfrak{M}_0$ uniformly with respect to $t \in [0,T]$, and satisfy the Hölder condition in t, with an exponent $v \in (0,1]$, uniformly with respect to $x \in K$. We assume also the uniform parabolicity condition $a_0(x,t) \geq \mu > 0$. Without loss of generality we may assume that $\alpha_{n+1} \stackrel{\text{def}}{=} \alpha(1-v) > \alpha_n$.

As in the Euclidean case [29, 97], the first step of the construction of a fundamental solution is to study the parametrized fundamental solution $Z(x,t,y,\theta)$ of the Cauchy problem for the equation

$$\frac{\partial u(x,t)}{\partial t} + a_0(y,\theta)(D^\alpha u)(x,t) = 0,$$

where $y \in K$ and $\theta > 0$ are parameters. We have (see (4.4) and (4.5))

$$Z(x,t,y,\theta) = \int_K \chi(-x\xi) \exp(-a_0(y,\theta)\|\xi\|^\alpha)\, d\xi, \tag{4.40}$$

and if $x \neq 0$, then

$$Z(x,t,y,\theta) = \sum_{m=1}^{\infty} \frac{(-1)^m}{m!} \cdot \frac{1-q^{\alpha m}}{1-q^{-\alpha m-1}} [a_0(y,\theta)t]^m \|x\|^{-\alpha m-1}. \tag{4.41}$$

By Lemma 4.1 and Lemma 4.2 we have the estimates

$$Z(x,t,y,\theta) \leq Ct(t^{1/\alpha} + \|x\|)^{-\alpha-1}, \quad x \in K,\ t > 0, \tag{4.42}$$

$$\left|\frac{\partial Z(x,t,y,\theta)}{\partial t}\right| \leq C(t^{1/\alpha} + \|x\|)^{-\alpha-1}, \tag{4.43}$$

$$|(D_x^\gamma Z)(x,t,y,\theta y,\theta)| \leq C(t^{1/\alpha} + \|x\|)^{-\gamma-1}, \quad 0 < \gamma \leq \alpha, \tag{4.44}$$

where the constants do not depend on y, θ.

As before, we get the identities

$$\int_K Z(x,t,y,\theta)\, dx = 1, \tag{4.45}$$

$$\frac{\partial Z(x,t,y,\theta)}{\partial t} = -a_0(y,\theta) \int_K \|\xi\|^\alpha \chi(-x\xi) \exp(-a_0(y,\theta)t\|\xi\|^\alpha)\, d\xi, \tag{4.46}$$

$$(D_x^\gamma Z)(x,t,y,\theta) = \int_K \chi(-x\eta)\|\eta\|^\gamma \exp(-a_0(y,\theta)t\|\eta\|^\alpha)\, d\eta, \quad x \neq 0, \tag{4.47}$$

$$\int_K (D_x^\gamma Z)(x,t,y,\theta)\, dx = 0. \tag{4.48}$$

One more estimate is given by the following lemma.

Lemma 4.5.

$$\left| \int_K \frac{\partial Z(x-y,t,y,\theta)}{\partial t}\, dy \right| \le C. \tag{4.49}$$

Proof. It follows from (4.45) and (4.46) that

$$\int_K \frac{\partial Z(x,t,y,\theta)}{\partial t}\, dx = 0. \tag{4.50}$$

Therefore

$$\int_K \frac{\partial Z(x-y,t,y,\theta)}{\partial t}\, dy = \int_K \left[\frac{\partial Z(x-y,t,y,\theta)}{\partial t}\, dy - \frac{\partial Z(x-y,t,x,\theta)}{\partial t} \right] dy. \tag{4.51}$$

The function $\dfrac{\partial Z(x,t,y,\theta)}{\partial t}$ belongs to the class $\mathfrak{M}_0$ in the variable y, and its local constancy exponent (which is equal to the one for a_0 by virtue of (4.46)) does not depend on the rest of variables. Thus the integral in the right-hand side of (4.51) is actually taken over the set $\{y \in K : \|y-x\| \ge q^{-N}\}$, where N does not depend on x, t, θ. Now the inequality (4.49) follows from (4.43). ∎

4.5.2. Heat potentials. Consider the heat potential

$$u(x,t,\tau) = \int_\tau^t d\theta \int_K Z(x-y,t-\theta,y,\theta) f(y,\theta)\, dy, \tag{4.52}$$

with $f \in \mathfrak{M}_\lambda$ $(0 \le \lambda < \alpha)$ uniformly with respect to θ, and f is continuous in (y,θ). Note the difference between (4.52) and the heat potential studied in Sect. 4.1 – the integral in (4.52) is not a convolution. In those cases where it is sufficient to use an estimate for Z or its derivatives (see (4.42)-(4.44)), the corresponding integrals are actually convolutions. Thus we obtain (as in (4.16)) the estimate

$$|u(x,t,\tau)| \le C(1 + \|x\|^\lambda), \quad x \in K, \tag{4.53}$$

which means that $u \in \mathfrak{M}_\lambda$.

In order to obtain the expressions for derivatives of $u(x,t,\tau)$ we repeat the arguments of Sect. 4.1, but in a modified way, taking into account the dependence of the kernel Z on its third and fourth variables, its local constancy with respect to its third variable, and the identities (4.46), (4.48). As a result we obtain the following identities:

$$\frac{\partial u(x,t,\tau)}{\partial t} = f(x,t) + \int_\tau^t d\theta \int_K \frac{\partial Z(x-y,t-\theta,y,\theta)}{\partial t}[f(y,\theta) - f(x,\theta)]\,dy$$
$$+ \int_\tau^t f(x,\theta)\,d\theta \int_K \frac{\partial Z(x-y,t-\theta,y,\theta)}{\partial t}\,dy; \quad (4.54)$$

$$(D^\gamma u)(x,t,\tau) = \int_\tau^t d\theta \int_K Z_\gamma(x-y,t-\theta,y,\theta) f(y,\theta)\,dy, \quad Z_\gamma = D_x^\gamma Z, \quad (4.55)$$

$\lambda < \gamma < \alpha$;

$$(D^\alpha u)(x,t,\tau) = \int_\tau^t d\theta \int_K Z_\alpha(x-y,t-\theta,y,\theta)[f(y,\theta) - f(x,\theta)]\,dy$$
$$+ \int_\tau^t f(x,\theta)\,d\theta \int_K [Z_\alpha(x-y,t-\theta,y,\theta) - Z_\alpha(x-y,t-\theta,x,\theta)]\,dy. \quad (4.56)$$

The formulas (4.54)-(4.56) remain valid also under slightly more general assumptions regarding the function f: it is sufficient to require that f be uniformly locally constant in $x \in K$, be continuous in (x,t), $t \in (0,T]$, and satisfy the bound

$$|f(x,t)| \le Ct^{-\rho}(1 + \|x\|^\lambda), \quad 0 \le \rho < 1. \quad (4.57)$$

4.5.3. Cauchy problem. Let us consider the Cauchy problem with the initial condition (4.2) for the equation (4.35). The conditions on f and φ are the same as in Sect. 4.1 (see Theorem 4.1): $\varphi \in \mathfrak{M}_\lambda$, $f \in \mathfrak{M}_\lambda$ uniformly with respect to t, and f is continuous in (x,t). Here we assume that $0 \le \lambda < \alpha_1$; if all the coefficients $a_1, \dots, a_n$ vanish identically, then we may assume $\lambda < \alpha$.

As above, we look for a solution $u(x,t)$, which is continuous on $K \times [0,T]$, continuously differentiable in t, and belongs to $\mathfrak{M}_\lambda$ uniformly with respect to t.

Theorem 4.6. *The Cauchy problem (4.35)-(4.2) possesses a solution*

$$u(x,t) = \int_0^t d\tau \int_K \Gamma(x,t,\xi,\tau) f(\xi,\tau)\, d\xi + \int_K \Gamma(x,t,\xi,0)\varphi(\xi)\, d\xi, \qquad (4.58)$$

where a fundamental solution $\Gamma(x,t,\xi,\tau)$, $x,\xi \in K$, $0 \le \tau < t \le T$, *is of the form*

$$\Gamma(x,t,\xi,\tau) = Z(x-\xi, t-\tau, \xi, \tau) + W(x,t,\xi,\tau), \qquad (4.59)$$

$$|W(x,t,\xi,\tau)| \le C\Big\{(t-\tau)^{1+\lambda}\left[(t-\tau)^{1/\alpha} + \|x-\xi\|\right]^{-\alpha-1}$$
$$+ (t-\tau)\sum_{k=1}^{n+1}\left[(t-\tau)^{1/\alpha} + \|x-\xi\|\right]^{-\alpha_k-1}\Big\}, \qquad (4.60)$$

while Z *satisfies the estimates (4.42), (4.43), (4.44), (4.49).*

Proof. Following the usual scheme of Levi's method [29, 97], we look for a fundamental solution of the Cauchy problem in the form (4.59), where

$$W(x,t,\xi,\tau) = \int_0^t d\theta \int_K Z(x-\eta, t-\theta, \eta, \theta)\Phi(\eta,\theta,\xi,\tau)\, d\eta, \qquad (4.61)$$

and Φ is determined from the integral equation

$$\Phi(x,t,\xi,\tau) = R(x,t,\xi,\tau) + \int_\tau^t d\theta \int_K R(x,t,\eta,\theta)\Phi(\eta,\theta,\xi,\tau)\, d\eta, \qquad (4.62)$$

in which

$$R(x,t,\xi,\tau) = [a_0(\xi,\tau) - a_0(x,t)]\, Z_\alpha(x-\xi, t-\tau, \xi, \tau)$$
$$- \sum_{k=1}^{n} a_k(x,t) Z_{\alpha_k}(x-\xi, t-\tau, \xi, \tau) - b(x,t) Z(x-\xi, t-\tau, \xi, \tau). \qquad (4.63)$$

Using (4.42) and (4.44) we obtain

$$|R(x,t,\xi,\tau)| \le C\sum_{k=1}^{n+1}\left[(t-\tau)^{1/\alpha} + \|x-\xi\|\right]^{-\alpha_k-1}. \qquad (4.64)$$

The integral equation (4.62) can be solved by the method of successive approximation:

$$\Phi(x,t,\xi,\tau) = \sum_{\nu=1}^{\infty} R_\nu(x,t,\xi,\tau), \tag{4.65}$$

where $R_1 \equiv R$ and

$$R_{\nu+1}(x,t,\xi,\tau) = \int_\tau^t d\theta \int_K R(x,t,\eta,\theta)R_\nu(\eta,\theta,\xi,\tau)\,d\eta, \quad \nu = 1,2,\ldots.$$

In order to prove the convergence of the series (4.65) we need the following lemma.

Lemma 4.6. *Let*

$$J(x,\xi,t,\tau) = \int_\tau^t (t-\mu)^{-\rho/\alpha}(\mu-\tau)^{-\sigma/\alpha}\,d\mu$$
$$\times \int_K \Big[(t-\mu)^{1/\alpha} + \|x-\eta\|\Big]^{-1-b_1} \Big[(\mu-\tau)^{1/\alpha} + \|\eta-\xi\|\Big]^{-1-b_2} d\eta, \tag{4.66}$$

where $0 \le \tau < t$, $b_1, b_2 > 0$, $\rho + b_1 < \alpha$, $\sigma + b_2 < \alpha$. *Then*

$$J(x,\xi,t,\tau) \le C\Big\{B(1-\frac{\rho+b_1}{\alpha}, 1-\frac{\sigma}{\alpha})(t-\tau)^{-(\rho+\sigma+b_1-\alpha)/\alpha}$$
$$\times \Big[(t-\tau)^{1/\alpha} + \|x-\xi\|\Big]^{-1-b_2} + B(1-\frac{\rho}{\alpha}, 1-\frac{\sigma+b_2}{\alpha})(t-\tau)^{-(\rho+\sigma+b_2-\alpha)/\alpha}$$
$$\times \Big[(t-\tau)^{1/\alpha} + \|x-\xi\|\Big]^{-1-b_1}\Big\}, \tag{4.67}$$

where C *depends only on* b_1, b_2; B *is the beta function.*

Proof. Let us decompose the domain of integration in (4.66) as $\Pi_1 \cup \Pi_2$,

$$\Pi_1 = \left(\tau, \frac{t+\tau}{2}\right) \times K, \quad \Pi_2 = \left[\frac{t+\tau}{2}, t\right) \times K.$$

Let also

$$\Pi_{11} = \Big\{\{\mu,\eta\} \in \Pi_1 : \ (\mu-\tau)^{1/\alpha} + \|\eta-\xi\| \le \frac{1}{2}\Big[(t-\tau)^{1/\alpha} + \|x-\xi\|\Big]\Big\},$$

$\Pi_{12} = \Pi_1 \setminus \Pi_{11}$,

$$\Pi_{21} = \left\{ \{\mu, \eta\} \in \Pi_2 : \ (t-\mu)^{1/\alpha} + \|x - \eta\| \le \frac{1}{2}\left[(t-\tau)^{1/\alpha} + \|x-\xi\|\right] \right\},$$

$\Pi_{22} = \Pi_2 \setminus \Pi_{21}$. Now $(\tau, t) \times K = \bigcup\limits_{j,k=1}^{2} \Pi_{jk}$ and $J(x, \xi, t, \tau) = \sum\limits_{j,k=1}^{2} J_{jk}$, where J_{jk} is the integral over Π_{jk}.

Let us use the elementary inequality $(u-v)^{\varkappa} \ge u^{\varkappa} - v^{\varkappa}$, where $0 \le v \le u/2$, $\varkappa \le 1$. If $\{\mu, \eta\} \in \Pi_{11}$, then (recall that $\alpha > 1$) $\mu - \tau \le (t-\tau)/2$,

$$\begin{aligned}(t-\mu)^{1/\alpha} + \|x-\eta\| &= [(t-\tau) - (\mu - \tau)]^{1/\alpha} + \|(x-\xi) - (\eta - \xi)\| \\ &\ge (t-\tau)^{1/\alpha} + \|x - \xi\| - \left[(\mu-\tau)^{1/\alpha} + \|\eta - \xi\|\right] \\ &\ge \frac{1}{2}\left[(t-\tau)^{1/\alpha} + \|x-\xi\|\right],\end{aligned}$$

whence

$$\begin{aligned}J_{11} &\le 2^{1+b_1}\left[(t-\tau)^{1/\alpha} + \|x-\xi\|\right]^{-1-b_1} \\ &\times \int\limits_{\tau}^{(t+\tau)/2} (t-\mu)^{-\rho/\alpha}(\mu-\tau)^{-\sigma/\alpha}\, d\mu \int\limits_{K} \left[(\mu-\tau)^{1/\alpha} + \|\eta\|\right]^{-1-b_2} d\eta \\ &\le C(b_1, b_2)\left[(t-\tau)^{1/\alpha} + \|x-\xi\|\right]^{-1-b_1} \\ &\times \int\limits_{\tau}^{(t+\tau)/2} (t-\mu)^{-\rho/\alpha}(\mu-\tau)^{-(\sigma+b_2)/\alpha}\, d\mu.\end{aligned}$$

After the change of variables $t - \mu = (t-\tau)\nu$ we get

$$\begin{aligned}J_{11} &\le C(b_1, b_2)\left[(t-\tau)^{1/\alpha} + \|x-\xi\|\right]^{-1-b_1} (t-\tau)^{-(\rho+\sigma+b_2-\alpha)/\alpha} \\ &\times \int\limits_{1/2}^{1} \nu^{-\rho/\alpha}(1-\nu)^{-(\sigma+b_2)/\alpha}\, d\nu \\ &\le C(b_1, b_2) B\left(1 - \frac{\rho}{\alpha}, 1 - \frac{\sigma + b_2}{\alpha}\right)(t-\tau)^{-(\rho+\sigma+b_2-\alpha)/\alpha} \\ &\times \left[(t-\tau)^{1/\alpha} + \|x-\xi\|\right]^{-1-b_1},\end{aligned}$$

Next,

$$J_{12} \le 2^{1+b_2} \left[(t-\tau)^{1/\alpha} + \|x-\xi\| \right]^{-1-b_2}$$

$$\times \int_{\tau}^{(t+\tau)/2} (t-\mu)^{-\rho/\alpha} (\mu-\tau)^{-\sigma/\alpha} \, d\mu \int_K \left[(t-\mu)^{1/\alpha} + \|\eta\| \right]^{-1-b_1} d\eta.$$

As before,

$$J_{12} \le C(b_1, b_2) B(1 - \frac{\rho + b_1}{\alpha}, 1 - \frac{\sigma}{\alpha})(t-\tau)^{-(\rho+\sigma+b_1-\alpha)/\alpha}$$

$$\times \left[(t-\tau)^{1/\alpha} + \|x-\xi\| \right]^{-1-b_2}.$$

After similar estimates for J_{21} and J_{22} we obtain the inequality (4.67).

■

Now we return to (4.65). From the estimate (4.64) and Lemma 4.6 we obtain by induction the estimates of iterated kernels:

$$|R_\nu(x, t, \xi, \tau)| \le CM^\nu \frac{[\Gamma(\lambda)]^\nu}{\Gamma(\nu\lambda)} \sum_{k=1}^{n+1} \left[(t-\tau)^{1/\alpha} + \|x-\xi\| \right]^{-\alpha_k - 1}.$$

It follows from the Stirling formula that the series (4.65) is convergent and

$$|\Phi(x, t, \xi, \tau)| \le C \sum_{k=1}^{n+1} \left[(t-\tau)^{1/\alpha} + \|x-\xi\| \right]^{-\alpha_k - 1}. \tag{4.68}$$

Now from (4.42), (4.61), (4.68), and Lemma 4.6 we obtain the estimate (4.60).

Denote by $u_1(x,t)$ and $u_2(x,t)$ the first and second summands in the right-hand side of (4.58). Substituting (4.59) and (4.61) into (4.58) we find that

$$u_1(x,t) = \int_0^t d\tau \int_K Z(x-\xi, t-\tau, \xi, \tau) f(\xi, \tau) \, d\xi$$

$$+ \int_0^t d\theta \int_K Z(x-\eta, t-\theta, \eta, \theta) F(\eta, \theta) \, d\eta,$$

$$u_2(x,t) = \int_K Z(x-\xi,t,\xi,0)\varphi(\xi)\,d\xi + \int_0^t d\theta \int_K Z(x-\eta,t-\theta,\eta,\theta)G(\eta,\theta)\,d\eta,$$

where

$$F(\eta,\theta) = \int_0^\theta d\tau \int_K \Phi(\eta,\theta,\xi,\tau)f(\xi,\tau)\,d\xi,$$

$$G(\eta,\theta) = \int_K \Phi(\eta,\theta,\xi,0)\varphi(\xi)\,d\xi.$$

It follows from the inequality (4.68) and Lemma 4.3 that

$$|F(\eta,\theta)| \le C, \quad |G(\eta,\theta)| \le C\theta^{-\alpha_{n+1}/\alpha}$$

for all $\eta \in K$, $\theta \in (0,T]$ (compare the last inequality with (4.57)). The functions F, G are uniformly locally constant. Indeed, let N be a local constancy exponent for all the functions on K, which appear in our Cauchy problem (4.35)-(4.2). The construction of the function P leads to the identity

$$R(x+\delta,t,\xi+\delta,\tau) = R(x,t,\xi,\tau),$$

if $\|\delta\| \le q^{-N}$. The we obtain recursively

$$R_{\nu+1}(x+\delta,t,\xi+\delta,\tau) = \int_\tau^t d\theta \int_K R(x+\delta,t,\eta,\theta)R_\nu(\eta,\theta,\xi+\delta,\tau)\,d\eta$$

$$= \int_\tau^t d\theta \int_K R(x+\delta,t,\zeta+\delta,\theta)R_\nu(\zeta+\delta,\theta,\xi+\delta,\tau)\,d\zeta = R_{\nu+1}(x,t,\xi,\tau),$$

so that $\Phi(x,t,\xi,\tau) = \Phi(x+\delta,t,\xi+\delta,\tau)$, whence

$$F(\eta+\delta,\theta) = \int_0^\theta d\tau \int_K \Phi(\eta+\delta,\theta,\xi+\delta,\tau)f(\xi+\delta,\tau)\,d\xi = F(\eta,\theta).$$

Similarly $G(\eta+\delta,\theta) = G(\eta,\theta)$, $\|\delta\| \le q^{-N}$.

Thus the potentials in the expressions for $u_1(x,t)$ and $u_2(x,t)$ satisfy the conditions, under which the differentiation formulas (4.54)-(4.56) were obtained. If we use those formulas, along with (4.46) and (4.47), we find after some simple transformations that the function $u(x,t) \in \mathfrak{M}_\lambda$ satisfies the equation (4.35).

Let us show that $u(x,t) \to \varphi(x)$ as $t \to 0$. Let

$$v(x,t) = \int_K Z(x-\xi,t,\xi,0)\varphi(\xi)\,d\xi.$$

Due to (4.59), (4.60), it is sufficient to verify that $v(x,t) \to \varphi(x)$ as $t \to 0$. By virtue of (4.45) we have

$$v(x,t) = \int_K [Z(x-\xi,t,\xi,0) - Z(x-\xi,t,x,0)]\varphi(\xi)\,d\xi$$
$$+ \int_K Z(x-\xi,t,x,0)[\varphi(\xi)-\varphi(x)]\,d\xi + \varphi(x).$$

Since Z (as a function of its third argument) and φ are locally constant, it follows that in both integrals the integration is actually performed over the set

$$\{\xi \in K:\ \|\xi - x\| \geq q^{-N}\}.$$

Applying the inequality (4.42) on this set, we see that both integrals tend to zero as $t \to 0$. ■

4.5.4. Non-negative solutions. Let us consider the situation, in which the conditions of Theorems 4.5 and 4.6 hold simultaneously.

Theorem 4.7. *If the coefficients $a_k(x,t)$ and $b(x,t)$ are nonnegative, then* $\Gamma(x,t,\xi,\tau) \geq 0$.

Proof. It is sufficient to show that the solution

$$u(x,t) = \int_K \Gamma(x,t,\xi,\tau)\varphi(\xi)\,d\xi$$

of the Cauchy problem for the equation (4.35) with $f(x,t) \equiv 0$ and the initial condition $u(x,\tau) = \varphi(x)$, $x \in K$, where $\varphi(x) \geq 0$, $\varphi \in \mathcal{D}(K)$, is nonnegative. It follows from (4.42), (4.59), (4.60) that $u(x,t) \to 0$ as

$\|x\| \to \infty$. Hence, if $u(x,t)$ fails to be nonnegative, then there exists a negative global minimum point:

$$u(x_0, t_0) = \inf_{x \in K, 0 \le t \le T} u(x,t) < 0. \tag{4.69}$$

As in the proof of Theorem 4.5, this can be shown to imply that at the point (x_0, t_0) all of the terms on the left in (4.35) are nonpositive, and hence are zero. Since $a_0(x_0, t_0) > 0$ by our uniform parabolicity assumption, we find that in particular $(D^\alpha u)(x_0, t_0) = 0$, which is possible only if $u(x, t_0) \equiv$const, so that $u(x, t_0) \equiv 0$. This contradicts (4.69). ■

The following relation is deduced from Theorem 4.5 in a standard way:

$$\Gamma(x,t,\xi,\tau) = \int_K \Gamma(x,t,y,\sigma)\Gamma(y,\sigma,\xi,\tau)\, dy, \quad \tau < \sigma < t \le T.$$

If $b(x,t) = 0$, then in addition

$$\int_K \Gamma(x,t,\xi,\tau)\, d\xi = 1.$$

Suppose now that the left-hand side of (4.35) is defined for all $t \ge 0$, and all the conditions, under which Theorems 4.5 and 4.6 hold, are fulfilled for any $T > 0$. As in Sect. 4.2, we come to the following probabilistic interpretation of the fundamental solution.

Corollary 4.1. *The fundamental solution $\Gamma(x,t,\xi,\tau)$ is the transition density of a bounded right-continuous strict Markov process without second kind discontinuities. If $b(x,t) \equiv 0$, then the process does not explode.*

4.6 Heat equation on a ball

4.6.1. The process and its generator. Let $\xi_\alpha(t)$ be the Markov process on a local field K constructed in Theorem 4.2. In this section we shall use $\xi_\alpha(t)$ in order to construct a natural Markov process on a ball

$$B_r = \{x \in K : \|x\| \le q^r\}.$$

Suppose that $\xi_\alpha(0) \in B_r$. Denote by $\xi_\alpha^{(r)}(t)$ the sum of all jumps of the process $\xi_\alpha(\tau)$, $\tau \in [0,t]$, whose absolute values exceed q^r. Since ξ_α is right continuous with left limits, $\xi_\alpha^{(r)}(t)$ is finite a.s., $\xi_\alpha^{(r)}(0) = 0$. Let us consider the process

$$\eta_\alpha(t) = \xi_\alpha(t) - \xi_\alpha^{(r)}(t).$$

Since the jumps of η_α never exceed q^r by absolute value, this process remains a.s. in B_r (due to the ultra-metric inequality). Such a property is of purely non-Archimedean nature; nothing of this kind can take place for real-valued stochastic processes.

Having constructed a process in the ball B_r we will now find its generator A_r^α. Since the operator $-D^\alpha$ is nonlocal we should not expect A_r^α to coincide with $-D^\alpha$ on locally constant functions concentrated on B_r. However it turns out to be connected with $-D^\alpha$ in a remarkably simple way.

Below we will often identify a function on B_r with its extension by zero onto the whole of K. Let $\mathcal{D}(B_r)$ consist of functions from $\mathcal{D}(K)$ supported in B_r. Then $\mathcal{D}(B_r)$ is dense in $L_2(B_r)$, and the operator D_r^α in $L_2(B_r)$ defined by restricting D^α to $\mathcal{D}(B_r)$ and considering $(D^\alpha\varphi)(x)$ only for $x \in B_r$, is essentially self-adjoint and positive (see Chapter 3, where this operator was denoted A_{B_r}. Denote by λ its smallest eigenvalue. Then (see Sect. 3.3)

$$\lambda = \frac{q-1}{q^{\alpha+1}-1} q^{\alpha(1-r)}.$$

Theorem 4.8. *If* $\eta_\alpha(0) = x$, *and* $\varphi \in \mathcal{D}(B_r)$ *then*

$$\frac{d}{dt}\mathbf{E}\varphi(\eta_\alpha(t))\bigg|_{t=0} = -\left(D_r^\alpha\varphi\right)(x) + \lambda\varphi(x). \tag{4.70}$$

Proof. It is sufficient to consider the case $x = 0$. By the general result about processes with independent increments [56] the processes $\eta_\alpha(t)$ and $\xi_\alpha^{(r)}(t)$ are independent. Thus

$$\mathbf{E}\chi(z\xi_\alpha(t)) = \mathbf{E}\chi(z\xi_\alpha^{(r)}(t)) \cdot \mathbf{E}\chi(z\eta_\alpha(t)) \tag{4.71}$$

for any $z \in K$. We know that $\mathbf{E}\chi(z\xi_\alpha(t)) = \exp\left(-t\|z\|^\alpha\right)$. The Lévy measure of the process ξ_α equals

$$\pi_\alpha(dx) = \frac{q^\alpha - 1}{1-q^{-\alpha-1}} \|x\|^{-\alpha-1}\, dx.$$

Since

$$\mathbf{E}\chi(z\xi_\alpha^{(r)}(t)) = \exp\left\{ t \int\limits_{K\setminus B_r} [\chi(zx) - 1]\pi_\alpha(dx) \right\}$$

(see [56]) we find from (4.71) that

$$\mathbf{E}\chi(z\eta_\alpha(t)) = \exp\left\{ t \int\limits_{B_r} [\chi(zx) - 1]\pi_\alpha(dx) \right\}.$$

This formula is a special case of a result from [32].

The character χ equals 1 on B_0. Hence $\mathbf{E}\chi(z\eta_\alpha(t)) = 1$ if $\|z\| \le q^{-r}$. Suppose that $\|z\| = q^k$, $k \ge -r+1$. Then

$$\int_{B_r} [\chi(zx) - 1]\pi_\alpha(dx) = \int_{q^{-k+1} \le \|x\| \le q^r} [\chi(zx) - 1]\pi_\alpha(dx).$$

Using the formula (1.28) for the Fourier transform of a radial function we find after elementary calculations that

$$\int_{B_r} [\chi(zx) - 1]\pi_\alpha(dx) = -\|z\|^\alpha + \lambda, \quad \|z\| > q^{-r},$$

so that

$$\mathbf{E}\chi(z\eta_\alpha(t)) = \begin{cases} 1, & \text{if } \|z\| \le q^{-r} \\ \exp(t(\lambda - \|z\|^\alpha)), & \text{if } \|z\| > q^{-r}. \end{cases} \tag{4.72}$$

If $\varphi = F\psi$, $\psi \in \mathcal{D}(K)$, then by virtue of (4.72)

$$\mathbf{E}\varphi(\eta_\alpha(t)) = \int_{\|z\| \le q^{-r}} \psi(z)\,dz + e^{\lambda t} \int_{\|z\| > q^{-r}} e^{-t\|z\|^\alpha} \psi(z)\,dz,$$

whence

$$\frac{d}{dt}\mathbf{E}\varphi(\eta_\alpha(t))\Big|_{t=0} = \lambda \int_{\|z\|>q^{-r}} \psi(z)\,dz - \int_{\|z\|>q^{-r}} \|z\|^\alpha \psi(z)\,dz. \tag{4.73}$$

By the Fourier inversion formula

$$\int_K \psi(z)\,dz = \varphi(0).$$

On the other hand, since $\operatorname{supp}\varphi \subset B_r$, we find that $\psi(z) = \psi(0)$ for $\|z\| \le q^{-r}$. Computing the integrals we obtain that

$$\int_{\|z\| \le q^{-r}} \|z\|^\alpha \psi(z)\,dz = \psi(0) \int_{\|z\| \le q^{-r}} \|z\|^\alpha\,dz$$

$$= \frac{1-q^{-1}}{1-q^{-\alpha-1}} q^{-(\alpha+1)r}\psi(0) = \lambda q^{-r}\psi(0) = \lambda \int_{\|z\| \le q^{-r}} \psi(z)\,dz,$$

so that (4.73) takes the form

$$\frac{d}{dt}\mathbf{E}\varphi(\eta_\alpha(t))\Big|_{t=0} = \lambda \int\limits_K \psi(z)\,dz - \int\limits_K \|z\|^\alpha \psi(z)\,dz = \lambda\varphi(0) - (D_r^\alpha \varphi)(0). \quad \blacksquare$$

4.6.2. Transition density. Consider the Cauchy problem

$$\frac{\partial u(x,t)}{\partial t} + (D_r^\alpha u)(x,t) - \lambda u(x,t) = 0, \quad x \in B_r,\ t > 0; \tag{4.74}$$

$$u(x,0) = \varphi(x), \quad x \in B_r, \tag{4.75}$$

where $\varphi \in \mathcal{D}(B_r)$.

A maximum principle argument (as in the proof of Theorem 4.5; note that a simple substitution reduces the equation (4.74) to a similar equation without the term λu) proves the uniqueness of the solution of (4.74)-(4.75). Thus the fundamental solution $Z_r(x-y,t)$ for the problem (4.74)-(4.75) can be identified with the transition density of the process η_α.

The next result gives a formula for the transition density; its structure is quite different from the ones appearing in conventional parabolic equations. Note in particular its dependence on the difference $x - y,\ x, y \in B_r$.

Theorem 4.9. *The solution of the problem (4.74)-(4.75) is given by the formula*

$$u(x,t) = \int\limits_{B_r} Z_r(x-y,t)\varphi(y)\,dy, \quad x \in B_r,\ t > 0, \tag{4.76}$$

where

$$Z_r(x,t) = e^{\lambda t} Z(x,t) + c(t), \quad x \in B_r,$$

$$c(t) = q^{-r} - q^{-r}(1-q^{-1})e^{\lambda t}\sum_{n=0}^{\infty} \frac{(-1)^n}{n!} t^n \frac{q^{-r\alpha n}}{1-q^{-\alpha n-1}},$$

and $Z(x,t)$ is given by (4.5). The function Z_r is non-negative, and

$$\int\limits_{B_r} Z_r(x,t)\,dx = 1. \tag{4.77}$$

Proof. Let us substitute (4.76) into (4.74) remembering that functions on B_r are meant to be extended onto K by zero. If $\theta_r(x)$, $x \in K$, is an indicator of the set B_r then we may write $u = u_1 + u_2$, where

$$u_1(x,t) = e^{\lambda t}\theta_r(x) \int\limits_{B_r} Z(x-y,t)\varphi(y)\,dy,$$

$$u_2(x,t) = c(t)\theta_r(x) \int\limits_{B_r} \varphi(y)\, dy, \quad x \in K,\ t > 0.$$

Introducing also the notation

$$w_1(x,t) = \theta_r(x) \int\limits_{B_r} Z(x-y,t)\varphi(y)\, dy,$$

$$w_2(x,t) = [1-\theta_r(x)] \int\limits_{B_r} Z(x-y,t)\varphi(y)\, dy, \quad x \in K,\ t > 0,$$

we obtain that

$$\left(\frac{\partial}{\partial t} + D^\alpha\right) w_1 = -\left(\frac{\partial}{\partial t} + D^\alpha\right) w_2$$

so that for $x \in B_r$

$$\frac{\partial w_1(x,t)}{\partial t} + (D^\alpha w_1)(x,t) = -(D^\alpha w_2)(x,t)$$

and

$$\frac{\partial u_1(x,t)}{\partial t} + (D_r^\alpha u_1)(x,t) - \lambda u_1(x,t) = -e^{\lambda t}(D^\alpha w_2)(x,t),$$

$x \in B_r$.

If $z \notin B_r$, then the ultra-metric property of the absolute value leads to the formula

$$w_2(z,t) = \int\limits_{B_r} \varphi(y)\, dy \sum_{n=1}^{\infty} \frac{(-1)^n}{n!} t^n \frac{1-q^{\alpha n}}{1-q^{-\alpha n-1}} \|z\|^{-\alpha n-1}.$$

Substituting this into the expression for D^α and calculating the integrals we find that for $x \in B_r$

$$(D^\alpha w_2)(x,t) = \frac{(1-q^\alpha)(1-q^{-1})}{1-q^{-\alpha-1}} \times \sum_{n=1}^{\infty} \frac{(-1)^n}{n!} t^n \frac{1-q^{\alpha n}}{1-q^{-\alpha n-1}} \cdot \frac{q^{-(r+1)(\alpha+\alpha n+1)}}{1-q^{-\alpha-\alpha n-1}}.$$

On the other hand,

$$\left(\left(\frac{\partial}{\partial t} + D_r^\alpha - \lambda\right) u_2\right)(x,t) = c'(t)\theta_r(x) \int\limits_{B_r} \varphi(y)\, dy$$

because θ_r is an eigenfunction of the operator D_r^α corresponding to the eigenvalue λ (see Sect. 3.3). Now it remains to compute $c'(t)$ and perform a simple series transformation, and we come to the equality (4.74).

The fact that the condition (4.75) is fulfilled follows from properties of Z (see Sect. 4.1), since $c(0) = 0$.

The inequality $Z_r(x,t) \geq 0$ and the identity (4.77) are evident from their probabilistic meaning. However it is not difficult to give direct proofs – the non-negativity may be proved by the maximum principle argument (as in the proof of Theorem 4.7), while (4.77) may be verified by an immediate calculation based on the representation

$$Z(x,t) = \int_K \chi(-xz) \exp\left(-t\|z\|^\alpha\right) dz. \qquad \blacksquare$$

It is well known (see e.g. [50]) that the unit ball $B_0^{(2)}$ in the 2-adic field $\mathbb{Q}_2$ is homeomorphic to the Cantor set $C \subset [0,1]$. Moreover the homeomorphism explicitly written in [50] transforms the Haar measure on $B_0^{(2)}$ into the Cantor staircase distribution on [0,1]. Thus the above results, together with general facts about transformations of Markov processes [27], lead to the explicit construction of the transition density of a Markov process on C, right continuous with left limits.

4.7 Heat equation on the group of units

Let U be the group of units of a local field K, A_U the operator over U obtained by restricting D^α to $\mathcal{D}(U)$ and considering $(D^\alpha\varphi)(x)$, $\varphi \in \mathcal{D}(U)$, only for $x \in U$. As we saw in Sect. 3.5, the operator A_U is connected with the multiplicative structure of the local field K. In particular, by Theorem 3.12, for any $\varphi \in \mathcal{D}(U)$

$$\left(A_U\varphi\right)(x) = \frac{1-q^\alpha}{1-q^{-\alpha-1}} \int_U \|1-y\|^{-\alpha-1}\left[\varphi(xy^{-1}) - \varphi(x)\right] dy + \lambda_0\varphi(x), \tag{4.78}$$

$x \in U$, where $\lambda_0 = \dfrac{q^{\alpha-1} - 2q^{-1} + 1}{1 - q^{-\alpha-1}}$ is the smallest eigenvalue of the closure $\overline{A_U}$ on $L_2(U)$.

It is natural to consider the Cauchy problem

$$\frac{\partial u(x,t)}{\partial t} + (A_U u)(x,t) - \lambda_0 u(x,t) = 0, \quad x \in U,\ t > 0; \tag{4.79}$$

$$u(x,0) = \varphi(x), \quad x \in U, \tag{4.80}$$

where $\varphi \in \mathcal{D}(U)$. As before, we shall understand $A_U u$ (where u need not belong to $\mathcal{D}(U)$) in the sense of the hypersingular integral in the right-hand side of (4.78).

The next theorem gives a fundamental solution for (4.79)-(4.80) (understood in the sense of the multiplicative structure of K and U). As in the previous section, it is obtained by a certain transformation of the fundamental solution (4.5) of D^α over K.

Theorem 4.10. *The solution of the problem (4.79)-(4.80) is given by the formula*

$$u(x,t) = \int_U Z_*(xy^{-1},t)\varphi(y)\,dy, \quad x \in U,\ t > 0, \tag{4.81}$$

where

$$Z_*(x,t) = e^{\lambda_0 t} Z(1-x,t) + c_0(t), \quad x \in U,$$

$$c_0(t) = \frac{q}{q-1}\left[1 - e^{\lambda_0 t}\sum_{n=0}^{\infty}\frac{(-1)^n}{n!}\frac{q^{\alpha n-1} - 2q^{-1} + 1}{1 - q^{-\alpha n-1}} t^n\right].$$

The function Z_* is non-negative, and

$$\int_U Z_*(x,t)\,dx = 1. \tag{4.82}$$

Proof. Let us substitute (4.81) into (4.79) remembering that functions on U are meant to be extended onto K by zero. If $\eta(x)$, $x \in K$, is an indicator of the set U, then we may write $u = u_1 + u_2$, where

$$u_1(x,t) = e^{\lambda_0 t}\eta(x)\int_U Z(1-xy^{-1},t)\varphi(y)\,dy$$

$$= e^{\lambda_0 t}\eta(x)\int_U Z(x-y,t)\varphi(y)\,dy\,,$$

$$u_2(x,t) = c_0(t)\eta(x) \int\limits_U \varphi(y)\, dy, \quad x \in K,\ t > 0.$$

Introducing also the notation

$$w_1(x,t) = \eta(x) \int\limits_U Z(x-y,t)\varphi(y)\, dy,$$

$$w_2(x,t) = [1-\eta(x)] \int\limits_U Z(x-y,t)\varphi(y)\, dy, \quad x \in K,\ t > 0,$$

we obtain that

$$\left(\frac{\partial}{\partial t} + D^\alpha\right) w_1 = -\left(\frac{\partial}{\partial t} + D^\alpha\right) w_2$$

so that for $x \in U$

$$\frac{\partial w_1(x,t)}{\partial t} + (D^\alpha w_1)(x,t) = -(D^\alpha w_2)(x,t)$$

and

$$\frac{\partial u_1(x,t)}{\partial t} + (A_U u_1)(x,t) - \lambda_0 u_1(x,t) = -e^{\lambda_0 t}(D^\alpha w_2)(x,t).$$

By the ultra-metric property of the absolute value

$$w_2(x,t) = \int\limits_U \varphi(y)\, dy \sum_{n=1}^{\infty} \frac{(-1)^n}{n!} \frac{1-q^{\alpha n}}{1-q^{-\alpha n-1}} t^n$$

if $\|x\| < 1$, and

$$w_2(x,t) = \int\limits_U \varphi(y)\, dy \sum_{n=1}^{\infty} \frac{(-1)^n}{n!} \frac{1-q^{\alpha n}}{1-q^{-\alpha n-1}} t^n \|x\|^{-\alpha n-1}$$

if $\|x\| > 1$. Remembering that $w_2(x,t) = 0$ if $\|x\| = 1$, substituting into (4.78) and calculating integrals we find that for any $x \in U$

$$(D^\alpha w_2)(x,t)$$
$$= \frac{1-q^\alpha}{1-q^{-\alpha-1}} \int\limits_U \varphi(y)\, dy \sum_{n=1}^{\infty} \frac{(-1)^n}{n!} t^n \frac{(1-q^{\alpha n})(q^{\alpha n+\alpha} - 2q^{-1} + 1)}{(1-q^{-\alpha n-1})(q^{\alpha n+\alpha+1} - 1)}.$$

On the other hand,

$$\left(\left(\frac{\partial}{\partial t} + A_U - \lambda_0\right) u_2\right)(x,t) = c_0'(t)\eta(x) \int\limits_U \varphi(y)\, dy$$

because η is an eigenfunction of the operator A_U corresponding to the eigenvalue λ_0. Now it remains to compute $c_0'(t)$, and after elementary (though somewhat tedious) transformations we come to the equality (4.79).

The fact that the condition (4.80) is fulfilled follows from properties of Z since $c_0(0) = 0$.

The inequality $Z_*(x,t) \geq 0$ can be proved by the maximum principle argument just as it was done in the preceding sections.

The identity (4.82) is verified by a direct calculation based on the representation

$$Z(s,t) = \int\limits_K \chi(-s\xi) \exp\left(-t\|\xi\|^\alpha\right) d\xi. \quad \blacksquare$$

The fundamental solution Z_* can be interpreted as a transition density of a non-exploding Markov process on U. It follows from the expression for Z_*, properties of Z and the general theory of Markov processes that the process is right-continuous, with left limits, without second kind discontinuities.

4.8 On parabolic pseudo-differential equations over $\mathbb{R}^n$

4.8.1. Motivation. It is interesting to compare the above results with those for parabolic pseudo-differential equations over $\mathbb{R}^n$. The main difference is the problem of convergence of hypersingular integrals. In the non-Archimedean case one has rich spaces of locally constant functions, on which hyper-singular integral operators exist due to the actual disappearance of singularities. The Euclidean case is much more complicated – rather subtle regularization procedures are necessary to guarantee the convergence. The regularisation depends crucially on the order of an equation. In particular, if the order is greater than 2, the regularized operator has a form preventing the use of the maximum principle argument used in the proof of Theorem 4.7 in order to show the non-negativity of the fundamental solution. Thus the problems of convergence and positivity in the Euclidean case are closely connected.

We shall consider the Cauchy problem

$$\frac{\partial u(x,t)}{\partial t} + (Au)(x,t) + \sum_{k=1}^{m}(A_k u)(x,t) = f(x,t), \quad x \in \mathbb{R}^n,\ t \in (0,T], \tag{4.83}$$

$$u(x,0) = \varphi(x), \tag{4.84}$$

where $A, A_1, \dots, A_m$ are pseudo-differential operators with the symbols $a(x,t,\xi), a_1(x,t,\xi), \dots, a_m(x,t,\xi)$, that is, e.g.

$$(Au)(x,t) = (2\pi)^{-n} \int_{\mathbb{R}^n} e^{ix\cdot\xi} a(x,t,\xi)\widehat{u}(\xi,t)\, d\xi, \tag{4.85}$$

where

$$\widehat{u}(\xi,t) = \int_{\mathbb{R}^n} e^{-iy\cdot\xi} u(y,t)\, dy.$$

The principal symbol $a(x,t,\xi)$ is assumed homogeneous of the order $\gamma \geq 1$ with respect to the variable ξ, and elliptic, the symbols $a_1(x,t,\xi), \dots, a_m(x,t,\xi)$ have the orders of homogeneity γ_k, $0 < \gamma_k < \gamma$, f and φ are bounded continuous functions. The expression (4.85) can be used only for smooth rapidly decreasing functions, and our first task is to reformulate it in terms of hyper-singular integrals.

4.8.2. Hyper-singular integrals. Suppose we have complex-valued bounded functions $f \in C(\mathbb{R}^n)$, $\Omega \in C(\mathbb{R}^n \times S^{n-1})$. The expression

$$\left(\mathbf{D}^{\alpha}_{\Omega} f\right)(x) = \frac{1}{d_{n,l}(\alpha)} \int_{\mathbb{R}^n} \Omega\left(x, \frac{h}{|h|}\right) \frac{\left(\Delta^l_h f\right)(x)}{|h|^{n+\alpha}}\, dh, \quad x \in \mathbb{R}^n, \tag{4.86}$$

where $\alpha > 0$, $\left(\Delta^l_h f\right)(x) = \sum\limits_{k=0}^{l} (-1)^k \binom{l}{k} f(x-kh)$, l is a natural number, $d_{n,l}(\alpha)$ is a normalization constant (its choice will be explained below), is called *a hyper-singular integral of the order α with the characteristic Ω*. We shall also use a restricted hyper-singular integral $\mathbf{D}^{\alpha}_{\Omega,\varepsilon} f$, which has the same form (4.86) except that the domain of integration is restricted to $\{h \in \mathbb{R}^n : |h| > \varepsilon\}$, $\varepsilon > 0$. The theory of hypersingular integrals was developed by Samko (see [130, 131]) for the case of a characteristic independent of x. The results we use below carry over virtually unchanged to the general situation.

Suppose first that the number α is not an integer. The integral (4.86) is absolutely convergent if $l > \alpha$, and the function f has bounded derivatives up to the order $[\alpha]+1$. The absolute convergence over the ball $\{|h| < \varepsilon\}$ is a consequence of the formula

$$\left(\Delta_h^l f\right)(x) = \sum_{|\varkappa|=\mu} \sum_{\nu=0}^{l} \frac{(-\nu h)^{\varkappa}(-1)^{\nu}}{\varkappa!} \binom{l}{\nu} (D^{\varkappa} f)(x - \theta_\nu \nu h), \tag{4.87}$$

where $0 < \theta_\nu < 1$, $l \geq \mu$, and the usual notation for operations with multi-indices is employed (of course, in this section, unlike the rest of the book, D denotes a usual differentiation). The formula (4.87) follows from the Taylor formula.

The restricted hyper-singular integral is absolutely convergent if, for example, f is bounded. The above convergence conditions can be easily extended to the case when $f(x)$ and its derivatives grow not too rapidly as $|x| \to \infty$. We shall use also conditionally convergent hyper-singular integrals of bounded functions:

$$\left(\mathbf{D}_\Omega^\alpha f\right)(x) = \lim_{\varepsilon \to 0} \left(\mathbf{D}_{\Omega,\varepsilon}^\alpha f\right)(x), \tag{4.88}$$

if the limit (4.88) exists for all $x \in \mathbb{R}^n$.

If the characterisic $\Omega(x, \sigma)$ is even in σ, that is $\Omega(x, -\sigma) = \Omega(x, \sigma)$, $x \in \mathbb{R}^n$, $\sigma \in S^{n-1}$, then the hyper-singular integral (4.86) makes sense also for $l > 2[\alpha/2]$, by the formula

$$\begin{aligned} \lim_{\varepsilon \to 0} \left(\mathbf{D}_{\Omega,\varepsilon}^\alpha f\right)(x) = &-\frac{1}{d_{n,l}(\alpha)} \sum_{k=1}^{(l-1)/2} \int_{\mathbb{R}^n} \Omega\left(x, \frac{h}{|h|}\right) \frac{\left(\Delta_h^{l+1} f\right)(x + kh)}{|h|^{n+\alpha}}\, dh \\ &- \frac{1}{2d_{n,l}(\alpha)} \int_{\mathbb{R}^n} \Omega\left(x, \frac{h}{|h|}\right) \frac{\left(\Delta_h^{l+1} f\right)\left(x + \frac{l+1}{2} h\right)}{|h|^{n+\alpha}}\, dh, \end{aligned} \tag{4.89}$$

where this time $l + 1 > \alpha$.

Next, let α be an integer. If α is even, then $(\mathbf{D}_\Omega^\alpha f)$ can be defined as before. In this case the hyper-singular integral operator $\mathbf{D}_\Omega^\alpha$ is actually a differential operator of the order α. If α is odd, then for $l > \alpha$ the integral in (4.86) vanishes identically for any function f. In this case $\mathbf{D}_\Omega^\alpha$ can be defined only for an even characteristic Ω by the formula (4.89) with $l = \alpha$.

Let $f \in C^l(\mathbb{R}^n)$, $|f(x)| \leq C(1 + |x|)^{-N_1}$, $|(D^{\varkappa} f)(x)| \leq C(1 + |x|)^{-N_2}$, $|\varkappa| = l$, where $N_1 > \alpha$, $N_2 > n$. Then

$$\left|\left(\mathbf{D}_\Omega^\alpha f\right)(x)\right| \leq C(1 + |x|)^{-\min(\alpha + N_1, N_2, \alpha + n)}. \tag{4.90}$$

Moreover, if Ω depends on a parameter t being bounded uniformly with respect to t, then the constant in (4.90) does not depend on t. The same estimate, with a constant independent of ε, holds for the restricted hyper-singular integral $\mathbf{D}^{\alpha}_{\Omega,\varepsilon} f$.

Let us consider the function $f_\xi(x) = e^{i\xi\cdot x}$. Obviously,

$$\left(\Delta^l_h f_\xi\right)(x) = e^{i\xi\cdot x} \sum_{k=0}^{l} (-1)^k \binom{l}{k} e^{-ik\xi\cdot h} = e^{i\xi\cdot x}\left(1 - e^{-ik\xi\cdot h}\right)^l,$$

whence

$$\left(\mathbf{D}^{\alpha}_{\Omega} f_\xi\right)(x) = \widetilde{\Omega}(x,\xi) f_\xi(x), \tag{4.91}$$

where

$$\widetilde{\Omega}(x,\xi) = \frac{1}{d_{n,l}(\alpha)} \int\limits_{\mathbb{R}^n} \frac{\left(1 - e^{-i\xi\cdot h}\right)^l}{|h|^{n+\alpha}} \Omega\left(x, \frac{h}{|h|}\right) dh.$$

The function $\widetilde{\Omega}$ is called *the symbol* of the hyper-singular integral $\mathbf{D}^{\alpha}_{\Omega} f$. The normalization constants $d_{n,l}(\alpha)$ are chosen in such a way that the symbol (hence the hyper-singular integral) does not depend on l. Their explicit form can be found in [131]. Here we only note that they do not depend on Ω, and $d_{n,l}(\alpha) > 0$ if $0 < \alpha < 2$. The symbol $\widetilde{\Omega}(x,\xi)$ is a homogeneous function of the degree α; for example, if $\Omega(x,\sigma) \equiv 1$, then $\widetilde{\Omega}(x,\xi) = |\xi|^\alpha$ (the corresponding hyper-singular integral operator is often called a Riesz derivative). The symbol can be represented as

$$\widetilde{\Omega}(x,\xi) = C(n,\alpha) \int\limits_{S^{n-1}} \Omega(x,\sigma)(i\xi\cdot\sigma)^\alpha \, d\sigma. \tag{4.92}$$

Note also that the symbol $\widetilde{\Omega}$ is even if and only if the characteristic Ω is even.

It follows from the formula (4.91) that on functions from the Schwartz space $\mathcal{S}(\mathbb{R}^n)$ (and more generally, on smooth functions, which decrease rapidly enough) the operator $\mathbf{D}^{\alpha}_{\Omega}$ coincides with the pseudo-differential operator with the symbol $\widetilde{\Omega}$. Indeed, if $f \in \mathcal{S}(\mathbb{R}^n)$, then

$$f(x) = (2\pi)^{-n} \int\limits_{\mathbb{R}^n} e^{ix\cdot\xi} \widehat{f}(\xi)\, d\xi,$$

and by the Fubini theorem

$$\left(\mathbf{D}^{\alpha}_{\Omega} f\right)(x) = (2\pi)^{-n} \int\limits_{\mathbb{R}^n} e^{ix\cdot\xi} \widetilde{\Omega}(x,\xi) \widehat{f}(\xi)\, d\xi.$$

In order to obtain a representation of a pseudo-differential operator in the form of a hyper-singular operator we shall need some results regarding spherical harmonics [112, 131].

A spherical harmonic of a degree $\nu \geq 0$ is a restriction of a homogeneous harmonic polynomial of of the degree $\nu \geq 0$ to the sphere S^{n-1}. The set of all spherical harmonics of a degree $\nu \geq 0$ is a finite-dimensional subspace H_ν of $L_2(S^{n-1})$. Let $\delta_\nu = \dim H_\nu$. Let us choose, for each ν, an orthonormal basis $\{Y_{\nu\mu}\}$, $\mu = 1, \ldots, \delta_\nu$. The function $Y_{\nu\mu}$ is even if ν is even; otherwise it is odd. $\{Y_{\nu\mu}\}$ is a complete system in $L_2(S^{n-1})$. If $\varphi \in C^{2r}(S^{n-1})$, then its Fourier coefficients

$$\varphi_{\nu\mu} = \int\limits_{S^{n-1}} \varphi(\sigma)\overline{Y_{\nu\mu}(\sigma)}\, d\sigma$$

satisfy the estimate

$$|\varphi_{\nu\mu}| \leq C_{n,r} M \nu^{-2r}, \tag{4.93}$$

where $M = \sup\limits_{\sigma\in S^{n-1}, |\varkappa|\leq 2r} |D^\varkappa \varphi(\sigma)$. It is also known that

$$\delta_\nu \leq C\nu^{n-2}, \quad |Y_{\nu\mu}(\sigma)| \leq C\nu^{(n-2)/2}. \tag{4.94}$$

Let $\widetilde{Y_{\nu\mu}}(\xi)$ be the symbol of the hyper-singular integral $\mathbf{D}^\alpha_{Y_{\nu\mu}} f$. Then if α is not an integer, or α is an odd integer and ν is even, then

$$\widetilde{Y_{\nu\mu}}(\xi) = |\xi|^\alpha \lambda(\nu, \alpha) Y_{\nu\mu}\left(\frac{\xi}{|\xi|}\right),$$

where

$$\lambda(\nu, \alpha) = E_{\nu,\alpha} i^\nu \frac{\Gamma\left(\frac{n+\alpha}{2}\right)\Gamma\left(1+\frac{\alpha}{2}\right)}{\Gamma\left(\frac{n+\alpha+\nu}{2}\right)\Gamma\left(1+\frac{\alpha-\nu}{2}\right)},$$

$$E_{\nu,\alpha} = (-1)^\nu \frac{\cos\frac{\alpha+\nu}{2}\pi}{\cos\frac{\alpha\pi}{2}} \quad \text{(if } \alpha \text{ is not an integer)},$$

$$E_{\nu,\alpha} = (-1)^{\nu/2} \quad \text{(if } \alpha \text{ is an odd integer)}.$$

Let us consider a pseudo-differential operator of the form (4.85), where the symbol $a(x,t,\xi)$ is continuous and homogeneous in ξ of the degree α, and either α is not an integer, or the symbol is even in ξ and α is an odd integer. Suppose that

$$\left|D^\varkappa_\xi a(x,t,\xi)\right| \leq C, \quad |\xi| = 1, x \in \mathbb{R}^n, t \geq 0, |\varkappa| \leq N,$$

where $N > 2n + \alpha - 1$. Expanding the symbol by spherical harmonics,

$$a(x,t,\xi) = \sum_{\nu=0}^{\infty} \sum_{\mu=1}^{\delta_\nu} c_{\nu\mu}(x,t) Y_{\nu\mu}(\xi), \quad |\xi| = 1,$$

we set

$$\Omega(x,t,\xi) = \sum_{\nu=0}^{\infty} \sum_{\mu=1}^{\delta_\nu} \frac{c_{\nu\mu}(x,t)}{\lambda(\nu,\alpha)} Y_{\nu\mu}(\xi). \tag{4.95}$$

According to (4.93), $|c_{\nu\mu}(x,t)| \le C\nu^{-N+1}$. Then, since

$$\left|\cos\left(\frac{\alpha+\nu}{2}\pi\right)\right| = \left|\cos\frac{\alpha\pi}{2}\right|, \quad \text{or} \quad \left|\cos\left(\frac{\alpha+\nu}{2}\pi\right)\right| = \left|\sin\frac{\alpha\pi}{2}\right|,$$

depending on whether ν is even or odd, we find that

$$\left|\frac{1}{\lambda(\nu,\alpha)}\right| \le C \left|\Gamma\left(\frac{n+\alpha+\nu}{2}\right)\Gamma\left(1 - \frac{\nu-\alpha}{2}\right)\right|.$$

If α is an odd integer, then $c_{\nu\mu}(x,t) = 0$ for odd ν. Let us use the identity [7]

$$\Gamma\left(1 - \frac{\nu-\alpha}{2}\right) = \frac{\pi}{\sin\left(\frac{\nu-\alpha}{2}\pi\right)\Gamma\left(\frac{\nu-\alpha}{2}\right)}.$$

Since $\left|\sin\left(\frac{\nu-\alpha}{2}\pi\right)\right|$ is equal to $\left|\cos\frac{\alpha\pi}{2}\right|$ or to $\left|\sin\frac{\alpha\pi}{2}\right|$ (in particular, it is equal to 1 if α is an odd integer and ν is even), and

$$\left|\frac{\Gamma\left(\frac{n+\alpha+\nu}{2}\right)}{\Gamma\left(\frac{\nu-\alpha}{2}\right)}\right| \le C\nu^{\frac{n}{2}+\alpha}$$

(see [7]), we have

$$\left|\frac{1}{\lambda(\nu,\alpha)}\right| \le C\nu^{\frac{n}{2}+\alpha},$$

which implies, together with (4.94), the estimate

$$\left|\sum_{\mu=1}^{\delta_\nu} \frac{c_{\nu\mu}(x,t)}{\lambda(\nu,\alpha)} Y_{\nu\mu}(\xi)\right| \le C\nu^{-N+2n+\alpha-2},$$

so that the series (4.95) is uniformly convergent.

Let us consider the hyper-singular integral operator $\mathbf{D}^{\alpha}_{\Omega}$. Its symbol is

$$\widetilde{\Omega}(x,t,\xi) = \sum_{\nu=0}^{\infty}\sum_{\mu=1}^{\delta_\nu} \frac{c_{\nu\mu}(x,t)}{\lambda(\nu,\alpha)}\widetilde{Y_{\nu\mu}}(\xi)$$

$$= |\xi|^{\alpha}\sum_{\nu=0}^{\infty}\sum_{\mu=1}^{\delta_\nu} c_{\nu\mu}(x,t)Y_{\nu\mu}\left(\frac{\xi}{|\xi|}\right) = a(x,t,\xi)$$

(it follows from (4.92) that the termwise transition from the characteristics to the symbols was legitimate). Thus a pseudo-differential operator A can be represented on $\mathcal{S}(\mathbb{R}^n)$ as the hyper-singular integral operator with the characteristic Ω.

4.8.3. Parametrix. Suppose that the principal symbol $a(x,t,\sigma)$ satisfies the following conditions:

a) $\operatorname{Re} a(x,t,\sigma) \geq a_0 > 0$, $x \in \mathbb{R}^n$, $t \in [0,T]$, $|\sigma| = 1$;

b) $a(x,t,\sigma)$ has N continuous derivatives in σ for $\sigma \neq 0$, and

$$|D^{\varkappa}_{\sigma} a(x,t,\sigma)| \leq C_N |\sigma|^{\gamma-|\varkappa|},$$

$$|D^{\varkappa}_{\sigma}[a(x,t,\sigma) - a(y,\tau,\sigma)]| \leq C_N\left(|x-y|^{\lambda} + |t-\tau|^{\lambda/\gamma}\right)|\sigma|^{\gamma-|\varkappa|}$$

for all $|\varkappa| \leq N$, $x,y,\sigma \in \mathbb{R}^n$ ($\sigma \neq 0$), $t,\tau \in [0,T]$. Here $\lambda \in (0,1)$ is a constant.

The parametrix of the problem (4.83)-(4.84) corresponding to "freezing" the symbol at the point $x = \xi$, $t = \theta$ is defined by

$$Z(x-y,t-\mu,\xi,\theta) = (2\pi)^{-n}\int_{\mathbb{R}^n} \exp[i(x-y)\sigma - a(\xi,\theta,\sigma)(t-\mu)]\,d\sigma, \tag{4.96}$$

$x,y \in \mathbb{R}^n$, $t > \mu$, $\theta \geq 0$.

Let us prove several estimates of oscillatory integrals, which will imply estimates of the function Z and its derivatives.

Let ν be a non-negative integer, $P_\nu(\sigma)$ a homogeneous polynomial of the degree ν,

$$\Phi_\nu(z,\xi,\theta) = \int_{\mathbb{R}^n} P_\nu(\sigma)\exp\{iz\cdot\sigma - a(\xi,\theta,\sigma)\}\,d\sigma.$$

Lemma 4.7. *If $N \geq 2n + \nu + [\gamma]$, then*

$$|\Phi_\nu(z,\xi,\theta)| \leq C(1+|z|)^{-n-\gamma-\nu}, \tag{4.97}$$

where C does not depend on ξ, θ.

Proof. Let $\eta_0 \in C_0^\infty(\mathbb{R}^n)$, $\eta_0(\sigma) = 1$ as $|\sigma| \leq 1$, $\eta_0(\sigma) = 0$ as $|\sigma| \geq 2$. Denote $\eta_1(\sigma) = 1 - \eta_0(\sigma)$,

$$\Phi_\nu^k(z,\xi,\theta) = \int_{\mathbb{R}^n} \eta_k(\sigma) P_\nu(\sigma) \exp\{iz\cdot\sigma - a(\xi,\theta,\sigma)\}\, d\sigma, \quad k = 0, 1.$$

Evidently, $\Phi_\nu = \Phi_\nu^0 + \Phi_\nu^1$.

Let us use the representation $e^{-t} = 1 - t + t^2 h(t)$, where $h(t) = t^{-2}(e^{-t} - 1 + t)$. It is easy to see that $h(t)$ has bounded derivatives of all orders on $[0,\infty)$. We have

$$\begin{aligned}\Phi_\nu^0(z,\xi,\theta) &= \int_{\mathbb{R}^n} \eta_0(\sigma) P_\nu(\sigma) e^{iz\cdot\sigma}\, d\sigma - \int_{\mathbb{R}^n} \eta_0(\sigma) P_\nu(\sigma) a(\xi,\theta,\sigma) e^{iz\cdot\sigma}\, d\sigma \\ &\quad + \int_{\mathbb{R}^n} \eta_0(\sigma) P_\nu(\sigma) [a(\xi,\theta,\sigma)]^2 h(a(\xi,\theta,\sigma)) e^{iz\cdot\sigma}\, d\sigma \\ &\stackrel{\text{def}}{=} \Phi_\nu^{0,1}(z) + \Phi_\nu^{0,2}(z,\xi,\theta) + \Phi_\nu^{0,3}(z,\xi,\theta).\end{aligned}$$

It is clear that $\Phi_\nu^{0,1} \in \mathcal{S}(\mathbb{R}^n)$. Let $a_\nu(\xi,\theta,\sigma) = P_\nu(\sigma) a(\xi,\theta,\sigma)$. Then a_ν is a homogeneous function of σ of the degree $\gamma + \nu$ of the same smoothness as a.

Let $\varphi \in C_0^\infty(\mathbb{R}^n)$, $\varphi(z) = 0$ as $|z| \leq 1$, $\psi(z) \equiv \varphi(-z)$. Then

$$\begin{aligned}\langle \Phi_\nu^{0,2}(\cdot,\xi,\theta), \varphi\rangle &= \int_{\mathbb{R}^n}\int_{\mathbb{R}^n} e^{iz\cdot\sigma} \eta_0(\sigma) a_\nu(\xi,\theta,\sigma) \varphi(z)\, d\sigma\, dz \\ &= \int_{\mathbb{R}^n} \eta_0(\sigma) a_\nu(\xi,\theta,\sigma) \widehat{\psi}(\sigma)\, d\sigma = \int_{\mathbb{R}^n} a_\nu(\xi,\theta,\sigma) \widehat{\psi}(\sigma)\, d\sigma \\ &\quad - \int_{\mathbb{R}^n} \eta_1(\sigma) a_\nu(\xi,\theta,\sigma) \widehat{\psi}(\sigma)\, d\sigma = \langle \widehat{a_\nu}(\xi,\theta,\cdot), \psi\rangle \\ &\quad - \int_{\mathbb{R}^n} \eta_1(\sigma) a_\nu(\xi,\theta,\sigma)\, d\sigma \int_{\mathbb{R}^n} \left(L^N e^{i\sigma\cdot z}\right) \varphi(z)\, dz,\end{aligned}$$

where $\widehat{a_\nu}$ is the Fourier transform of a_ν in the sense of distributions from $\mathcal{S}'(\mathbb{R}^n)$, $L = -i|z|^{-2}\sum\limits_{k=1}^{n} z_k \dfrac{\partial}{\partial\sigma_k}$, $Le^{i\sigma\cdot z} = e^{i\sigma\cdot z}$.

Let us expand the function $a_\nu(\xi,\theta,\zeta)$, $|\zeta| = 1$, by spherical harmonics:

$$a_\nu(\xi,\theta,\zeta) = \sum_{l=0}^{\infty}\sum_{\mu=1}^{\delta_l} g_{l\mu}(\xi,\theta)Y_{l\mu}(\zeta).$$

Then for all $\sigma \in \mathbb{R}^n$, $\sigma \neq 0$,

$$a_\nu(\xi,\theta,\sigma) = |\sigma|^{\gamma+\nu}\sum_{l=0}^{\infty}\sum_{\mu=1}^{\delta_l} g_{l\mu}(\xi,\theta)Y_{l\mu}\left(\frac{\sigma}{|\sigma|}\right).$$

It is known [100] that the Fourier transform of the function $|\sigma|^{\gamma+\nu}Y_{l\mu}\left(\frac{\sigma}{|\sigma|}\right)$ equals

$$(-i)^l 2^{n+\gamma+\nu}\pi^{n/2}\frac{\Gamma\left(\frac{n+l+\gamma+\nu}{2}\right)}{\Gamma\left(\frac{l-\gamma-\nu}{2}\right)}\cdot\frac{Y_{l\mu}\left(\frac{z}{|z|}\right)}{|z|^{n+\gamma+\nu}}, \tag{4.98}$$

if $l-\gamma-\nu \neq 0,-2,-4,\ldots$ (see [40] regarding the interpretation of the homogeneous function (4.98) as an element of $\mathcal{S}'(\mathbb{R}^n)$). Otherwise (that is possible only for a finite number of values of l) the above Fourier transform is a distribution concentrated at the origin. If $|z| \geq 1$, we have

$$\widehat{a_\nu}(\xi,\theta,z) = \sum_{l=l_0}^{\infty}\sum_{\mu=1}^{\delta_l}\Bigg\{g_{l\mu}(\xi,\theta)(-i)^l 2^{n+\gamma+\nu}\pi^{n/2} \times\frac{\Gamma\left(\frac{n+l+\gamma+\nu}{2}\right)}{\Gamma\left(\frac{l-\gamma-\nu}{2}\right)}\cdot\frac{Y_{l\mu}\left(\frac{z}{|z|}\right)}{|z|^{n+\gamma+\nu}}\Bigg\},\quad l_0 \geq 0. \tag{4.99}$$

The series (4.99) is a priori convergent in the distribution sense. However by virtue of (4.93), (4.94) it converges uniformly for $|z| = 1$, so that for $|z| \geq 1$

$$|\widehat{a_\nu}(\xi,\theta,z)| \leq C|z|^{-n-\gamma-\nu}, \tag{4.100}$$

where the constant does not depend on ξ,θ.

Next,

$$\int_{\mathbb{R}^n}(L^N e^{i\sigma\cdot z})\,\varphi(z)\,dz = (-i)^N\sum_{|\varkappa|=N} D_\sigma^\varkappa\int_{\mathbb{R}^n}|z|^{-2N}z^\varkappa e^{i\sigma\cdot z}\varphi(z)\,dz$$

(recall that $\varphi(z) = 0$ for $|z| \leq 1$). Taking into account that $\eta_1(\sigma) = 0$ for $|\sigma| \leq 1$ we find after integration by parts that

$$\int_{\mathbb{R}^n} \eta_1(\sigma) a_\nu(\xi, \theta, \sigma) \int_{\mathbb{R}^n} \left(L^N e^{i\sigma\cdot z}\right) \varphi(z)\, dz$$

$$= i^N \sum_{|\varkappa|=N} \int_{\mathbb{R}^n} D_\sigma^\varkappa [\eta_1(\sigma) a_\nu(\xi, \theta, \sigma)]\, d\sigma \int_{\mathbb{R}^n} |z|^{-2N} z^\varkappa e^{i\sigma\cdot z} \varphi(z)\, dz.$$

Obviously,

$$D_\sigma^\varkappa [\eta_1(\sigma) a_\nu(\xi, \theta, \sigma)] = \eta_1(\sigma) D_\sigma^\varkappa a_\nu(\xi, \theta, \sigma) + R(\xi, \theta, \sigma),$$

where $R(\xi, \theta, \sigma) = 0$ for $|\sigma| \leq 1$ and $|\sigma| \geq 2$. The function $D_\sigma^\varkappa a_\nu(\xi, \theta, \sigma)$ is homogeneous in σ of the degree $\gamma + \nu - N < -n$. By the Fubini theorem

$$\langle \Phi_\nu^{0,2}(\cdot, \xi, \theta), \varphi \rangle = \int_{\mathbb{R}^n} \widehat{a_\nu}(\xi, \theta, -z) \varphi(z)\, dz$$

$$- i^N \sum_{|\varkappa|=N} \int_{\mathbb{R}^n} \left\{ \int_{\mathbb{R}^n} [\eta_1(\sigma) D_\sigma^\varkappa a_\nu(\xi, \theta, \sigma)\right.$$

$$\left. + R(\xi, \theta, \sigma)]\, e^{i\sigma\cdot z}\, d\sigma \right\} |z|^{-2N} z^\varkappa \varphi(z)\, dz,$$

whence for $|z| \geq 1$

$$\Phi_\nu^{0,2}(z, \xi, \theta) = \widehat{a_\nu}(\xi, \theta, -z)$$

$$- i^N |z|^{-2N} \sum_{|\varkappa|=N} z^\varkappa \int_{\mathbb{R}^n} [\eta_1(\sigma) D_\sigma^\varkappa a_\nu(\xi, \theta, \sigma) + R(\xi, \theta, \sigma)]\, e^{i\sigma\cdot z}\, d\sigma,$$

$$\left|\Phi_\nu^{0,2}(z, \xi, \theta)\right| \leq |\widehat{a_\nu}(\xi, \theta, -z)| + C|z|^{-N},$$

and due to (4.100)

$$\left|\Phi_\nu^{0,2}(z, \xi, \theta)\right| \leq C|z|^{-n-\gamma-\nu}, \quad |z| \geq 1, \tag{4.101}$$

where C does not depend on ξ, θ.

Let us find a bound for $\Phi_\nu^{0,3}(z, \xi, \theta)$, $|z| \geq 1$. Let $z = (z_1, \dots, z_n)$; consider such a number j that $|z_j| \geq |z_k|$, $k = 1, \dots, n$. Then $|z| \leq \sqrt{n}|z_j|$. Integrating $N_1 = n + \nu + [\gamma] + 1$ times by parts with respect to the variable

σ_j we find that

$$\Phi_\nu^{0,3}(z,\xi,\theta) = \frac{(-1)^{N_1}}{(iz_j)^{N_1}} \int\limits_{\mathbb{R}^n} e^{i\sigma\cdot z} \frac{\partial^{N_1}}{\partial\sigma_j^{N_1}} \left\{\eta_0(\sigma)P_\nu(\sigma)[a(\xi,\theta,\sigma)]^2 h(a(\xi,\theta,\sigma))\right\} d\sigma.$$

Since $\eta_0(\sigma) = 1$ as $|\sigma| \le 1$, all the terms emerging in the process of differentiating, which contain derivatives of η_0, are continuous and have compact supports. Next, computing the derivative

$$\frac{\partial^{N_1}}{\partial\sigma_j^{N_1}} \left\{\eta_0(\sigma)P_\nu(\sigma)[a(\xi,\theta,\sigma)]^2 h(a(\xi,\theta,\sigma))\right\}$$

with the use of the Leibnitz formula and the Faà di Bruno formula for higher derivatives of a superposition of functions [47], it is not hard to calculate that the maximal order of singularity at the point $\sigma = 0$ does not exceed $n + [\gamma] + 1 - 3\gamma < n$. Therefore for $|z| \ge 1$ we have

$$\left|\Phi_\nu^{0,3}(z,\xi,\theta)\right| \le C|z_j|^{-N_1} \le C|z|^{-n-\gamma-\nu}, \quad |z| \ge 1,$$

that is $\left|\Phi_\nu^0(z,\xi,\theta)\right| \le C|z|^{-n-\gamma-\nu}$.

The function $\Phi_\nu^1(z,\xi,\theta)$ is estimated in the same way as $\Phi_\nu^{0,3}(z,\xi,\theta)$, due to the fact that $\eta_1(\sigma) = 0$ as $|\sigma| \le 1$. We get

$$\left|\Phi_\nu^1(z,\xi,\theta)\right| \le C|z|^{-n-\gamma-\nu}, \quad |z| \ge 1,$$

so that $|\Phi_\nu(z,\xi,\theta)| \le C|z|^{-n-\gamma-\nu}$, $|z| \ge 1$. In order to prove (4.97), it remains to notice that

$$\left|\Phi_\nu^1(z,\xi,\theta)\right| \le \int\limits_{\mathbb{R}^n} |P_\nu(\sigma)| e^{-a_0|\sigma|^\gamma}\, d\sigma$$

as $|z| \le 1$. ■

Let $Q_\alpha(r,\sigma)$ be a homogeneous function in σ of the degree $\alpha > 0$ depending on a parameter r of an arbitrary nature, which has, as $\sigma \ne 0$, N continuous derivatives in σ, and

$$|D_\sigma^\varkappa Q_\alpha(r,\sigma)| \le C B_N(r)|\sigma|^{\alpha-|\varkappa|},$$

whenever $|\varkappa| \le N$, $\sigma \ne 0$, for all values of r. Let us consider the function

$$\Psi_\alpha(z,\xi,\theta,r) = \int\limits_{\mathbb{R}^n} Q_\alpha(r,\sigma)\exp\{iz\cdot\sigma - a(\xi,\theta,\sigma)\}\, d\sigma.$$

Lemma 4.8. *If $N \geq 2n + [\alpha] + [\gamma] + 1$, then*

$$|\Psi_\alpha(z, \xi, \theta, r)| \leq CB(r)(1 + |z|)^{-n-\alpha}, \tag{4.102}$$

where C does not depend on z, ξ, θ, r.

The proof consists essentially of repetition of the arguments from the proof of Lemma 4.7. The only new point that appears is the following. In the decomposition of the integral Ψ_α the term

$$\Psi_\alpha^{0,1}(z, r) = \int_{\mathbb{R}^n} \eta_0(\sigma) Q_\alpha(r, \sigma) \exp\{iz \cdot \sigma\} \, d\sigma$$

(in contrast to the function $\Phi_\nu^{0,1}$ from the proof of Lemma 4.7) need not belong to $\mathcal{S}(\mathbb{R}^n)$. The techniques of Lemma 4.7 gives the estimate

$$\left|\Psi_\alpha^{0,1}(z, r)\right| \leq CB(r)|z|^{-n-\alpha}, \quad |z| \geq 1,$$

which leads to the inequality (4.102). ■

Let us consider the function

$$\Phi_0(z, \xi, \theta) = \int_{\mathbb{R}^n} \exp\{iz \cdot \sigma - a(\xi, \theta, \sigma)\} \, d\sigma.$$

Lemma 4.9. *If $N \geq 2n + [\gamma]$, then for any $z, \xi_1, \xi_2 \in \mathbb{R}^n$, $\theta > 0$,*

$$|\Phi_0(z, \xi_1, \theta) - \Phi_0(z, \xi_2, \theta)| \leq C|\xi_1 - \xi_2|^\lambda (1 + |z|)^{-n-\gamma}.$$

The proof is similar to that of Lemma 4.7. Some technical differences are caused by the necessity to obtain estimates of expressions of the form

$$\frac{\partial^l}{\partial \sigma_j^l}[h(a(\xi_1, \theta, \sigma)) - h(a(\xi_2, \theta, \sigma))]$$

in two situations: for $h(s) = e^{-s}$, $|\sigma| \geq 1$, and for a function h about which we know only that it has bounded derivatives of any order, for $|\sigma| \leq 2$.

By the Faà di Bruno formula

$$\frac{\partial^l}{\partial\sigma_j^l}[h(a(\xi_1,\theta,\sigma)) - h(a(\xi_2,\theta,\sigma))] = \sum C_{m_1\dots m_\mu}\Bigg\{ h^{(k)}(a(\xi_1,\theta,\sigma))$$
$$\times \left[\frac{\partial a(\xi_1,\theta,\sigma)}{\partial\sigma_j}\right]^{m_1}\left[\frac{\partial^2 a(\xi_1,\theta,\sigma)}{\partial\sigma_j^2}\right]^{m_2}\cdots\left[\frac{\partial^\mu a(\xi_1,\theta,\sigma)}{\partial\sigma_j^\mu}\right]^{m_\mu}$$
$$- h^{(k)}(a(\xi_2,\theta,\sigma))\left[\frac{\partial a(\xi_2,\theta,\sigma)}{\partial\sigma_j}\right]^{m_1}\left[\frac{\partial^2 a(\xi_2,\theta,\sigma)}{\partial\sigma_j^2}\right]^{m_2}\cdots$$
$$\times\left[\frac{\partial^\mu a(\xi_2,\theta,\sigma)}{\partial\sigma_j^\mu}\right]^{m_\mu}\Bigg\}, \qquad (4.103)$$

where the summation is over all tuples $(m_1,\dots,m_\mu)$ of natural numbers, such that $m_1 + 2m_2 + \cdots + \mu m_\mu = l$ while μ varies from 1 to l; here $k = m_1 + \cdots + m_\mu$. Let us use the elementary formulas

$$b_1\cdots b_\nu - b_1'\cdots b_\nu'$$
$$= (b_1 - b_1')b_2\cdots b_\nu + b_1'(b_2 - b_2')b_3\cdots b_\nu + \cdots + b_1'b_2'\cdots b_{\nu-1}'(b_\nu - b_\nu');$$
$$b_\nu - (b')^\nu = (b-b')(b^{\nu-1} + b^{\nu-2}b' + \cdots + (b')^{\nu-1}.$$

Each summand in (4.103) is a sum of expressions containing one of the differences $h^{(k)}(a(\xi_1,\theta,\sigma)) - h^{(k)}(a(\xi_2,\theta,\sigma))$ or

$$\left[\frac{\partial^\nu a(\xi_1,\theta,\sigma)}{\partial\sigma_j^\nu}\right]^{m_\nu} - \left[\frac{\partial^\nu a(\xi_2,\theta,\sigma)}{\partial\sigma_j^\nu}\right]^{m_\nu}$$

multiplied by a bounded function and by a homogeneous function of the degree $m_1(\gamma-1) + m_2(\gamma-2) + \cdots + m_\nu(\gamma-\nu) = \gamma k - l$ (in the first case), or $\gamma k - l - m_\nu(\gamma-\nu)$ (in the second case). Further,

$$\left|h^{(k)}(a(\xi_1,\theta,\sigma)) - h^{(k)}(a(\xi_2,\theta,\sigma))\right| \le C|a(\xi_1,\theta,\sigma) - a(\xi_2,\theta,\sigma)|$$
$$\le C|\xi_1 - \xi_2|^\lambda|\sigma|^\gamma;$$

$$\left|\left[\frac{\partial^\nu a(\xi_1,\theta,\sigma)}{\partial\sigma_j^\nu}\right]^{m_\nu} - \left[\frac{\partial^\nu a(\xi_2,\theta,\sigma)}{\partial\sigma_j^\nu}\right]^{m_\nu}\right| \le C|\xi_1-\xi_2|^\lambda|\sigma|^{m_\nu(\gamma-\nu)}.$$

Thus for $|\sigma| \le 2$

$$\left|\frac{\partial^l}{\partial\sigma_j^l}[h(a(\xi_1,\theta,\sigma)) - h(a(\xi_2,\theta,\sigma))]\right| \le C|\xi_1-\xi_2|^\lambda|\sigma|^{\gamma-l};$$

for $h(s) = e^{-s}$, $|\sigma| \geq 1$, we get

$$\left| \frac{\partial^l}{\partial \sigma_j^l} [e^{-a(\xi_1,\theta,\sigma)} - e^{-a(\xi_2,\theta,\sigma)}] \right| \leq C|\xi_1 - \xi_2|^\lambda |\sigma|^{\gamma+(\gamma-l)l} e^{-a_0|\sigma|^\gamma}.$$

With these estimates taken into account the proof of Lemma 4.9 goes just like that of Lemma 4.7. ■

The next lemma can be proved in a similar way.

Lemma 4.10. *Assume the conditions of Lemma 4.8. Then*

$$|\Psi_\alpha(z, \xi_1, \theta, r) - \Psi_\alpha(z, \xi_2, \theta, r)| \leq CB(r)|\xi_1 - \xi_2|^\lambda (1 + |z|)^{-n-\gamma-\alpha},$$

where C does not depend on ξ_1, ξ_2, θ, r.

Under the above conditions a), b) we obtain from Lemmas 4.7 – 4.10 the following estimates of the parametrized fundamental solution Z (uniformly with respect to the parameters):

$$|D_x^\varkappa Z(x - y, t - \mu, \xi, \theta)| \leq C_\varkappa(t - \mu) \left[(t - \mu)^{1/\gamma} + |x - y|\right]^{-n-\gamma-|\varkappa|}, \tag{4.104}$$

$|\varkappa| \leq N - 2n - [\gamma]$,

$$\left| \frac{\partial}{\partial t} Z(x - y, t - \mu, \xi, \theta) \right| \leq C \left[(t - \mu)^{1/\gamma} + |x - y|\right]^{-n-\gamma}; \tag{4.105}$$

$$\begin{aligned} &|Z(x - y, t - \mu, \xi_1, \theta) - Z(x - y, t - \mu, \xi_2, \theta)| \\ &\qquad \leq C(t - \mu)|\xi_1 - \xi_2|^\lambda \left[(t - \mu)^{1/\gamma} + |x - y|\right]^{-n-\gamma}; \end{aligned} \tag{4.106}$$

$$\begin{aligned} &\left| \frac{\partial}{\partial t} Z(x - y, t - \mu, \xi_1, \theta) - \frac{\partial}{\partial t} Z(x - y, t - \mu, \xi_2, \theta) \right| \\ &\qquad \leq C(t - \mu)|\xi_1 - \xi_2|^\lambda \left[(t - \mu)^{1/\gamma} + |x - y|\right]^{-n-2\gamma}. \end{aligned} \tag{4.107}$$

We have the equation

$$\int_{\mathbb{R}^n} Z(x - y, t - \mu, \xi, \theta)\, dy = 1, \quad z, \xi \in \mathbb{R}^n,\ t > \mu,\ \theta \geq 0, \tag{4.108}$$

which follows from (4.96) and the Fourier inversion. Using (4.105) we get from (4.108) that

$$\int_{\mathbb{R}^n} \frac{\partial Z}{\partial t}(x-y, t-\mu, \xi, \theta)\, dy = 0. \tag{4.109}$$

The inequality

$$\left| \int_{\mathbb{R}^n} \frac{\partial Z}{\partial t}(x-y, t-\mu, y, \mu)\, dy \right| \le C(t-\mu)^{-1+\lambda/\gamma}. \tag{4.110}$$

4.8.4. Heat potential. Let us consider the heat potential

$$u(x,t,\tau) = \int_{\tau}^{t} d\mu \int_{\mathbb{R}^n} Z(x-y, t-\mu, y, \mu) f(y,\mu)\, dy. \tag{4.111}$$

We assume the above conditions a), b) with $N \ge 2n + 2[\gamma] + 1$. Suppose also that

$$|f(y,\mu)| \le C(\mu-\tau)^{-\rho}, \quad |f(x,\mu) - f(y,\mu)| \le C|x-y|^{\lambda}(\mu-\tau)^{-\rho},$$

where $0 \le \rho < 1$, $0 < \lambda \le 1$. The integral in (4.111) is absolutely convergent due to (4.104). It also follows from (4.104) that the function $u(x,t,\tau)$ has continuous derivatives in x of any order $< \gamma$, and they can be obtained by differentiating under the sign of integral.

Just as in the non-Archimedean case (see Sect. 4.1 and 4.5) we obtain formulas for $\frac{\partial u(x,t,\tau)}{\partial t}$ and $(\mathbf{D}_{\Omega}^{\alpha} u)(x,t,\tau)$, $\alpha < \gamma$, which are similar to (4.18) and (4.19). Note only that the calculation of $(\mathbf{D}_{\Omega}^{\alpha} u)(x,t,\tau)$ is based on the inequalities

$$\begin{aligned} &\left|(\Delta_h^l Z)(x-y, t-\mu, y, \mu)\right| \\ &\le C|h|^{[\alpha]+1}(t-\mu) \sum_{\nu=0}^{l} \left[(t-\mu)^{1/\gamma} + |x - \theta_\nu \nu h - y|\right]^{-n-\gamma-[\alpha]-1} \end{aligned} \tag{4.112}$$

$(0 < \theta_\nu < 1)$ if $|h| \le (t-\mu)^{1/\gamma}$, and

$$\begin{aligned} &\left|(\Delta_h^l Z)(x-y, t-\mu, y, \mu)\right| \\ &\qquad \le C(t-\mu) \sum_{\nu=0}^{l} \left[(t-\mu)^{1/\gamma} + |x - y - \nu h|\right]^{-n-\gamma}, \end{aligned} \tag{4.113}$$

if $|h| \geq (t-\mu)^{1/\gamma}$, both of which are consequences of (4.87) and (4.104).

The case $\alpha = \gamma$ is much more complicated. Assume that the symbol $\widetilde{\Omega}(x,\sigma)$ has $N \geq 2n + 2[\gamma] + 1$ continuous derivatives in $\sigma \neq 0$, and that

$$\left|D_\sigma^\varkappa \widetilde{\Omega}(x,\sigma)\right| \leq C_N |\sigma|^{\gamma-|\varkappa|}, \quad \left|D_\sigma^\varkappa \left[\widetilde{\Omega}(x,\sigma) - \widetilde{\Omega}(y,\sigma)\right]\right| \leq C_N |x-y|^\lambda |\sigma|^{\gamma-|\varkappa|}, \quad |\varkappa| \leq N.$$

Let also $\widetilde{\Omega}(x,\sigma) \neq 0$ if $\sigma \neq 0$. If γ is an integer, and the symbol $\widetilde{\Omega}(x,\sigma)$ is not a polynomial in σ (which is possible only for an odd γ and an even characteristic Ω), we assume in addition that in the expansion

$$\left[\widetilde{\Omega}(x,\sigma)\right]^{-1} = \sum_{\nu=0}^{\infty} \sum_{\mu=1}^{\delta_{2\nu}} c_{2\nu,\mu} Y_{2\nu,\mu}(\sigma), \quad |\sigma| = 1,$$

$c_{2\nu,\mu}(x) = 0$ if $\gamma = n + 2\nu + 2k$, $k = 0,1,2,\ldots$. We shall not consider the case when $\widetilde{\Omega}$ is a polynomial in σ because it is covered by the classical theory of parabolic differential equations.

Lemma 4.11. *Under the above conditions the hyper-singular integral* $\mathbf{D}_\Omega^\gamma u$ *exists in the sense of the conditional convergence (4.88), and*

$$(\mathbf{D}_\Omega^\gamma u)(x,t,\tau) = \int_\tau^t d\mu \int_{\mathbb{R}^n} Z_\Omega(x, x-y, t-\mu, y, \mu)[f(y,\mu) - f(x,\mu)]\, dy$$
$$+ \int_\tau^t f(x,\mu)\, d\mu \int_{\mathbb{R}^n} [Z_\Omega(x, x-y, t-\mu, y, \mu) - Z_\Omega(x, x-y, t-\mu, x, \mu)]\, dy, \tag{4.114}$$

where $Z_\Omega = \mathbf{D}_\Omega^\gamma Z$, *that is*

$$Z_\Omega(x,z,t,\xi,\theta) = (2\pi)^{-n} \int_{\mathbb{R}^n} \widetilde{\Omega}(x,\sigma) \exp[iz\cdot\sigma - a(\xi,\theta,\sigma)t]\, d\sigma,$$

and $\widetilde{\Omega}(x,\sigma)$ *is the symbol of the hyper-singular integral operator* $\mathbf{D}_\Omega^\gamma$.

Proof. Denote

$$u_s(x,t,\tau) = \int_\tau^{t-s} d\mu \int_{\mathbb{R}^n} Z(x-y, t-\mu, y, \mu) f(y,\mu)\, dy, \quad 0 < s < t-\tau.$$

Using (4.112) and (4.113) we prove the absolute convergence of the hyper-singular integral $\mathbf{D}_\Omega^\gamma u_s$, and the formula

$$\left(\mathbf{D}_\Omega^\gamma u_s\right)(x,t,\tau) = \int_\tau^{t-s} d\mu \int_{\mathbb{R}^n} Z_\Omega(x, x-y, t-\mu, y, \mu) f(y,\mu)\, dy.$$

We shall need the following properties of the function Z_Ω:

$$\int_{\mathbb{R}^n} Z_\Omega(x, z, t, \xi, \theta)\, dz = 0; \tag{4.115}$$

$$|Z_\Omega(x, x-y, t-\mu, y, \mu)| \le C\left[(t-\mu)^{1/\gamma} + |x-y|\right]^{-n-\gamma}; \tag{4.116}$$

$$\begin{aligned} &|Z_\Omega(x, x-y, t-\mu, y, \mu) - Z_\Omega(x, x-y, t-\mu, x, \mu)| \\ &\qquad \le C|x-y|^\lambda (t-\mu)\left[(t-\mu)^{1/\gamma} + |x-y|\right]^{-n-2\gamma}. \end{aligned} \tag{4.117}$$

The formula (4.115) is a consequence of (4.108); we use (4.112), (4.113), and the Fubini theorem. The inequality (4.116) follows from Lemma 4.8, while (4.117) is a consequence of Lemma 4.10.

According to (4.115)

$$\begin{aligned} \left(\mathbf{D}_\Omega^\gamma u_s\right)(x,t,\tau) &= \int_\tau^{t-s} d\mu \int_{\mathbb{R}^n} Z_\Omega(x, x-y, t-\mu, y, \mu)[f(y,\mu) - f(x,\mu)]\, dy \\ &+ \int_\tau^{t-s} f(x,\mu)\, d\mu \int_{\mathbb{R}^n} [Z_\Omega(x, x-y, t-\mu, y, \mu) \\ &\qquad - Z_\Omega(x, x-y, t-\mu, x, \mu)]\, dy, \end{aligned} \tag{4.118}$$

Let $\Phi(x,t,\tau)$ be the function in the right-hand side of (4.114) (the integrals converge due to (4.116), (4.117)). Taking into account (4.116), (4.117) we see from (4.118) that uniformly with respect to $x \in \mathbb{R}^n$

$$\lim_{s\to 0} \left(\mathbf{D}_\Omega^\gamma u_s\right)(x,t,\tau) = \Phi(x,t,\tau). \tag{4.119}$$

Let us consider the restricted hyper-singular integral

$$\left(\mathbf{D}_{\Omega,\varepsilon}^\gamma u_s\right)(x,t,\tau) = \int_\tau^{t-s} d\mu \int_{\mathbb{R}^n} Z_{\Omega,\varepsilon}(x, x-y, t-\mu, y, \mu) f(y,\mu)\, dy, \tag{4.120}$$

where $\varepsilon > 0$, $Z_{\Omega,\varepsilon} = \mathbf{D}^{\gamma}_{\Omega,\varepsilon} Z$. Our immediate goal is to obtain an integral representation expressing $\mathbf{D}^{\gamma}_{\Omega,\varepsilon} u_s$ in terms of $\mathbf{D}^{\gamma}_{\Omega} u_s$. We introduce the constant characteristic $\Omega_\xi(\sigma) = \Omega(\xi, \sigma)$ depending on $\xi \in \mathbb{R}^n$ as a parameter. Let $\widetilde{\Omega_\xi}$ be the corresponding symbol. Fixing the rest of the variables we denote $F(x) = Z(x - y, t - \mu, y, \mu)$ and express F via $\mathbf{D}^{\gamma}_{\Omega_\xi} F$. Since the Fourier transform of the function $G_\xi = \mathbf{D}^{\gamma}_{\Omega_\xi} F$ equals $\widetilde{\Omega_\xi}(\zeta)\widehat{F}(\zeta)$, we find that

$$F(x) = (2\pi)^{-n} \int_{\mathbb{R}^n} \frac{e^{i\zeta\cdot x}}{\widetilde{\Omega_\xi}(\zeta)} \widehat{G_\xi}(\zeta)\, d\zeta = (2\pi)^{-n} \int_{\mathbb{R}^n} \frac{\widehat{G_{x,\xi}}(\zeta)}{\widetilde{\Omega_\xi}(\zeta)}\, d\zeta,$$

where $G_{x,\xi}(z) = \left(\mathbf{D}^{\gamma}_{\Omega_\xi} F\right)(x + z)$.

Let us expand the homogeneous function $1/\widetilde{\Omega_\xi}$ in a series of spherical harmonics:

$$\frac{1}{\widetilde{\Omega_\xi}(\zeta)} = |\zeta|^{-\gamma} \sum_{\nu=0}^{\infty} \sum_{\mu=1}^{\delta_\nu} g_{\nu\mu}(\xi) Y_{\nu\mu}\left(\frac{\zeta}{|\zeta|}\right).$$

For each fixed $\xi \in \mathbb{R}^n$ the series converges uniformly with respect to ζ, so that

$$F(x) = (2\pi)^{-n} \sum_{\nu=0}^{\infty} \sum_{\mu=1}^{\delta_\nu} g_{\nu\mu}(\xi) \int_{\mathbb{R}^n} \widehat{G_{x,\xi}}(\zeta) Y_{\nu\mu}\left(\frac{\zeta}{|\zeta|}\right) |\zeta|^{-\gamma}\, d\zeta. \tag{4.121}$$

The Fourier transform (in the sense of the space $\mathcal{S}'(\mathbb{R}^n)$) of the distribution $|\zeta|^{-\gamma} Y_{\nu\mu}\left(\frac{\zeta}{|\zeta|}\right)$ equals

$$i^\nu \pi^{n/2} 2^{n-\gamma} \frac{\Gamma((n + \nu - \gamma)/2)}{\Gamma((\nu + \gamma)/2)} |z|^{\gamma - n} \check{Y}_{\nu\mu}\left(\frac{z}{|z|}\right)$$

(see [100]), where $\check{Y}_{\nu\mu}(\sigma) = \check{Y}_{\nu\mu}(-\sigma)$. Strictly speaking, we cannot use this to transform the integral in (4.121), since $G_{x,\xi} \notin \mathcal{S}(\mathbb{R}^n)$. However if $n/2 < \operatorname{Re}\lambda < n$, then the formula

$$\int_{\mathbb{R}^n} \widehat{G_{x,\xi}}(\zeta) Y_{\nu\mu}\left(\frac{\zeta}{|\zeta|}\right) |\zeta|^{-\lambda}\, d\zeta$$

$$= i^\nu \pi^{n/2} 2^{n-\lambda} \frac{\Gamma((n + \nu - \lambda)/2)}{\Gamma((\nu + \lambda)/2)} \int_{\mathbb{R}^n} G_{x,\xi}(z) |z|^{\lambda - n} \check{Y}_{\nu\mu}\left(\frac{z}{|z|}\right) dz \tag{4.122}$$

is nevertheless valid, since in this case the function $|\zeta|^{-\lambda} Y_{\nu\mu}\left(\frac{\zeta}{|\zeta|}\right)$ can be represented as a sum of functions from $L_1(\mathbb{R}^n)$ and $L_2(\mathbb{R}^n)$. An explicit form of the function $\widehat{G_{x,\xi}}$ and the estimate for the function $G_{x,\xi}$ (see [100]) enable us to perform in the standard way [40, 100] the analytic continuation in λ and to prove (4.122) for $\lambda = \gamma \neq n + \nu + 2k$, $k = 0, 1, 2, \ldots$.

Denote

$$H(\xi, z) = \pi^{n/2} 2^{n-\gamma} |z|^{\gamma-n} \frac{\Gamma((n+\nu-\gamma)/2)}{\Gamma((\nu+\gamma)/2)} \times \sum_{\nu=0}^{\infty} \sum_{\mu=1}^{\delta_\nu} i^\nu \frac{\Gamma((n+\nu-\gamma)/2)}{\Gamma((\nu+\gamma)/2)} g_{\nu\mu}(\xi) Y_{\nu\mu}\left(\frac{z}{|z|}\right).$$

The series converges uniformly with respect to z. Moreover, it follows from (4.93), (4.94) and the inequality

$$\left| D^{\varkappa} Y_{\nu\mu}\left(\frac{z}{|z|}\right)\right| \leq C \nu^{|\varkappa| + (n-2)/2} |z|^{-|\varkappa|}$$

that the function $H(\xi, z)$ is smooth in z for $z \neq 0$: $H(\xi, \cdot) \in C^{[\gamma]+1}(S^{n-1})$. Now

$$F(x) = (2\pi)^{-n} \int\limits_{\mathbb{R}^n} H(\xi, z) \left(\mathbf{D}^{\gamma}_{\Omega_\xi} F\right)(x+z)\, dz,$$

whence

$$\left(\mathbf{D}^{\gamma}_{\Omega_\xi,\varepsilon} F\right)(x) = (2\pi)^{-n} \left[d_{n,l}(\gamma)\right]^{-1} \int\limits_{\mathbb{R}^n} \left(\mathbf{D}^{\gamma}_{\Omega_\xi} F\right)(\zeta)\, d\zeta \times \int\limits_{|h|>\varepsilon} \frac{\Delta^l_h \check{H}(\xi, x-\zeta)}{|h|^{n+\gamma}} \Omega_\xi\left(\frac{h}{|h|}\right) dh.$$

Writing out the expression for the difference Δ^l_h in detail and making the substitution $h = |\zeta| y$, $\zeta = \varepsilon\eta$, after elementary transformations we get

$$\left(\mathbf{D}^{\gamma}_{\Omega_\xi,\varepsilon} F\right)(x) = \int\limits_{\mathbb{R}^n} K(\xi, \eta) \left(\mathbf{D}^{\gamma}_{\Omega_\xi} F\right)(x - \varepsilon\eta)\, d\eta, \tag{4.123}$$

where

$$K(\xi,\eta) = (2\pi)^{-n}\left[d_{n,l}(\gamma)\right]^{-1}|\eta|^{-n}\int\limits_{|y|>|\eta|^{-1}}\frac{\left(\Delta_y^l H_1\right)\left(\xi,\frac{\eta}{|\eta|}\right)}{|y|^{n+\gamma}}$$

$$\times\,\Omega_\xi\left(\frac{y}{|y|}\right)dy, \quad H_1(\xi,\sigma) = |\sigma|^{n-\gamma}\check{H}(\xi,\sigma).$$

The function $K(\xi,\eta)$ is integrable with respect to η, and

$$\int\limits_{\mathbb{R}^n} K(\xi,\eta)\,d\eta = 1. \tag{4.124}$$

For the case $\gamma < n$ this was proved by Samko [130] in connection with the inversion problem of generalized Riesz potentials. The proof in the general case is a word-for word repetition of the arguments in [130].

Setting $\xi = x$ in (4.123) and returning to the previous notation we find that

$$Z_{\Omega,\varepsilon}(x, x-y, t-\mu, y, \mu) = \int\limits_{\mathbb{R}^n} K(x,\eta)Z_\Omega(x, x-\varepsilon\eta-y, t-\mu, y, \mu)\,d\eta.$$

Substituting in (4.120) we get that

$$\left(\mathbf{D}^\gamma_{\Omega,\varepsilon}u_s\right)(x,t,\tau) = \int\limits_{\mathbb{R}^n} K(x,\eta)\left(\mathbf{D}^\gamma_\Omega u_s\right)(x-\varepsilon\eta,t,\tau)\,d\eta. \tag{4.125}$$

As $s \to 0$, we have that $u_s(x,t,\tau) \to u(x,t,\tau)$, unfiformly with respect to $x \in \mathbb{R}^n$. Hence for $s \to 0$

$$\left(\mathbf{D}^\gamma_{\Omega,\varepsilon}u_s\right)(x,t,\tau) \longrightarrow \left(\mathbf{D}^\gamma_{\Omega,\varepsilon}u\right)(x,t,\tau).$$

Using (4.119) we pass to the limit in (4.125) as $s \to 0$. Then

$$\left(\mathbf{D}^\gamma_{\Omega,\varepsilon}u\right)(x,t,\tau) = \int\limits_{\mathbb{R}^n} K(x,\eta)\Phi(x-\varepsilon\eta,t,\tau)\,d\eta. \tag{4.126}$$

It follows from the uniform convergence (4.119) that the function $\Phi(x,t,\tau)$ is continuous in x. Now from (4.126), (4.124), and the dominated convergence theorem we get the existence of the limit

$$\left(\mathbf{D}^\gamma_\Omega u\right)(x,t,\tau) = \lim_{\varepsilon\to 0}\left(\mathbf{D}^\gamma_{\Omega,\varepsilon}u\right)(x,t,\tau) = \Phi(x,t,\tau). \quad \blacksquare$$

4.8.5. Cauchy problem. Now we are ready to consider the Cauchy problem (4.83), (4.84). Let us list the conditions imposed upon the symbols of pseudo-differential operators from (4.83). As before, the principal symbol $a(x,t,\xi)$ is a homogeneous function with respect to ξ of the degree $\gamma \geq 1$. The symbols $a_k(x,t,\xi)$ are homogeneous in ξ with the degrees γ_k, $0 < \gamma_k < \gamma$. Fixing a natural number $N \geq 2n + 2[\gamma] + 1$ we assume the conditions a), b) from Sect. 4.8.3 with regard to the principal symbol. Other symbols should be such that if $\xi \neq 0$, $|\varkappa| \leq N$, $x, y \in \mathbb{R}^n$, $t, \tau \in [0,T]$, then

$$\left|D_\xi^\varkappa a_k(x,t,\xi)\right| \leq C|\xi|^{\gamma_k - |\varkappa|},$$

$$\left|D_\xi^\varkappa [a_k(x,t,\xi) - a_k(y,\tau,\xi)]\right| \leq C\left(|x-y|^\lambda + |t-\tau|^{\lambda/\gamma}\right)|\xi|^{\gamma_k - |\varkappa|},$$

where C does not depend on x, y, ξ, t, τ.

Some additional requirements (needed in order to represent pseudo-differential operators as hyper singular ones) must be imposed on the symbols of an integer order. Namely, a symbol of an even order is assumed to be a polynomial in ξ, and a symbol of an odd order is assumed to be either a polynomial or an even function of ξ. Finally, if γ is an odd integer, we assume that in the expansion of the function $[a(x,t,\xi)]^{-1}$, $|\xi| =!$, in spherical harmonics the coefficients at $Y_{2\nu,\mu}$ with $\gamma = n + 2\nu + 2k$, $k = 0, 1, 2, \ldots$, are equal to zero. This condition is automatically satisfied if $\gamma < n$, or if n is even.

The initial function ϕ is assumed to be bounded and continuous, while the function f is bounded, jointly continuous, and

$$|f(x,t) - f(y,t)| \leq C|x-y|^\lambda,$$

where C does not depend on x, y, t. We may assume without loss of generality that $\max\limits_{1\leq k\leq m} \gamma_k < \gamma - \lambda$.

By a solution of the problem (4.83), (4.84) we mean a bounded function $u(x,t)$, jointly continuous on $\mathbb{R}^n \times [0,T]$, which satisfies (4.84), and satisfies the equation (4.83) with A and A_k replaced by the appropriate conditionally convergent hyper-singular integral operators $\mathbf{D}_\Omega^\gamma$ and $\mathbf{D}_{\Omega_k}^\gamma$, or by differential operators understood in the classical sense, if corresponding symbols are polynomials in ξ.

Theorem 4.11. *The Cauchy problem (4.83)-(4.84) possesses a solution*

$$u(x,t) = \int_0^t d\tau \int_{\mathbb{R}^n} \Gamma(x,t,\xi,\tau) f(\xi,\tau)\, d\xi + \int_{\mathbb{R}^n} \Gamma(x,t,\xi,0)\varphi(\xi)\, d\xi,$$

where a fundamental solution $\Gamma(x,t,\xi,\tau)$, $x,\xi\in\mathbb{R}^n$, $0\le\tau<t\le T$, *is of the form*

$$\Gamma(x,t,\xi,\tau)=Z(x-\xi,t-\tau,\xi,\tau)+W(x,t,\xi,\tau),$$

$$|W(x,t,\xi,\tau)|\le C\left\{(t-\tau)^{(\gamma+\lambda)/\gamma}\left[(t-\tau)^{1/\gamma}+|x-\xi|\right]^{-n-\gamma}\right.$$
$$\left.+(t-\tau)\sum_{k=1}^{n+1}\left[(t-\tau)^{1/\gamma}+|x-\xi|\right]^{-n-\gamma_k}\right\},\quad \gamma_{m+1}=\gamma-\lambda,$$

while Z *satisfies the estimates (4.104) - (4.107).*

The proof is analogous to that of Theorem 4.6. Note in particular that Lemma 4.6 carries over to this case virtually unchanged. ■

Analogs of Theorem 4.5 and 4.7 also hold in this situation (see [74]). In particular, we have the following property, which is basic for the probabilistic interpretation of the equation (4.83).

Theorem 4.12. *If* $\gamma\le 2$, *and the characteristics* Ω *and* Ω_k *are even and non-negative, then* $\Gamma(x,t,\xi,\tau)\ge 0$ *for all* $x,\xi\in\mathbb{R}^n$, $t>\tau$.

4.9 Comments

The formula (4.5) for the heat kernel, as well as the fact of its positivity, were found independently by several authors; see [52, 60, 76, 53, 154, 16, 149]. A systematic treatment of the Cauchy problem (4.1)-(4.2) and more general parabolic equations over $\mathbb{Q}_p$ was given by the author [76]. Our exposition of the theory of parabolic pseudo-differential equations over $\mathbb{R}^n$ follows [74], where references to some earlier papers (by Eidelman, Drin', Fedoryuk and others) can be found.

For a generalization to some equations on vector spaces over local fields and division rings see [149]; that paper, as well as [60, 16], contains various versions of the Feynman-Kac formula. An extensive exposition of the potential theory based on the heat kernel (4.5) was given by Haran [53]. The results of Sect. 4.6 and 4.7 regarding heat equations over balls and the group of units are taken from [83, 85].

Chapter 5

Construction of Processes

5.1 Notations and preliminaries

Let $\xi_\alpha(t)$ be the non-Archimedean analog of the symmetric stable process introduced in Sect. 4.2. It is a stochastic process with independent increments on a local field K. We fix the origin as the starting point of ξ_α. As we know, ξ_α is stochastically continuous, and we assume its paths to be almost surely right-continuous with finite left limits. If Γ is a Borel subset of $K \setminus \{0\}$, denote by $q_\alpha(t,\Gamma)$ the number of jumps of the process ξ_α belonging to Γ, on the interval $[0,t)$. If Γ is contained in the complement $K \setminus B_n$ of a ball $B_n = \{x \in K : \ \|x\| \le q^n\}$ $(n \in \mathbb{Z})$ then $q_\alpha(t,\Gamma) < \infty$.

For a fixed Γ $q_\alpha(t,\Gamma)$ is a Poisson process with the parameter $t\pi_\alpha(\Gamma)$ (see [56]), that is

$$\mathbf{E}q_\alpha(t,\Gamma) = t\pi_\alpha(\Gamma),$$

$$\mathbf{P}\{q_\alpha(t,\Gamma) = l\} = \frac{[t\pi_\alpha(\Gamma)]^l}{l!}e^{-t\pi_\alpha(\Gamma)}, \quad l = 0,1,2,\dots .$$

As before, π_α is the Lévy measure,

$$\pi_\alpha(dx) = \frac{q^\alpha - 1}{1 - q^{-\alpha-1}}\|x\|^{-\alpha-1}\,dx.$$

For any fixed t $q_\alpha(t,\Gamma)$ is a random non-negative countably additive integer-valued measure on the Borel σ-algebra $\mathcal{B}(K \setminus \{0\})$. It may also be interpreted as a finitely additive K-valued random measure. If $\Gamma_1 \cap \Gamma_2 = \varnothing$, then the random variables $q_\alpha(t,\Gamma_1)$ and $q_\alpha(t,\Gamma_2)$ are independent. The characteristic function can be easily calculated:

$$\mathbf{E}\chi(zq_\alpha(t,\Gamma)) = \exp(\pi(t,\Gamma)[\chi(z) - 1]) \ , \quad z \in K.$$

For an interval $\Delta = [t', t'') \subset [0, T]$ we shall write $q_\alpha(\Delta, \Gamma) = q_\alpha(t'', \Gamma) - q_\alpha(t', \Gamma)$. The random variables $q_\alpha(\Delta, \Gamma)$ and $q_\alpha(t'' - t', \Gamma)$ have the same distribution. We also use the notation

$$\pi_\alpha(dt, dx) = \frac{q^\alpha - 1}{1 - q^{-\alpha-1}} \|x\|^{-\alpha-1} dt\, dx \ , \quad \pi_\alpha(\Delta, \Gamma) = \int_\Delta \int_\Gamma \pi_\alpha(dt, dx).$$

The theory of integration of K-valued functions on a compact set $\Sigma \subset K$ with respect to a K-valued measure possesses some features different from the conventional integration theory (see [71, 134, 127]). It is sufficient for our purposes to consider the case of an integer-valued finitely additive measure $r(dx)$. In this case the integral $\int_\Sigma f(x) r(dx)$ exists in the Riemann sense for every continuous function $f : \Sigma \to K$ and

$$\|I\| \leq \sup_{x \in \Sigma} \|f(x)\|. \tag{5.1}$$

Note that the measure need not be countably additive (with respect to the topology of K).

Below we shall use several Banach spaces over the field K consisting of K-valued functions. The space $C(\Sigma, K)$ consists of all continuous K-valued functions defined on a compact set Σ; it is endowed with the supremum norm. If $(\Omega, \mathcal{E}, \mu)$ is a measurable space,we define $L_\gamma(\Omega, K)$, $\gamma \geq 1$, as the set of all functions $f : \Omega \to K$ measurable with respect to the σ-algebra $\mathcal{E}$ and the Borel σ-algebra $\mathcal{B}(K)$ such that

$$\|f\|_{L_\gamma} = \left\{ \int_\Omega \|f(x)\|^\gamma \, d\mu(x) \right\}^{1/\gamma} < \infty.$$

Completeness of this space can be proved in a standard way (for the properties of measurable functions with values in a general metric space see [113]).

If X is a Banach space over K, then $D([0, T], X)$ consists of functions on the interval $[0, T]$, with values in X, which are right-continuous and have left limits in X. Well-known properties of the space D summarized in [15, 105] carry over to this situation (in fact they are valid for an arbitrary metric space X). Note that functions from $D([0, T], K)$ are constant between their jumps; that follows from total disconnectedness of K.

5.2 Stochastic integrals of deterministic functions

Consider a function

$$f(t,u) = \sum_{k=1}^{\infty} a_k(t)u^k \ , \quad 0 \le t \le T < \infty, \ u \in K, \tag{5.2}$$

where $a_k \in D([0,T],K)$,

$$\lim_{k\to\infty} \sup_{0\le t\le T} \|a_k(t)\|^{1/k} = 0. \tag{5.3}$$

The condition (5.3) guarantees convergence of the series (5.2) for every $u \in K$, uniformly with respect to $t \in [0,T]$ (see [71, 134]). In fact, the convergence is uniform on each sphere $S_n = \{x \in K : \|x\| = q^n\}$, $n \in \mathbb{Z}$. Thus $f \in D([0,T], C(S_n, K))$, for each $n \in \mathbb{Z}$.

Our task is to define the stochastic integral

$$I = \int_0^T \int_K f(t,u) q_\alpha(dt,du). \tag{5.4}$$

We assume that K is a *local field of characteristic zero*, that is a finite extension of $\mathbb{Q}_p$. Let us begin with the definition of integrals

$$I_n = \int_0^T \int_{S_n} f(t,u) q_\alpha(dt,du), \quad n \in \mathbb{Z}. \tag{5.5}$$

Considering f as a vector-function from $D([0,T], C(S_n, K))$ we can approximate it uniformly by step vector-functions $f_m(t,u)$ (see [15]). The function f_m corresponds to such a partition

$$[0,T] = \bigcup_{i=1}^{N_m} \Delta_i^{(m)}$$

by non-intersecting intervals that the oscillation of the vector-function f on each of the subintervals does not exceed ε_m, where $\varepsilon_m \to 0$ for $m \to \infty$. The vector-function f_m is constant on $\Delta_i^{(m)}$: $f_m(t,u) = f_{m,i}(u)$, $u \in S_n$ (it can be obtained by fixing the value of $f(t,u)$ at some point from $\Delta_i^{(m)}$), and we may set

$$I_{n,m} = \int_0^T \int_{S_n} f_m(t,u) q(dt,du) \stackrel{\text{def}}{=} \sum_{i=1}^{N_m} \int_{S_n} f_{m,i}(u) q_\alpha(\Delta_i^{(m)}, du),$$

where the integrals are understood as in Sect. 5.1.

It follows from the estimate (5.1) and the ultra-metric inequality that the sequence $\{I_{n,m}\}_{m=1}^{\infty}$ is fundamental in K almost surely, and its limit does not depend on the choice of the approximations $\{f_m\}$. We define the integral (5.5) as

$$I_n = \lim_{m\to\infty} I_{n,m}.$$

Using (5.1) again we obtain with probability 1 the inequality

$$\|I_n\| \le \sup_{0\le t\le T, u\in S_n} \|f(t,u)\|. \tag{5.6}$$

Now we define the stochastic integral (5.4) by the series

$$I = \sum_{n=-\infty}^{\infty} I_n. \tag{5.7}$$

Theorem 5.1. *The series (5.7) is convergent in probability.*

Proof. By virtue of the ultra-metric inequality it is sufficient to prove that $I_n \to 0$ in probability when $n \to \pm\infty$. If $n < 0$ then according to (5.3) and (5.6) we have with probability 1

$$\|I_n\| \le \sup_{k\ge 1} q^{nk} \sup_{0\le t\le T} \|a_k(t)\| \le Cq^n$$

(as before, we denote by C various positive constants) so that almost surely $I_n \to 0$ when $n \to -\infty$.

Let $u \in S_n$, $n \ge 1$, $\|v\| \le 1$. Then

$$\begin{aligned}\|f(t,u+v) - f(t,u)\| &= \left\| \sum_{k=1}^{\infty} a_k(t) \sum_{\nu=0}^{k-1} u^{\nu} v^{k-\nu} C_k^{\nu} \right\| \\ &\le \sup_{t,k,\nu} \|a_k(t)\| \cdot \|u\|^{\nu} \cdot \|v\|^{k-\nu} \le \|v\| \sup_{t,k} \left[q^{n(k-1)} \|a_k(t)\| \right] \\ &= q^{-n} \|v\| \sup_{t,k} \left[\|a_k(t)\|^{1/k} q^n \right]^k .\end{aligned}$$

Using (5.3) we conclude that there exists such a natural number $l(n)$ that

$$\|f(t,u+v) - f(t,u)\| \le q^{-n}$$

as soon as $\|v\| \le q^{-l(n)-1}$.

Let $N(n) \geq 0$ be such an integer that

$$\sup_{0 \leq t \leq T, u \in S_n} \|f(t,u)\| \leq q^{N(n)}.$$

We can write the canonical representation of the element $f(t,u) \in K$ as follows:

$$f(t,u) = \varphi_n(t,u) + \psi_n(t,u), \tag{5.8}$$

where

$$\varphi_n(t,u) = \beta^{-N(n)} \left(\xi_0 + \xi_1 \beta + \cdots + \xi_{N(n)+n-1} \beta^{N(n)+n-1} \right),$$

$$\psi_n(t,u) = \beta^{-N(n)} \left(\xi_{N(n)+n} \beta^{N(n)+n} + \cdots \right),$$

$\xi_j = \xi_j(t,u) \in S$ (as before, S is a complete set of representatives in the ring of integers $O \subset K$ of classes from the residue field O/P). Evidently,

$$\|\psi_n(t,u)\| \leq q^{-n}, \quad u \in S_n. \tag{5.9}$$

If $u = \beta^{-n}(u_0 + u_1\beta + \cdots)$ is the canonical representation of u, then the function $\varphi_n(t,u)$ depends only on components u_j with $j \leq l(n)$. Indeed, if $\|v\| \leq q^{-l(n)-1}$ then

$$\begin{aligned}
\|\varphi_n(t,u+v) - \varphi_n(t,u)\| &= \|[\varphi_n(t,u+v) - f(t,u+v)] \\
&\quad + [f(t,u+v) - f(t,u)] + [f(t,u) - \varphi_n(t,u)]\| \\
&\leq \max\{\|\psi_n(t,u+v)\|, \|f(t,u+v) - f(t,u)\|, \|\psi_n(t,u)\|\} \leq q^{-n}.
\end{aligned}$$

On the other hand,

$$\varphi_n(t,u+v) - \varphi_n(t,u) = \beta^{-N(n)} \sum_{j=0}^{N(n)+n-1} [\xi_j(t,u+v) - \xi_j(t,u)]\beta^j.$$

If $\varphi_n(t,u+v) \neq \varphi_n(t,u)$ then

$$\|\varphi_n(t,u+v) - \varphi_n(t,u)\| \geq q^{-n+1}$$

and we come to a contradiction.

It is clear that $\varphi_n, \psi_n \in D([0,T], C(S_n, K))$. We may write $I_n = I'_n + I''_n$, where

$$I'_n = \int_0^T \int_{S_n} \varphi_n(t,u) q_\alpha(dt, du);$$

$$I''_n = \int_0^T \int_{S_n} \psi_n(t,u) q_\alpha(dt, du).$$

We have $\|I''_n\| \le q^{-n} \to 0$ almost surely due to (5.6), (5.9). It remains to consider I'_n.

Since $f \in D([0,T], C(S_n, K))$, each of the functions ξ_j may have only a finite number of jumps with respect to the variable t, and the jump points do not depend on u. The same conclusion is valid for the function φ_n. Taking into account the behaviour of φ_n with respect to u described above we obtain that

$$\begin{aligned} I'_n = \beta^{-N(n)} \sum_{i=1}^{m_n} \sum_{u_0 \in S, u_0 \notin P} \sum_{u_1 \in S} \cdots \sum_{u_{l(n)} \in S} \Big\{ \xi_0(t_i^{(n)}, u_0 + u_1\beta + \cdots \\ + u_{l(n)}\beta^{l(n)}) + \xi_1(t_i^{(n)}, u_0 + u_1\beta + \cdots + u_{l(n)}\beta^{l(n)})\beta \\ + \cdots + \xi_{N(n)+n-1}(t_i^{(n)}, u_0 + u_1\beta + \cdots + u_{l(n)}\beta^{l(n)})\beta^{N(n)+n-1}\Big\} \\ \times q_\alpha(\Delta_i^{(n)}, B(u_0, u_1, \cdots, u_{l(n)})) \end{aligned}$$

where the intervals $\Delta_i^{(n)}$, $i = 1, \ldots, m_n$, form a partition of the interval $[0,T]$, $t_i^{(n)} \in \Delta_i^{(n)}$, the set $B(u_0, u_1, \ldots, u_{l(n)})$ consists of all elements $u \in S_n$ with fixed coefficients $u_0, \ldots, u_{l(n)}$ in the canonical representation

$$u = \beta^{-n} \sum_{j=0}^{\infty} u_j \beta^j.$$

Now we get with probability 1 the inequality

$$\|I'_n\| \le q^{N(n)} \max_i \max_{u_0, \ldots, u_{l(n)}} \|q_\alpha(\Delta_i^{(n)}, B(u_0, u_1, \ldots, u_{l(n)}))\|,$$

whence for any $\varepsilon > 0$

$$\mathbf{P}\{\|I'_n\| \ge \varepsilon\} \le \sum_{i=1}^{m_n} \sum_{u_0 \in S, u_0 \notin P} \sum_{u_1 \in S} \cdots \sum_{u_{l(n)} \in S} \mathbf{P}\Big\{\|q_\alpha(\Delta_i^{(n)}, B(u_0, u_1, \ldots, u_{l(n)}))\| \ge \varepsilon q^{-N(n)}\Big\}. \quad (5.10)$$

By our assumption, the field K is a finite extension of $\mathbb{Q}_p$. Let $M = \deg K/\mathbb{Q}_p$. Since $q_\alpha(\cdot,\cdot)$ is integer-valued, we have $\|q_\alpha(\cdot,\cdot)\| = |q_\alpha(\cdot,\cdot)|_p^M$. On the other hand (see Sect. 1.3), $M = ef$, $q = p^f$, where e is the ramification index, f is the index of inertia of the extension $K/\mathbb{Q}_p$. We rewrite the estimate (5.10) as

$$\mathbf{P}\{\|I_n'\| \geq \varepsilon\} \leq \sum_{i=1}^{m_n} \sum_{u_0 \in S, u_0 \notin P} \sum_{u_1 \in S} \cdots \sum_{u_{l(n)} \in S} \mathbf{P}\Big\{|q_\alpha(\Delta_i^{(n)}, B(u_0, u_1, \ldots, u_{l(n)}))|_p \geq \varepsilon^{1/M} p^{-e^{-1}N(n)}\Big\},$$

so that

$$\mathbf{P}\{\|I_n'\| \geq \varepsilon\} \leq \sum_{i=1}^{m_n} \sum_{u_0 \in S, u_0 \notin P} \sum_{u_1 \in S} \cdots \sum_{u_{l(n)} \in S} \mathbf{P}\Big\{|q_\alpha(\Delta_i^{(n)}, B(u_0, u_1, \ldots, u_{l(n)}))|_p \geq \varepsilon^{1/M} p^{-N_1(n)}\Big\}, \quad (5.11)$$

where $N_1(n) = [e^{-1}N(n)] + 1$.

Our aim is to prove that $P\{\|I_n'\| \geq \varepsilon\} \to 0$ as $n \to \infty$. It is sufficient to consider $\varepsilon = p^{-M\nu}$, $\nu \in \mathbb{Z}$, $\nu > 0$. Using the notation $a \mid b$ if a is a divisor of b,and $a \nmid b$ in the opposite case,we find that

$$\begin{aligned}
&\mathbf{P}\left\{|q_\alpha(\Delta_i^{(n)}, B(u_0, u_1, \ldots, u_{l(n)}))|_p \geq \varepsilon p^{-N_1(n)}\right\} \\
&\quad \leq \sum_{j=0}^{N_1(n)+\nu} \mathbf{P}\left\{q_\alpha(\Delta_i^{(n)}, B(u_0, u_1, \ldots, u_{l(n)})) = k,\ p^j \mid k,\ p^{j+1} \nmid k\right\} \\
&\quad = \mathbf{P}\left\{q_\alpha(\Delta_i^{(n)}, B(u_0, u_1, \ldots, u_{l(n)})) = k,\ p^{N_1(n)+\nu+1} \nmid k\right\} \\
&\quad = 1 - \mathbf{P}\left\{q_\alpha(\Delta_i^{(n)}, B(u_0, u_1, \ldots, u_{l(n)})) = p^{N_1(n)+\nu+1} j,\ j = 0, 1, \ldots\right\} \\
&\quad = 1 - \sum_{j=0}^{\infty} \mathbf{P}\left\{q_\alpha(\Delta_i^{(n)}, B(u_0, u_1, \ldots, u_{l(n)})) = p^{N_1(n)+\nu+1} j\right\}.
\end{aligned}$$

Denoting by $|\Delta|$ the length of an interval Δ and using the formula for distribution of the Poisson random measure we obtain that

$$\begin{aligned}
&\mathbf{P}\left\{|q_\alpha(\Delta_i^{(n)}, B(u_0, u_1, \ldots, u_{l(n)}))|_p \geq \varepsilon p^{-N_1(n)}\right\} \\
&\quad \leq 1 - \exp\left(-c|\Delta_i^{(n)}| q^{-\alpha n - l(n)}\right) \sum_{j=0}^{\infty} \frac{[c|\Delta_i^{(n)}| q^{-\alpha n - l(n)}]^{jp^{N_1(n)+\nu+1}}}{(jp^{N_1(n)+\nu+1})!},
\end{aligned}$$

where $c > 0$. We have used the fact that due to invariance of the measure

$$\pi_\alpha(\Delta_i^{(n)}, B(u_0, u_1, \dots, u_{l(n)}))$$
$$= \frac{q^\alpha - 1}{1 - q^{-\alpha-1}} |\Delta_i^{(n)}| \int\limits_{B(u_0,u_1,\dots,u_{l(n)})} \|u\|^{-\alpha-1}\, du$$
$$= \frac{q^\alpha - 1}{1 - q^{-\alpha-1}} |\Delta_i^{(n)}| q^{-(\alpha-1)n}(q-1)^{-1} q^{-l(n)} \int\limits_{S_n} du$$
$$= c|\Delta_i^{(n)}| q^{-\alpha n - l(n)}.$$

Substituting into (5.11) we come to the estimate

$$\mathbf{P}\{|I_n'|_p \geq \varepsilon\} \leq S_n^{(1)} + S_n^{(2)}$$

where

$$S_n^{(1)} = q^{l(n)+1} \sum_{i=1}^{m_n} \left[1 - \exp\left(-c|\Delta_i^{(n)}| q^{-\alpha n - l(n)}\right)\right],$$

$$S_n^{(2)} = q^{l(n)+1} \sum_{i=1}^{m_n} \exp\left(-c|\Delta_i^{(n)}| q^{-\alpha n - l(n)}\right) \times$$
$$\times \sum_{j=1}^{\infty} \frac{\left[c|\Delta_i^{(n)}| q^{-\alpha n - l(n)}\right]^{jp^{N_1(n)+\nu+1}}}{(jp^{N_1(n)+\nu+1})!}.$$

It follows from the inequality $e^{-x} \geq 1 - x,\ x \geq 0$, that

$$S_n^{(1)} \leq c q^{1-\alpha n} \sum_{i=1}^{m_n} |\Delta_i^{(n)}| = c q^{1-\alpha n} T \to 0\,, \quad n \to \infty.$$

The estimate for $S_n^{(2)}$ is based on the elementary inequality

$$x^k e^{-x} \leq k^k e^{-k}\,,\ x \geq 0\,,\ k \geq 1.$$

We obtain that

$$\left[c|\Delta_i^{(n)}| q^{-\alpha n - l(n)}\right]^{jp^{N_1(n)+\nu+1}-1} \exp\left(-c|\Delta_i^{(n)}| q^{-\alpha n - l(n)}\right)$$
$$\leq \left(jp^{N_1(n)+\nu+1} - 1\right)^{jp^{N_1(n)+\nu+1}-1} \exp\left(-jp^{N_1(n)+\nu+1} + 1\right),$$

whence

$$S_n^{(2)} \le cTeq^{1-\alpha n} \sum_{j=1}^{\infty} \frac{(jp^{N_1(n)+\nu+1}-1)^{jp^{N_1(n)+\nu+1}-1} e^{-jp^{N_1(n)+\nu+1}}}{(jp^{N_1(n)+\nu+1})!}.$$

By the Stirling formula

$$\left(jp^{N_1(n)+\nu+1}-1\right)! \ge \sqrt{2\pi}\left(jp^{N_1(n)+\nu+1}\right)^{1/2} \times$$
$$\times \left(jp^{N_1(n)+\nu+1}-1\right)^{jp^{N_1(n)+\nu+1}-1} e^{-jp^{N_1(n)+\nu+1}},$$

so that

$$S_n^{(2)} \le \frac{cTe}{\sqrt{2\pi}} q^{1-\alpha n} \sum_{j=1}^{\infty} j^{-3/2} \to 0\,, \quad n \to \infty. \quad \blacksquare$$

Now we can use our stochastic integral to obtain a representation for the process ξ_α. Note that a similar representation for real-valued processes is not available; one can construct only a representation for parts of the processes "separated from the origin"; see [142]. We shall write $\xi \approx \eta$ if the processes ξ, η have the same finite dimensional distributions. If the function f does not depend on t we write $\int\limits_K f(u)q_\alpha(t,du)$ instead of $\int\limits_0^t \int\limits_K f(u)q_\alpha(ds,du)$.

Theorem 5.2. *The process ξ_α admits a representation*

$$\xi_\alpha(t) \approx \int\limits_K uq(t,du)\,, \quad 0 \le t < \infty.$$

Proof. Since $\xi_\alpha(t)$ and $\int\limits_{Q_p} uq(t,du)$ are both processes with independent increments it is sufficient to prove coincidence, for every fixed t, of their characteristic functions. Thus we have to prove that for any $\eta \in K$

$$\mathbf{E}\chi(\eta\xi_\alpha(t)) \overset{?}{=} \mathbf{E}\chi\left(\eta \int\limits_K uq_\alpha(t,du)\right). \tag{5.12}$$

The left-hand side of (5.12) is equal to $\exp(-t\|\eta\|^\alpha)$. By Theorem 5.1

$$\mathbf{E}\chi\left(\eta \int\limits_K uq_\alpha(t,du)\right) = \lim_{N\to\infty} \prod_{n=-N}^{N} \mathbf{E}\chi\left(\eta \int\limits_{S_n} uq_\alpha(t,du)\right).$$

Approximating the integral over S_n by Riemann sums, using independence properties of q_α and the expression for the characteristic function we find that

$$\mathbf{E}\chi\left(\eta\int_{S_n} uq_\alpha(t,du)\right) = \exp\left\{\int_{S_n}[\chi(\eta u)-1]\pi_\alpha(t,du)\right\},$$

whence

$$\mathbf{E}\chi\left(\eta\int_{K} uq_\alpha(t,du)\right) = \exp\left\{\int_{K}[\chi(\eta u)-1]\pi_\alpha(t,du)\right\}$$
$$= \exp\left(-t\|\eta\|^\alpha\right). \quad \blacksquare$$

5.3 Stochastic integrals of random functions

5.3.1. The main lemma. Our construction of the stochastic integral will be based on the following general fact.

Lemma 5.1. *Let ζ be a non-negative integer-valued random variable having the Poisson distribution with the mean value $a > 0$:*

$$\mathbf{P}\{\zeta = k\} = e^{-a}\frac{a^k}{k!}.$$

Then for any prime number p, and any $\gamma \in [1,\infty)$,

$$\mathbf{E}|\zeta|_p^\gamma \le \frac{p^\gamma}{p^\gamma - 1}a. \tag{5.13}$$

Proof. The inequality $|\zeta|_p < p^{-n}$ $(n \in \mathbb{Z})$ holds if and only if $\zeta = mp^{n+1}$, $m \in \{0,1,\dots\}$. Therefore

$$\mathbf{P}\{|\zeta|_p = p^{-n}\} \le \mathbf{P}\{|\zeta|_p \ge p^{-n}\} = 1 - \sum_{m=0}^{\infty}\mathbf{P}\{\zeta = mp^{n+1}\}$$
$$= 1 - e^{-a}\sum_{m=0}^{\infty}\frac{a^{mp^{n+1}}}{(mp^{n+1})!},$$

so that

$$\mathbf{E}|\zeta|_p^\gamma = \sum_{n=0}^{\infty} p^{-\gamma n}\mathbf{P}\{|\zeta|_p = p^{-n}\}$$

$$\leq \frac{p^\gamma}{p^\gamma - 1} - e^{-a}\sum_{n=0}^{\infty} p^{-\gamma n}\sum_{m=0}^{\infty}\frac{a^{mp^{n+1}}}{(mp^{n+1})!}.$$

Considering the function

$$\varphi(x) = \frac{p^\gamma}{p^\gamma - 1} - e^{-x}\sum_{n=0}^{\infty} p^{-\gamma n}\sum_{m=0}^{\infty}\frac{x^{mp^{n+1}}}{(mp^{n+1})!} - \frac{p^\gamma}{p^\gamma - 1}x\,, \quad x \geq 0,$$

we see that $\varphi(0) = 0$,

$$\varphi'(x) = e^{-x}\left[\sum_{n=0}^{\infty} p^{-\gamma n}\sum_{m=0}^{\infty}\frac{x^{mp^{n+1}}}{(mp^{n+1})!} - \sum_{n=0}^{\infty} p^{-\gamma n}\sum_{m=1}^{\infty}\frac{x^{mp^{n+1}-1}}{(mp^{n+1}-1)!}\right] - \frac{p^\gamma}{p^\gamma - 1}$$

$$\leq e^{-x}\sum_{n=0}^{\infty} p^{-\gamma n}\sum_{k=0}^{\infty}\frac{x^k}{k!} - \frac{p^\gamma}{p^\gamma - 1} = 0$$

which results in (5.13). ■

Returning to our stable-like process $\xi_\alpha(t)$ with values in a local field K, a finite extension of $\mathbb{Q}_p$, we obtain the following estimate.

Corollary 5.1. *Let Δ be an interval on the real axis, $\Gamma \subset K$ a Borel set. For any $\gamma \in [1, \infty)$*

$$\mathbf{E}\|q_\alpha(\Delta, \Gamma)\|^\gamma \leq C_{\gamma,K}\pi_\alpha(\Delta, \Gamma), \quad C_{\gamma,K} > 0. \tag{5.14}$$

5.3.2. Definition and properties. Let the probability space $(\Omega, \mathcal{F}, \mathbf{P})$ be equipped with a right-continuous increasing family $\mathcal{F}_t$ $(0 \leq t \leq T)$ of sub-σ-algebras of $\mathcal{F}$ containing all $\mathbf{P}$-zero sets. Suppose that for any $t \in [0, T]$ and any compact set $\Sigma \subset K\backslash\{0\}$ the random variable $q_\alpha(t, \Sigma)$ is $\mathcal{F}_t$-measurable. We assume also that the random variables $q_\alpha(s_1, \Sigma_1)$, $\ldots$, $q_\alpha(s_n, \Sigma_n)$, where $s_1, \ldots, s_n \in [t, T]$, $\Sigma_1, \ldots, \Sigma_n$ are compact subsets of $K \setminus \{0\}$, are jointly independent of any event from $\mathcal{F}_t$.

We call a random function $f(t, u)$ defined on $[0, T] \times K$, with values in K, a random step function if there exists a non-random partition $0 =$

$t_0 < t_1 < \ldots < t_n = T$ and non-random Borel sets $\Gamma_1, \ldots, \Gamma_m \subset K$, $\bigcup_{j=1}^m \Gamma_j = K$, such that $f(t,u)$ takes a constant (random) value on every set $[t_k, t_{k+1}) \times \Gamma_j$, $k = 0, 1, \ldots, n-1$; $j = 1, \ldots, m$.

Denote by $M(\mathcal{F}_t)$ the set of all $\mathcal{F}_t$-measurable random step functions f such that $f(t,u) = 0$ for $u \in B_k$ where $k \in \mathbb{Z}$ depends on f.

For any function $f \in M(\mathcal{F}_t)$ we define the stochastic integral I_f as

$$I_f \equiv \int_0^T \int_K f(t,u) q_\alpha(dt, du) = \sum_{k=0}^{n-1} \sum_{j=1}^{m} f(t_k, u_j) q_\alpha([t_k, t_{k+1}), \Gamma_j),$$

where $u_j \in \Gamma_j$ are arbitrary points. It follows from the inequality (5.14) and the independence assumptions that

$$\begin{aligned} \mathbf{E}\|I_f\|^\gamma &\le E \max_{k,j} \|f(t_k, u_j)\|^\gamma \|q_\alpha([t_k, t_{k+1}), \Gamma_j)\|^\gamma \\ &\le C_{\gamma,K} \sum_{k,j} \mathbf{E}\|f(t_k, u_j)\|^\gamma \pi_\alpha([t_k, t_{k+1}), \Gamma_j) \\ &= C_{\gamma,K} \int_0^T \int_K \mathbf{E}\|f(t,u)\|^\gamma \pi_\alpha(dt, du). \end{aligned}$$

Denote by $M_\gamma(\mathcal{F}_t)$ the completion of $M(\mathcal{F}_t)$ with respect to the norm

$$\|f\|_\gamma = \left\{ \int_0^T \int_K \mathbf{E}\|f(t,u)\|^\gamma \pi_\alpha(dt, du) \right\}^{1/\gamma}.$$

We have proved that $f \mapsto I_f$ is a linear continuous operator from $M_\gamma(\mathcal{F}_t)$ to $L_\gamma(\Omega, K)$ defined on a dense linear manifold $M(\mathcal{F}_t)$. Its extension I_f by continuity onto the whole space $M_\gamma(\mathcal{F}_t)$ will be called a stochastic integral, now defined for any $f \in M_\gamma(\mathcal{F}_t)$. We have the estimate

$$\mathbf{E}\|I_f\|^\gamma \le C_{\gamma,K} \int_0^T \int_K \mathbf{E}\|f(t,u)\|^\gamma \pi_\alpha(dt, du). \tag{5.15}$$

Let us describe explicitly a subset of $M_\gamma(\mathcal{F}_t)$.

Proposition 5.1. *Suppose that a random function $f(t,u)$, $t \in [0,T]$, $u \in K$, with values in K, satisfies the following conditions:*

1) $f(t,u)$ is $\mathcal{F}_t$-measurable for any t, u;

2) $f \in D([0,T], C(\Sigma, L_\gamma(\Omega, K)))$ *for any compact set* $\Sigma \subset K \setminus \{0\}$*;*

3) $u \mapsto f(t,u)$ *is a continuous mapping from* K *to* $L_\gamma(\Omega, K)$*, uniformly with respect to* $t \in [0,T]$*;*

4) $\|f\|_\gamma < \infty$.

Then $f \in M_\gamma(\mathcal{F}_t)$.

Proof. It is sufficient to consider functions $f(t,u)$ vanishing for $\|u\| \leq q^m$ and for $\|u\| \geq q^N$, $m < N$. Denote

$$B_{m,N} = \left\{u \in K : \ q^{m+1} \leq \|u\| \leq q^{N-1}\right\}.$$

It follows from the condition 2 (see [15]) that, given $\varepsilon > 0$, there exists a partition $0 = t_0 < t_1 < \cdots < t_r = T$ for which

$$\sup_{t_i \leq t, s < t_{i+1}} \sup_{u \in B_{m,N}} \mathbf{E}\|f(s,u) - f(t,u)\|^\gamma < \varepsilon \ , \quad i = 0, 1, \ldots, r-1. \quad (5.16)$$

Using the condition 3 we can cover $B_{m,N}$ by non-intersecting balls $V_1, \ldots, V_l$ in such a way that

$$\sup_{t \in [0,T]} \sup_{u,v \in V_j} \mathbf{E}\|f(t,u) - f(t,v)\|^\gamma < \varepsilon \ , \quad j = 1, \ldots, l. \quad (5.17)$$

Choose arbitrary points $u_j \in V_j$, $j = 1, \ldots, l$, and set

$$f_\varepsilon(t,u) = \begin{cases} f(t_{i-1}, u_j), & \text{if } t \in [t_{i-1}, t_i), \ u \in V_j \\ 0, & \text{if } u \notin B_{m,N}. \end{cases}$$

It follows from the condition 1 that $f_\varepsilon \in M(\mathcal{F}_t)$, and from (5.16),(5.17) that

$$\|f - f_\varepsilon\|_\gamma^\gamma < 2\varepsilon \int_0^T \int_{B_{m,N}} \pi_\alpha(dt, du).$$

The right-hand side becomes arbitrarily small when $\varepsilon \to 0$. ■

5.4 Stochastic differential equations

5.4.1. Existence and uniqueness. In this section we consider an equation

$$x(t) = x_0 + \int_0^t \int_K F(s, x(s), u) q_\alpha(ds, du) \ , \quad 0 \leq t \leq T, \quad (5.18)$$

with a random K-valued function F; a more special case leading to Markov processes will be discussed below. Our presentation follows the general lines of [141]. However there are some features here different from the stochastic analysis in the real domain. First of all our equation does not contain diffusion terms (since there are no continuous processes on a totally disconnected space K) or drift terms (since one cannot integrate a K-valued function with respect to the Lebesgue measure). We systematically consider stochastic processes (assumed to be continuous in the mean) as functions with values in $L_\gamma(\Omega, K)$. Thus we are unable to deal with individual paths; we shall not distinguish between stochastically equivalent processes, and the equality sign in the equation (5.18) means in fact stochastic equivalence.

The equation (5.18) will be considered under the following assumptions.

(A_1) *The random variable x_0 with values in K is $\mathcal{F}_0$-measurable and* $\mathbf{E}\|x_0\|^\gamma < \infty$.

(A_2) *For each $x, u \in K$, $t \in [0,T]$, the function $F(t,x,u)$ is $\mathcal{F}_t$-measurable.*

(A_3) $F(\cdot, x, u) \in D([0,T], C_b(K \times S_n, L_\gamma(\Omega, K)))$, *for every $n \in \mathbb{Z}$; here C_b is the space of bounded continuous vector functions with* sup *norm.*

(A_4) *The mapping $u \mapsto F(t,x,u)$ from K to $L_\gamma(\Omega, K)$ is continuous,uniformly with respect to $t \in [0,T]$, $x \in K$.*

(A_5) $\int_0^T \int_K \mathbf{E}\left[\sup_{x\in K} \|F(t,x,u)\|^\gamma\right] \|u\|^{-\alpha-1}\, du < \infty.$

(A_6) *For each $n \in \mathbb{Z}$ there exists such a constant $L_n > 0$ that almost surely for all $t \in [0,T]$, $x, y \in K$, $u \in S_n$*

$$\|F(t,x,u) - F(t,y,u)\| \le L_n \|x-y\|.$$

(A_7) $\sum_{n=-\infty}^{\infty} L_n q^{-n(\alpha+1)} < \infty.$

Denote by X the Banach space of (classes of stochastically equivalent) processes $x(t)$, $t \in [0,T]$, with values in K, $\mathcal{F}_t$-measurable for each t, continuous as functions on $[0,T]$ with values in $L_\gamma(\Omega, K)$, equipped with the norm $\|x\|_X = \sup_{0\le t\le T} \{\mathbf{E}\|x(t)\|^\gamma\}^{1/\gamma}$.

Theorem 5.3. *Under the assumptions (A_1) – (A_7) the equation (5.18) possesses a unique solution $x \in X$.*

Proof. Consider the operator

$$(\mathfrak{S}x)(t) = x_0 + \int_0^t \int_K F(s, x(s), u) q_\alpha(ds, du) \ , \quad x \in X. \tag{5.19}$$

The stochastic integral in (5.19) exists by Proposition 5.1 – its condition 1 follows from (A_2) and the inclusion $x \in X$, the condition 2 is a consequence of (A_3) and (A_6), and the conditions 3 and 4 follow from (A_4) and (A_5) respectively.

From (A_5) and the inequality (5.15) we obtain that for any $t', t'' \in [0, T]$

$$\mathbf{E}\|(\mathfrak{S}x)(t') - (\mathfrak{S}x)(t'')\|^\gamma = \mathbf{E}\left\| \int_{t'}^{t''} \int_K F(s, x(s), u) q_\alpha(ds, du) \right\|^\gamma$$

$$\leq C \int_{t'}^{t''} \int_K \mathbf{E}\|F(s, x(s), u)\|^\gamma \|u\|^{-\alpha-1} ds\, du.$$

Thus (A_1),(A_2) and (A_5) guarantee that $\mathfrak{S}x \in X$.

If $x_1, x_2 \in X$ then

$$\mathbf{E}\,\|(\mathfrak{S}x_1)(t) - (\mathfrak{S}x_2)(t)\|^\gamma$$

$$\leq C_{\gamma,K} \int_0^t \int_K \mathbf{E}\|F(s, x_1(s), u) - F(s, x_2(s), u)\|^\gamma \pi_\alpha(ds, du)$$

$$\leq C \int_0^t \sum_{n=-\infty}^{\infty} L_n \mathbf{E}\|x_1(s) - x_2(s)\|^\gamma \int_{S_n} \|u\|^{-\alpha-1}\, du\, ds$$

$$\leq \mu \int_0^t \mathbf{E}\|x_1(s) - x_2(s)\|^\gamma\, ds$$

with some $\mu > 0$; we have used (5.15), (A_6) and (A_7). Iterating we find that

$$\mathbf{E}\,\|(\mathfrak{S}^n x_1)(t) - (\mathfrak{S}^n x_2)(t)\|^\gamma \leq \frac{\mu^n t^{n-1}}{(n-1)!} \|x_1 - x_2\|_X,$$

whence

$$\|\mathfrak{S}^n x_1 - \mathfrak{S}^n x_2\|_X \leq \frac{\mu^n T^{n-1}}{(n-1)!} \|x_1 - x_2\|_X, \tag{5.20}$$

so that $\mathfrak{S}^n$ is a contraction for sufficiently large $n \geq n_0$.

Let $\overline{x} \in X$ be the unique fixed point of $\mathfrak{S}^{n_0}$. Its uniqueness implies that $\mathfrak{S}$ cannot possess more than one fixed point. On the other hand, we have $\mathfrak{S}^{kn_0}\overline{x} = \overline{x}$ for any $k = 1, 2, \ldots$ whence

$$\|\overline{x} - \mathfrak{S}\overline{x}\|_X = \|\mathfrak{S}^{kn_0}\overline{x} - \mathfrak{S}^{kn_0+1}\overline{x}\|_X \leq \frac{\mu^{kn_0}T^{kn_0-1}}{(kn_0-1)!}\|\overline{x} - \mathfrak{S}\overline{x}\|_X$$

in accordance with (5.20). Passing to the limit for $k \to \infty$ we come to the equality $\overline{x} = \mathfrak{S}\overline{x}$. ■

Using the inequality (5.15), assumptions (A_6), (A_7) and the Gronwall inequality we obtain the following estimate for the dependence of a solution upon its initial value x_0. If $x'(t), x''(t)$ are solutions of the equation (5.18) corresponding to the initial values x_0', x_0'' then

$$\mathbf{E}\|x'(t) - x''(t)\|^\gamma \leq e^{Ct}\mathbf{E}\|x_0' - x_0''\|^\gamma. \tag{5.21}$$

5.4.2. Markov processes. Suppose that F is a non-random function satisfying (with obvious simplifications) the conditions (A_1)-(A_7). Below we assume $\gamma = 1$. Let $x_{s,z}(t)$ be a solution of the equation

$$x(t) = z + \int_s^t \int_K F(\sigma, x(\sigma), u) q_\alpha(d\sigma, du), \tag{5.22}$$

$z \in K$, $s \geq 0$.

It can be shown in a standard way that $x_{s,z}(t)$ is a Markov process on K.

Denote by $H(K)$ the set of all real-valued bounded functions on K satisfying the Lipschitz condition

$$|\varphi(z') - \varphi(z'')| \leq C\|z' - z''\|.$$

It follows from the estimate (5.21) that the stochastic evolution defined by the equation (5.18) preserves the space $H(K)$: if $\varphi \in H(K)$ then

$$|\mathbf{E}\varphi(x_{s,z'}(t)) - \mathbf{E}\varphi(x_{s,z''}(t))| \leq C\|z' - z''\|, \quad s \leq t \leq T.$$

In our next result we show that the generator of our Markov process acts on functions from $H(K)$ as a hyper-singular integral operator. Here we need instead of (A_5) a stronger assumption:

(A_5') $\int_K \sup_{x \in K, t \in [0,T]} \|F(t,x,u)\| \cdot \|u\|^{-\alpha-1} du < \infty.$

Proposition 5.2. *If $\varphi \in H(K)$ then*

$$\lim_{t\to s+0} \frac{\mathbf{E}\varphi(x_{s,z}(t)) - \varphi(z)}{t-s} = \frac{q^\alpha - 1}{1 - q^{-\alpha-1}} \int_K [\varphi(z + F(s,z,u)) - \varphi(z)] \|u\|^{-\alpha-1} du, \quad (5.23)$$

uniformly with respect to $z \in K$.

Proof. Consider the process

$$\widetilde{x}_{s,z}(t) = z + \int_s^t \int_K F(\sigma, z, u) q_\alpha(d\sigma, du)$$

and set $y(t) = x_{s,z}(t) - \widetilde{x}_{s,z}(t)$. We have

$$y(t) = \int_s^t \int_K [F(\sigma, x_{s,z}(\sigma), u) - F(\sigma, z, u)]\, q_\alpha(d\sigma, du),$$

whence

$$\mathbf{E}\|y(t)\| \le C \int_s^t \int_K \mathbf{E}\|F(\sigma, x_{s,z}(\sigma), u) - F(\sigma, z, u)\| \pi_\alpha(d\sigma, du)$$
$$\le C \int_s^t \mathbf{E}\|x_{s,z}(\sigma) - z\|\, d\sigma.$$

This means that

$$|\mathbf{E}\varphi(x_{s,z}(t)) - \mathbf{E}\varphi(\widetilde{x}_{s,z}(t))| \le C\mathbf{E}\|y(t)\|$$
$$\le C \int_s^t d\sigma \int_s^\sigma \int_K \mathbf{E}\|F(\tau, x_{s,z}(\tau), u)\| \pi_\alpha(d\tau, du)$$
$$\le C \int_s^t d\sigma \int_s^\sigma d\tau \int_K \sup_{\zeta \in K} \|F(\tau, \zeta, u)\| \cdot \|u\|^{-\alpha-1} du = o(t-s),$$

uniformly with respect to $z \in K$.

Hence the right-hand side of (5.23) can be written as

$$\lim_{t\to s+0} \frac{\mathbf{E}\varphi(\widetilde{x}_{s,z}(t)) - \varphi(z)}{t-s}, \tag{5.24}$$

and the transition to (5.24) is uniform with respect to z.

Approximate, for each fixed z,the function $F(\sigma, z, u)$ by step functions $F_n(\sigma, u)$, $n = 1, 2, \dots$ so that $F_n \to F$ in $M_1(\mathcal{F}_t)$,

$$\sup_{s\le t\le T} \mathbf{E}\left\| \int_s^t \int_K F_n(\sigma, u) q_\alpha(d\sigma, du) - \int_s^t \int_K F(\sigma, z, u) q_\alpha(d\sigma, du) \right\| \longrightarrow 0, \quad n \to \infty.$$

We may assume that $F_n(\sigma, u) = 0$ for $\|u\| \le q^{-n}$, $n = 1, 2, \dots$. If

$$x^{(n)}(t) = z + \int_s^t \int_K F_n(\sigma, u) q_\alpha(d\sigma, du)$$

then

$$\sup_{s\le t\le T} \mathbf{E}\|x^{(n)}(t) - \widetilde{x}_{s,z}(t)\| \longrightarrow 0, \quad n \to \infty. \tag{5.25}$$

Fixing n denote by $\{\sigma_j\}_{j=1,\dots,N}$ such jump times of our basic process ξ_α that

$$\|\xi_\alpha(\sigma_j) - \xi_\alpha(\sigma_j - 0)\| > q^{-n}.$$

Let the sequence $\{\sigma_j\}$ be arranged at increasing order. Then

$$x^{(n)}(t) = z + \sum_{\sigma_m \le t} F_n(\sigma_m, q_m)$$

with probability 1. Here $q_m = \xi_\alpha(\sigma_m) - \xi_\alpha(\sigma_m - 0)$. Since the function $x^{(n)}(t)$ has almost surely a finite number of jumps on the interval $[0, t]$ being constant between them we can express an increment of $\varphi(x^{(n)}(t))$ as a sum of corresponding jumps. Thus we come to the representation

$$\varphi(x^{(n)}(t)) - \varphi(z) = \int_s^t \int_K \left[\varphi(x^{(n)}(\sigma - 0) + F_n(\sigma, u)) - \varphi(x^{(n)}(\sigma - 0))\right] q_\alpha(d\sigma, du) \tag{5.26}$$

(compare to a similar reasoning in the proof of the Ito formula for real stochastic integrals [58]).

The real-valued integrand in (5.26) is bounded, $\mathcal{F}_\sigma$-measurable and vanishes for $||u|| \leq q^{-n}$. We can insert expectation inside the integral in (5.26) which results in

$$\mathbf{E}\varphi(x^{(n)}(t)) - \varphi(z)$$
$$= \int_s^t \int_K \mathbf{E}\left[\varphi(x^{(n)}(\sigma - 0) + F_n(\sigma, u)) - \varphi(x^{(n)}(\sigma - 0))\right] \pi_\alpha(d\sigma, du).$$

Since $x^{(n)}$ is continuous in the mean, and φ is Lipschitz we find that $\mathbf{E}\varphi(x^{(n)}(\cdot))$ is a continuous (real-valued) function so that

$$\mathbf{E}\varphi(x^{(n)}(t)) - \varphi(z)$$
$$= \int_s^t \int_K \mathbf{E}\left[\varphi(x^{(n)}(\sigma) + F_n(\sigma, u)) - \varphi(x^{(n)}(\sigma))\right] \pi_\alpha(d\sigma, du). \quad (5.27)$$

Our next task is passing to the limit in (5.27) for $n \to \infty$. Note that F_n is constructed from F by fixing its values on sets of small oscillation. This process can be performed in such a way that

$$||F_n(\sigma, u)|| \leq ||F(\sigma, z, u)||$$

(one should choose the values or their left limits at minimum points). Using the relation (5.25),the Lipschitz property of φ and the assumption (A_5) (or (A_5')) we apply the dominated convergence theorem and obtain the formula

$$\mathbf{E}\varphi(\widetilde{x}_{s,z}(t)) - \varphi(z)$$
$$= \int_s^t \int_K \mathbf{E}\left[\varphi(\widetilde{x}_{s,z}(\sigma) + F(\sigma, z, u)) - \varphi(\widetilde{x}_{s,z}(\sigma))\right] \pi_\alpha(d\sigma, du). \quad (5.28)$$

The internal integral in (5.28) is right continuous as a function of σ at the point $\sigma = s$, uniformly with respect to z, due to the assumptions (A_3) and (A_5'). Dividing both sides of (5.28) by $t - s$ and passing to the limit for $t \to s$ we obtain the desired equality (5.23). ■

5.5 Rotation-invariant processes with independent increments

Let K be a finite extension of $\mathbb{Q}_p$. In this section we give a description and explicit construction of all rotation-invariant temporally and spatially homogeneous processes with independent increments on K.

Let $Y(t)$ be a temporally and spatially homogeneous stochastically continuous process with independent increments on K, $Y(0) = 0$, μ_t is the distribution of the random variable $Y(t)$, $\widehat{\mu_t}(z)$ is the corresponding characteristic function, that is

$$\widehat{\mu_t}(z) = \int_K \chi(zx)\mu_t(dx), \quad z \in K.$$

Suppose that $Y(t)$ is symmetric, that is all the measures μ_t are invariant with respect to the reflection: $\mu_t(\Gamma) = \mu_t(-\Gamma)$ for any Borel set $\Gamma \subset K$. It is well-known (see e.g. [10, 56]) that

$$\widehat{\mu_t}(z) = \exp\left[t \int_{K\setminus\{0\}} (\chi(xz) - 1)F(dx)\right], \quad z \in K, \tag{5.29}$$

where χ is a rank zero additive character, F is a symmetric σ-finite measure on $K \setminus \{0\}$ with finite mass outside any neighbourhood of the origin. Such measures (called *Lévy measures*) are in one-to-one correspondence with processes of the above class.

Let $Q_t(x, dy)$ be the transition probability of the process Y_t. Denote by $B_m(x)$ the ball $\{y \in K : \ \|y - z\| \leq q^m\}$.

Lemma 5.2. *The following conditions are equivalent.*

(i) *For each* $m \in \mathbb{Z}$ *and* $t \geq 0$

$$h_{m,t}(x) = Q_t(x, B_m(0)) = Q_t(0, B_m(-x))$$

depends only on the absolute value $\|x\|$.

(ii) $\|Y(t)\|$ *is a Markov process on* $\|K\| = \{\|x\|, x \in K\}$, *that is, for any real function* g *on* K, *such that the value* $g(x)$ *depends only on* $\|x\|$, *it follows that*

$$\mathbf{E}g(Y(t) + x) = \mathbf{E}g(Y(t) + x'),$$

whenever $\|x\| = \|x'\|$.

(iii) *The Lévy measure F is such that*

$$F(\Gamma) = F(u\Gamma) \tag{5.30}$$

for any Borel set Γ if $\|u\| = 1$.

Proof. (i)⇒(ii). If g is such as in (ii), then

$$g = \sum_{m \in \mathbb{Z}} b_m \left(\omega_{B_m(0)} - \omega_{B_{m-1}(0)}\right) + b_{-\infty}\omega_{\{0\}},$$

for some $b_m, b_{-\infty} \in \mathbb{R}$, where ω_Γ is the indicator of a set Γ. Then

$$\mathbf{E}g(Y(t) + x) = \sum_{m \in \mathbb{Z}} b_m \left(h_{m,t}(x) - h_{m-1,t}(x)\right) + b_{-\infty} \lim_{m \to -\infty} h_{m,t}(x),$$

which depends only on $\|x\|$.

(ii)⇒(i). Let $g = \omega_{B_m(0)}$. If $\|x\| = \|x'\|$, then

$$h_{m,t}(x) = \mathbf{E}g(Y(t) + x) = \mathbf{E}g(Y(t) + x') = h_{m,t}(x').$$

(i)⇒(iii). Let $u \in K$, $\|u\| = 1$. Then

$$\widehat{Q_t}(u^{-1}z) = \int_K \chi(zx)Q_t(u\,dx) = \int_K \chi(zx)Q_t(dx) = \widehat{Q_t}(z),$$

whence

$$\exp\left[t \int_{K\setminus\{0\}} (\chi(xz) - 1)F(dx)\right] = \exp\left[t \int_{K\setminus\{0\}} (\chi(xz) - 1)F(u\,dx)\right]$$

for all t, so that $F(\Gamma) = F(u\Gamma)$ for any Borel set Γ.

(iii)⇒(i). Put $Q_t^{(u)}(\Gamma) = Q_t(u\Gamma)$, $\|u\| = 1$, and let F be the Lévy measure of Q_t. Then for any $z \in K$

$$\widehat{Q_t^{(u)}}(z) = \exp\left[t \int_{K\setminus\{0\}} (\chi(xz) - 1)F(u\,dx)\right]$$

$$= \exp\left[t \int_{K\setminus\{0\}} (\chi(xz) - 1)F(dx)\right] = \widehat{Q_t}(z),$$

whence $Q_t^{(u)} = Q_t$. ■

Now, if (Y_t, Q_t) is any rotation-invariant process on K, then by Lemma 5.2, its Lévy measure must satisfy the condition (5.30). Set

$$a(m) = F(K \setminus B_m(0)).$$

Then the sequence $\{a(m)\}_{m\in\mathbb{Z}}$ is such that

$$a(m+1) \le a(m), \quad m \in \mathbb{Z}, \tag{5.31}$$

$$\lim_{m\to+\infty} a(m) = 0. \tag{5.32}$$

Conversely, for a given sequence $\{a(m)\}_{m\in\mathbb{Z}}$ of non-negative numbers with the properties (5.31) and (5.32), there exists a unique measure F on $K \setminus \{0\}$ satisfying (5.30) and

$$F(K \setminus B_m(0)) = a(m). \tag{5.33}$$

Indeed, every ball V in K, which does not contain 0, lies within a certain sphere S_N centered at 0, and any other ball within the same sphere is obtained from V by a rotation. Thus $F(V)$ can be recovered from $F(S_N)$, while the latter is determined by (5.33).

In other words, a sequence $\{a(m)\}_{m\in\mathbb{Z}}$ satisfying (5.31) and (5.32) corresponds in the one-two-one way to a rotation-invariant process (Y_t, Q_t), whose Lévy measure is given by (5.33). In fact the transition probability can be written explicitly. Denote

$$P_m(t) = (1-q^{-1}) \sum_{i=0}^{\infty} q^{-i} \exp\left[-(q-1)^{-1}(qa(m+i) - a(m+i+1))t\right],$$

$t \ge 0$, and let

$$\begin{aligned} P_t(x, B_m(y)) &= P_m(t), \text{ if } \ \|x-y\| \le q^m, \\ P_t(x, B_m(y)) &= (q-1)^{-1} q^{1-\nu} (P_{m+\nu}(t) - P_{m+\nu-1}(t)), \\ &\qquad \text{if } \ \|x-y\| = q^{m+\nu}, \nu \ge 1. \end{aligned} \tag{5.34}$$

Theorem 5.4. *P_t defines the transition probability of a rotation-invariant non-exploding process X_t, whose Lévy measure satisfies (5.33).*

Proof. It follows from (5.31) that $P_t(x, B_m(y)) \ge 0$. Since every open subset Γ of K is a disjoint union of balls, we can define $P_t(x, \Gamma)$ in an obvious way obtaining a Borel measure on K. By (5.32) $P_m(t) \to 1$ as $m \to \infty$, so that $P_t(x, K) = 1$. It is clear that $P_m(0) = 1$, whence $P_0(x, K \setminus \{x\}) = 0$.

Let us consider the family of measures $\Phi_t(\Gamma)$ defined initially on balls as

$$\Phi_t(B_m(y)) = P_m(t), \text{ if } \|y\| \le q^m,$$
$$\Phi_t(B_m(y)) = (q-1)^{-1} q^{1-\nu} (P_{m+\nu}(t) - P_{m+\nu-1}(t)), \quad (5.35)$$
$$\text{if } \|y\| = q^{m+\nu}, \nu \ge 1.$$

It is clear that $P_t(x, \Gamma) = \Phi_t(\Gamma - x)$ (of course, this is just a relation between the transition probability of a process with independent increments, and the associated semigroup of measures; see e.g. [43]). The semigroup property of P_t is equivalent to the fact that Φ_t is a semigroup of measures. The latter is equivalent to the property

$$\widehat{\Phi_{t+s}} = \widehat{\Phi_t}\widehat{\Phi_s}. \quad (5.36)$$

of the characteristic function

$$\widehat{\Phi_t}(z) = \int_K \chi(xz) \Phi_t(dx).$$

Below we shall find $\widehat{\Phi_t}$ explicitly, so that the identity (5.36) will be evident. However we have first to study the family of measures (5.35).

Lemma 5.3. *If* $\lim_{m \to -\infty} a(m) = \infty$, *then for each* $t > 0$ *the measure* Φ_t *is absolutely continuous with respect to the Haar measure. If* $\lim_{m \to -\infty} a(m) < \infty$, *then for each* $t > 0$ *the measure* Φ_t *is absolutely continuous with respect to the Haar measure everywhere except the origin. In both cases its density outside the origin is given by*

$$\varphi_t(y) = (q-1)^{-1} q^{1-m} (P_m(t) - P_{m-1}(t)) \quad \text{if } \|y\| = q^m.$$

Proof. Suppose that $a(m) \to \infty$ as $m \to -\infty$. Then $P_m(t) \to 0$ as $m \to -\infty$, for any $t > 0$. If $\|z\| \le q^m$, then

$$\int_{B_m(z)} \varphi_t(y)\, dy = \int_{B_m(0)} \varphi_t(y)\, dy = P_m(t) = \Phi_t(B_m(z)).$$

If $\|z\| = q^{m+\nu}$, $\nu \ge 1$, then $\|y - z\| \le q^m$ implies $\|y\| = q^{m+\nu}$. Therefore

$$\int_{B_m(z)} \varphi_t(y)\, dy = (q-1)^{-1} q^{1-\nu} (P_{m+\nu}(t) - P_{m+\nu-1}(t)) = \Phi_t(B_m(z)), \quad (5.37)$$

and this is what was required.

Suppose now that $a(m) \to W < \infty$ as $m \to -\infty$. In this case

$$\Phi_t(\{0\}) = \lim_{m\to-\infty} P_m(t) = e^{-Wt} > 0,$$

so that Φ_t is not absolutely continuous at the origin. At the same time, (5.37) holds in this case too. ■

Lemma 5.4. *The characteristic function* $\widehat{\Phi_t}$ *is given by*

$$\widehat{\Phi_t}(z) = \exp\left[-(q-1)^{-1}(qa(-n) - a(-n+1))t\right], \tag{5.38}$$

where $n = \log_q \|z\|$.

Proof. If $\|z\| = q^n$, then

$$\widehat{\Phi_t}(z) = \int\limits_{B_{-n}(0)} \chi(xz)\Phi_t(dx) + \sum_{m=-n+1}^{\infty} \int\limits_{\|x\|=q^m} \chi(xz)\varphi_t(x)\,dx$$

$$= P_{-n}(t) + (q-1)^{-1} \sum_{m=-n+1}^{\infty} q^{1-m}(P_m(t) - P_{m-1}(t)) \int\limits_{\|x\|=q^m} \chi(xz)\,dx.$$

By (1.27) we find that

$$\int\limits_{\|x\|=q^m} \chi(xz)\,dx = \begin{cases} -q^{m-1}, & \text{if } m = -n+1, \\ 0, & \text{if } m > -n+1, \end{cases}$$

so that

$$\widehat{\Phi_t}(z) = P_{-n}(t) - (q-1)^{-1}\left(P_{-n+1}(t) - P_{-n}(t)\right),$$

and after simple transformations we come to (5.38). ■

Now we can go on with *the proof of Theorem 5.4*. The expression (5.38) immediately implies (5.36), so that Φ_t is a semigroup of measures, while P_t is a transition probability. It remains to prove the equality (5.33) for the corresponding Lévy measure.

Let F be the measure defined by (5.30) and (5.33). In view of Lemma 5.4 and the uniqueness of the Lévy measure for symmetric processes, it is sufficient to show that

$$\int\limits_K (\chi(xz) - 1)F(dx) = -(q-1)^{-1}(qa(-n) - a(-n+1)) \tag{5.39}$$

for $\|z\| = q^n$. Since F is rotation-invariant, we may assume that $z = \beta^{-n}$.

The integral on the left in (5.39) is actually taken over those x, for which $\|x\| \geq q^{-n+1}$. If $\|x\| = q^m$, $m \geq -n+1$, we write a canonical representation

$$x = \beta^{-m}(x_0 + x_1\beta + \cdots), \quad x_j \in S, x_0 \notin P,$$

so that

$$\int_{\|x\|=q^m} (\chi(xz) - 1)F(dx)$$
$$= \sum_{(x_0,x_1,\dots,x_{m+n-1})} \Big\{ \left(\chi\left(\beta^{-m-n}\left(x_0 + x_1\beta + \cdots + x_{m+n-1}\beta^{m+n-1}\right)\right) - 1\right)$$
$$\times \int_{B_{-n}(\beta^{-m}(x_0+x_1\beta+\cdots+x_{m+n-1}\beta^{m+n-1}))} F(dy) \Big\}.$$

The sphere $S_m = \{x \in K : \|x\| = q^m\}$ can be decomposed into the union of $(q-1)q^{m+n-1}$ "closed" balls of the radius q^{-n}. By the rotation invariance

$$\int_{B_{-n}(\beta^{-m}(x_0+x_1\beta+\cdots+x_{m+n-1}\beta^{m+n-1}))} F(dy) = (q-1)^{-1}q^{-m-n+1}F(S_m)$$
$$= (q-1)^{-1}q^{-m-n+1}[a(m-1) - a(m)].$$

On the other hand,

$$\sum_{(x_0,x_1,\dots,x_{m+n-1})} \chi\left(\beta^{-m-n}\left(x_0 + x_1\beta + \cdots + x_{m+n-1}\beta^{m+n-1}\right)\right)$$
$$= q^n \sum_{(x_0,x_1,\dots,x_{m+n-1})} \int_{B_{-n}(\beta^{-m}(x_0+x_1\beta+\cdots+x_{m+n-1}\beta^{m+n-1}))} \chi(\beta^{-n}y)\,dy$$
$$= q^n \int_{\|y\|=q^m} \chi(\beta^{-n}y)\,dy = \begin{cases} -1, & \text{if } m = -n+1, \\ 0, & \text{if } m > -n+1. \end{cases}$$

As a result we get that

$$\int_K (\chi(xz) - 1)F(dx) = -q(q-1)^{-1}[a(-n) - a(-n+1)]$$

$$- \sum_{m=-n+2}^{\infty} [a(m-1) - a(m)] = -(q-1)^{-1}[qa(-n) - a(-n+1)],$$

as desired. ■

Example. For the process $\xi_\alpha(t)$ considered above, we have

$$F(dx) = \pi_\alpha(dx) = \frac{q^\alpha - 1}{1 - q^{-\alpha-1}} \|x\|^{-\alpha-1} dx,$$

so that

$$a(m) = \frac{q^\alpha - 1}{1 - q^{-\alpha-1}} \int_{\|x\| > q^m} \|x\|^{-\alpha-1} dx = \frac{1 - q^{-1}}{1 - q^{-\alpha-1}} q^{-\alpha m}. \tag{5.40}$$

5.6 The generator and its spectrum

Let T_t be the semigroup of operators on $L_2(K)$ determined by the transition probability $P_t(x, \Gamma)$. It can be verified by a direct calculation that

$$\int_K g(x)\, dx \int_K P_t(x, dy) f(y) = \int_K f(y)\, dy \int_K g(x) P_t(y, dx) \tag{5.41}$$

for any $f, g \in L_2(K)$ (note that it is sufficient to prove (5.41) for the case when f and g are indicators of balls). It follows from (5.41) that the operators T_t are self-adjoint.

The semigroup T_t is strongly continuous. Indeed, let $f \in L_2(K)$ be of the form $f = \sum_{i=1}^{n} c_i \omega_i$, where ω_i is the indicator of a ball V_i of a radius q^m, and the balls are disjoint. Using (5.41) and the semigroup property, we find that

$$\|T_t f - f\|^2_{L_2(K)} = \sum_{i,j=1}^{n} c_i c_j \left[q^m P_{2t}(V_i, V_j) - 2q^m P_t(V_i, V_j)\right] + q^m \sum_{i=1}^{n} c_i^2,$$

where

$$P_t(V_i, V_j) = q^{-m} \int_{V_i} P_t(x, V_j)\, dx = P_t(V_j, V_i).$$

As $t \to 0$,

$$\|T_t f - f\|^2_{L_2(K)} \longrightarrow q^m \left[\sum_{i,j=1}^{n} c_i c_j (\delta_{ij} - 2\delta_{ij}) + \sum_{i=1}^{n} c_i^2 \right] = 0.$$

Together with the contractivity of T_t, this implies $T_t f \to f$ for any $f \in L_2(K)$.

Now $T_t = e^{-Ht}$, $t \geq 0$, where H is a self-adjoint non-negative operator (the generator of T_t). By the definition,

$$(Hf)(x) = \lim_{t \downarrow 0} t^{-1} \left[f(x) - \int_K f(y) P_t(x, dy) \right],$$

whenever the strong limit exists. In particular, the domain $\mathfrak{D}(H)$ contains $\mathcal{D}(K)$.

Indeed, $\mathcal{D}(K)$ is spanned by indicators of all balls in K. Hence it is sufficient to consider the indicator function $\omega_{z,m}$ of a ball $B_m(z)$, $m \in \mathbb{Z}$, $z \in K$. We have for $x \in B_m(z)$

$$\begin{aligned}(H\omega_{z,m})(x) &= \lim_{t \downarrow 0} t^{-1} [1 - P_t(x, B_m(z))] \\ &= \sum_{i=0}^{\infty} q^{-i} [a(m+i) - q^{-1} a(m+i+1)] = a(m).\end{aligned}$$

If $\operatorname{dist}(x, B_m(z)) = q^{m+\nu}$, $\nu = 1, 2, \ldots$, then

$$\begin{aligned}(H\omega_{z,m})(x) &= q^{-\nu} \left\{ \sum_{i=0}^{\infty} q^{-i} [a(m+\nu+i) - q^{-1} a(m+\nu+i-1)] \right. \\ &\quad \left. - q(q-1)^{-1} a(m+\nu-1) + (q-1)^{-1} a(m+\nu) \right\} \\ &= -q^{-\nu+1} (q-1)^{-1} [a(m+\nu-1) - a(m+\nu)].\end{aligned}$$

It is easy to see that the above limits are the limits in the sense of $L_2(K)$. Thus we have found that

$$(H\omega_{z,m})(x) = \begin{cases} a(m), & \text{if } x \in B_m(z), \\ -q^{-\nu+1}(q-1)^{-1}[a(m+\nu-1) & \\ -a(m+\nu)], & \text{if } \|x - z\| = q^{m+\nu}. \end{cases} \tag{5.42}$$

Due to the linearity the formula (5.42) determines H on $\mathcal{D}(K)$, so that $\mathcal{D}(K) \subset \mathfrak{D}(H)$.

Theorem 5.5. *The operator H is essentially self-adjoint on $\mathcal{D}(K)$.*

Proof. Since H is non-negative, it will be sufficient to prove that for some $\lambda < 0$ the set $(H - \lambda I)\mathcal{D}(K)$ is dense in $L_2(K)$.

Let $f \perp (H - \lambda I)\mathcal{D}(K)$, and $f \neq 0$ on a set of a positive measure. There exist such $z \in K$, $m \in \mathbb{Z}$, that

$$\gamma = \int_{B_m(z)} f(x)\,dx \neq 0.$$

We may assume that f is real, $\gamma > 0$, and

$$\gamma \geq \sup_{\nu \geq 1} \left| \int_{\|x-z\|=q^{m+\nu}} f(x)\,dx \right|.$$

Using (5.42) we find that

$$(f, H\omega_{z,m})_{L_2(K)} = a(m)\gamma + \sum_{\nu=1}^{\infty} \int_{\|x-z\|=q^{m+\nu}} f(x)(H\omega_{z,m})(x)\,dx$$

$$= a(m)\gamma - (q-1)^{-1} \sum_{\nu=1}^{\infty} q^{-\nu+1}[a(m+\nu-1) - a(m+\nu)] \int_{\|x-z\|=q^{m+\nu}} f(x)\,dx$$

$$\geq a(m)\gamma - \gamma \sum_{\nu=1}^{\infty} [a(m+\nu-1) - a(m+\nu)]$$

$$= \gamma \lim_{\nu\to\infty} a(m+\nu) = 0,$$

whence

$$(f, (H - \lambda I)\omega_{z,m})_{L_2(K)} \geq -\lambda\,(f, \omega_{z,m})_{L_2(K)} = -\lambda\gamma > 0$$

for $\lambda < 0$, and we have arrived at a contradiction. ∎

Let us study the spectrum of H. It is clear from (5.42), (5.31), and (5.32), that the indicator function of a ball $B_m(z)$ is an eigenfunction of H if and only if $a(m) = 0$, in which case the eigenvalue is zero. However we shall see that certain linear combinations of the indicator functions give all the eigenfunctions of H.

For any element $z \in K$ with the canonical decomposition $z = \sum \zeta_j \beta^j$, $\zeta_j \in S$, and for any $m \in \mathbb{Z}$ we can write

$$z = \sum_{j=-\nu}^{\infty} \zeta_{-m+j} \beta_{-m+j}, \quad \nu \geq 0.$$

Then the ball $B_m(z)$ is determined by the elements $\zeta_{-m-\nu}, \ldots, \zeta_{-m-1}$, and we shall use the notation $\{\zeta_{-m-\nu}, \ldots, \zeta_{-m-1}\}$ instead of $B_m(z)$.

Theorem 5.6. *For any $m \in \mathbb{Z}$ and $z = \sum_{j=-m-\nu}^{\infty} \zeta_j \beta^j \in K$, there is an eigenvalue λ_m of the operator H given by*

$$\lambda_m = a(m) + (q-1)^{-1}(a(m) - a(m+1)),$$

and a $(q-1)$-dimensional eigenspace spanned by vectors of the form

$$e_{z,m} = \sum_{\zeta_{-m-1} \in S} b_{\zeta_{-m-1}} \omega_{\{\zeta_{-m-\nu}, \ldots, \zeta_{-m-1}\}}, \tag{5.43}$$

where $\sum_{\zeta_{-m-1} \in S} b_{\zeta_{-m-1}} = 0$, and $\omega_{\{\zeta_{-m-\nu}, \ldots, \zeta_{-m-1}\}}$ is the indicator function of the ball $\{\zeta_{-m-\nu}, \ldots, \zeta_{-m-1}\}$.

The linear hull spanned by the functions $e_{z,m}$, $m \in \mathbb{Z}$, $z \in K$, is dense in $L_2(K)$.

Proof. Let us consider the function (5.43). It is evident that $\operatorname{supp} e_{z,m} \subset \{\zeta_{-m-\nu}, \ldots, \zeta_{-m-2}\}$. Note also that the assumption $\sum b_{\zeta_{-m-1}} = 0$ implies the inclusion

$$\operatorname{supp} H e_{z,m} \subset \{\zeta_{-m-\nu}, \ldots, \zeta_{-m-2}\}.$$

Indeed, if $x \notin \{\zeta_{-m-\nu}, \ldots, \zeta_{-m-2}\}$, then

$$\operatorname{dist}(x, \{\zeta_{-m-\nu}, \ldots, \zeta_{-m-2}, \zeta_{-m-1}\}) = q^{m+n+1},$$

where $n \geq 1$ does not depend on ζ_{-m-1}. Now by (5.42)

$$(H e_{z,m})(x) = -q^{-n+1}(q-1)^{-1}[a(m+n) - a(m+n+1)] \sum b_{\zeta_{-m-1}} = 0.$$

This observation enables us to reduce the condition $H e_{z,m} = h e_{z,m}$, $h \in \mathbb{C}$, to the system of algebraic equations

$$\sum_{\zeta \in S} a_{\eta\zeta} b_\zeta = 0, \quad \eta \in S, \tag{5.44}$$

$$\sum_{\zeta \in S} b_\zeta = 0, \tag{5.45}$$

where $a_{\zeta\zeta} = a(m) - h$, $\zeta \in S$,

$$a_{\eta\zeta} = -(q-1)^{-1}[a(m) - a(m+1)], \quad \eta \neq \zeta.$$

By elementary algebra (for the necessary determinant see [98]) we show that the system (5.44)-(5.45) has a non-trivial solution if and only if $h = \lambda_m$, and in this case the system is equivalent to the single equation (5.45). Therefore we obtain a $(q-1)$-dimensional eigenspace corresponding to the eigenvalue λ_m.

In order to prove the second assertion, it is sufficient to show that for any $z \in K$ and $m \in \mathbb{Z}$, the indicator function $\omega_{z,m}$ can be approximated by linear combinations of the eigenfunctions $e_{z,n}$. We shall give a construction of the approximation for $z = 0$ (other cases can be obtained by appropriate shifts).

We have

$$e_{0,n} = \sum_{\zeta_{-n-1} \in S} b_{\zeta_{-n-1}} \omega_{\{\zeta_{-n-1}\}}.$$

We know that the condition (5.45) determines $q-1$ linearly independent vectors $e_{0,n}^k$ $(k = 0, 1, \ldots, q-2)$, which can be obtained by setting

$$e_{0,n}^k = \sum_{\zeta_{-n-1} \in S} b_{\zeta_{-n-1}}^k \omega_{\{\zeta_{-n-1}\}},$$

where the coefficients are defined as follows. Let us enumerate S identifying the values of ζ_{-n-1} with $\zeta = 0, 1, \ldots, q-1$, in such a way that the element from P is identified with $\zeta = 0$, and set

$$b_\zeta^k = \begin{cases} 1, & \zeta = 0, 1, \ldots, k, \\ -(k+1), & \zeta = k+1, \\ 0, & \zeta = k+2, \ldots, q-1, \end{cases} \quad , \quad k = 0, 1, \ldots, q-2.$$

With this choice of b_ζ^k (5.45) is satisfied for each k, and the functions $e_{0,n}^k$ are pairwise orthogonal in $L_2(K)$. We also have

$$\left\| e_{0,n}^k \right\|_{L_2(K)}^2 = q^n[k + 1 + (k+1)^2] = q^n(k+1)(k+2).$$

The theorem will be proved if we show that for any $m \in \mathbb{Z}$

$$q^m = \|\omega_{0,m}\|_{L_2(K)}^2 = \sum_{n=m}^{\infty} \sum_{k=0}^{q-2} \left| \left(\omega_{0,m}, e_{0,n}^k\right)_{L_2(K)} \left\| e_{0,n}^k \right\|_{L_2(K)}^{-1} \right|^2.$$

For $n \geq m$ we have $(\omega_{0,m}, e^k_{0,n})_{L_2(K)} = q^m$ since $B_m(0) \cap \{\zeta_{-n+1}\} = \varnothing$ for $\zeta_{-n-1} \notin P$, and $b^k_0 = 1$ for all k. Hence

$$\sum_{n=m}^{\infty} \sum_{k=0}^{q-2} \left| (\omega_{0,m}, e^k_{0,n})_{L_2(K)} \left\| e^k_{0,n} \right\|^{-1}_{L_2(K)} \right|^2$$
$$= q^m \sum_{n=m}^{\infty} q^{m-n} \sum_{k=0}^{q-2} \frac{1}{(k+1)(k+2)} = q^m (q-1) q^{-1} \sum_{i=0}^{\infty} q^{-i} = q^m.$$

We have used that $\sum\limits_{k=1}^{l} \frac{1}{k(k+1)} = \frac{l}{l+1}$, which can be easily verified by induction. This completes the proof. ■

It follows from Theorem 5.6 that the operator H has a pure point spectrum of infinite multiplicity. Note that an explicit construction of a complete orthogonal system of eigenfunctions is known at present (see Chapter 3) only for the operator D^α, which (as we saw above) corresponds to the case

$$a(m) = \frac{1 - q^{-1}}{1 - q^{-\alpha-1}} q^{-\alpha m}, \quad \alpha > 0.$$

5.7 Recurrence and hitting probabilities

In this section we continue the investigation of the rotation-invariant processes (X_t, P_t) described in Theorem 5.4. We shall give criteria for such a process to be recurrent or to hit a single point with a positive probability.

Theorem 5.7. *The process X_t is recurrent if and only if*

$$\sum_{i=1}^{\infty} \frac{q^{-i}}{a(i)} = \infty. \tag{5.46}$$

Proof. For $m \in \mathbb{Z}$

$$\int_0^{\infty} P_t(x, B_m(x))\, dt = q^{-1}(q-1)^2 \sum_{i=0}^{\infty} q^{-i} [q a(m+i) - a(m+i+1)]^{-1}$$
$$= q^{m-1}(q-1)^2 \sum_{i=m}^{\infty} q^{-i} [q a(i) - a(i+1)]^{-1},$$

which diverges for any m if and only if

$$\sum_{i=1}^{\infty} q^{-i}[qa(i) - a(i+1)]^{-1} = \infty.$$

Since $(q-1)a(i) \le qa(i) - a(i+1) \le qa(i)$, we come to (5.46). ■

Denote

$$\tau_x = \inf\{t > 0 : \ X_t = x\}, \quad x \in K.$$

Let $C^\lambda = C^\lambda\{x\}$ $(\lambda > 0)$ be the λ-capacity of a one-point set. General properties of these objects are summarized in the following well-known proposition, which holds for any spatially homogeneous standard Markov process on K.

Proposition 5.3. (i) *Suppose that for any $t > 0$ $P_t(x, dy)$ has a density $p_t(x, y)$ with respect to the Haar measure. Then $C^\lambda > 0$ if and only if the function*

$$g^\lambda(x) = \int_0^\infty \exp(-\lambda t) p_t(0, x)\, dt$$

is bounded.

(ii) *If for some $\lambda > 0$ $C^\lambda = 0$, then $C^\lambda = 0$ for all $\lambda > 0$ and $\mathbf{P}(\tau_x < \infty) = 0$ for almost all x.*

(iii) *If P_t has a density and if $g^\lambda(x)$ is bounded and continuous, then*

$$\mathbf{E}[\exp(-\lambda \tau_x) : \ \tau_x < \infty] = \frac{g^\lambda(x)}{g^\lambda(0)}$$

for any $x \in K$.

For *the proof* see e.g. [116]. ■

Applying Proposition 5.3 to our process (X_t, P_t) we obtain the following criterion regarding λ-capacities.

Theorem 5.8. *Suppose that $a(m) \to \infty$ as $m \to -\infty$. Then $C^\lambda > 0$ if and only if*

$$\sum_{i=-\infty}^{0} \frac{q^{-i}}{1 + a(i)} < \infty. \tag{5.47}$$

Proof. By Proposition 5.3, $C^\lambda > 0$ if and only if $g^\lambda(x)$ is bounded. If $\|x\| = q^m$, then

$$g^\lambda(x) = (1 - q^{-1}) \sum_{i=m}^{\infty} \frac{q^{-i}}{\lambda + (q-1)^{-1}(qa(i) - a(i+1))} - \frac{q^{-m}}{\lambda + (q-1)^{-1}(qa(m-1) - a(m))}. \quad (5.48)$$

Since the right-hand side tends to zero as $m \to +\infty$, $g^\lambda(x)$ is bounded for large $\|x\|$. Therefore for g^λ to be bounded, it is necessary and sufficient that the right-hand side of (5.48) converges as $m \to -\infty$. It remains to note that

$$(q-1)a(i) \leq qa(i) - a(i+1) \leq qa(i),$$

and we come to the condition (5.47). ■

Let us consider the probabilities of hitting single points.

Theorem 5.9. (i) *If* $\lim_{m \to -\infty} a(m) = W < \infty$, *then*

$$\mathbf{P}(\tau_x < \infty) = 0, \quad x \neq 0,$$
$$\mathbf{P}(\tau_0 < \infty) = 1.$$

(ii) *If* $\lim_{m \to -\infty} a(m) = \infty$ *and* $\sum_{i=-\infty}^{0} \frac{q^{-i}}{1 + a(i)} = \infty$, *then*

$$\mathbf{P}(\tau_x < \infty) = 0$$

for any $x \in K$.

(iii) *If* $\sum_{i=-\infty}^{0} \frac{q^{-i}}{1 + a(i)} < \infty$, *then*

$$\mathbf{P}(\tau_x < \infty) = \lim_{\lambda \downarrow 0} \frac{g^\lambda(x)}{g^\lambda(0)}$$

for any $x \in K$.

Proof. (i) Let $W_t^{(m)}$ be the number of jumps of X_t on the time interval $(0, t]$ with the absolute values exceeding q^m. As we know (see [56]; in fact we have used similar results for the process ξ_α), $W_t^{(m)}$ has a Poisson distribution with the mean $a(m)t$. If $\theta = \inf\{t > 0 : \ X_t \neq 0\}$, then

$$\mathbf{P}(\theta \geq t) = \mathbf{P}(\text{no jumps occur on } (0,t]) = \lim_{m\to\infty} \mathbf{P}(W_t^{(m)} = 0) = \exp(-Wt).$$

Therefore

$$\mathbf{P}(\theta > 0) = \lim_{t\to 0} \exp(-Wt) = 1,$$

so that $\mathbf{P}(\tau_0 < \infty) = 1$. Now let $x \neq 0$. Denote $J_t = X_t - X_{t-0}$. For each $t > 0$, $\|J_s\| > 0$ for only a finite number of values of $s \in (0,t]$, since

$$\mathbf{E}(\text{card}\{s \leq t : \ \|J_s\| > 0\}) = \lim_{m\to -\infty} a(m)t < \infty.$$

Let t_k be the time of the k-th jump. We can write

$$X_t = \sum_{s \leq t} J_s,$$

whence

$$\mathbf{P}(\tau_x < \infty) \leq \sum_{k=1}^{\infty} \mathbf{P}(J_{t_1} + \cdots + J_{t_k} = x). \tag{5.49}$$

On the other hand,

$$\mathbf{P}(J_{t_i} \in \Gamma) = \frac{F(\Gamma \setminus \{0\})}{F(K \setminus \{0\})},$$

and $F|_{K\setminus\{0\}}$ has no atoms (due to the rotation-invariance), so are the distributions of J_{t_i}. Since J_{t_i} are independent,

$$\mathbf{P}(J_{t_1} + \cdots + J_{t_k} = x) = 0$$

for each k. Together with (5.49) this shows that $\mathbf{P}(\tau_x < \infty) = 0$.

(ii) By Proposition 5.3 and Theorem 5.8 we have

$$\mathbf{P}(\tau_x < \infty) = 0 \tag{5.50}$$

for almost all x. On the other hand, for each $m \in \mathbb{Z}$ the sphere $S_m = \{x \in K : \ \|x\| = q^m\}$ has a positive Haar measure, and $P(\tau_x < \infty)$ is the same for all $x \in S_m$, since X_t is rotation-invariant. Hence (5.50) holds for all $x \neq 0$. As we know, in the present case $P_t(x, dy)$ is absolutely continuous, and we can see (just as for processes on $\mathbb{R}^n$; see [67]) that (5.50) holds for $x = 0$ too.

(iii) By our assumption, g^λ is bounded, and we can easily see that it is continuous. By Proposition 5.3

$$\mathbf{E}\left[\exp(-\lambda\tau_x) : \tau_x < \infty\right] = \frac{g^\lambda(x)}{g^\lambda(0)}.$$

Letting $\lambda \to 0$ gives our assertion. ■

Example. Consider the process $\xi_\alpha(t)$, $\alpha > 0$. As we know, in this case $a(m) = c_\alpha q^{-\alpha m}$, $c_\alpha =$ const. We have

$$\lim_{m \to -\infty} a(m) = \infty.$$

The process is recurrent if and only if $\alpha \geq 1$. The capacities C^λ are positive if and only if $\alpha > 1$. If $\alpha \leq 1$, then $\mathbf{P}(\tau_x < \infty) = 0$.

Let us consider the hitting probabilities for the case $\alpha \geq 1$. In our situation

$$p_t(0, x) = \int_K \chi(-x\xi) e^{-t\|\xi\|^\alpha}\, d\xi,$$

so that

$$g^\lambda(x) = \int_K \frac{\chi(-x\xi)}{\lambda + \|\xi\|^\alpha}\, d\xi = \sum_{n=-\infty}^{\infty} \frac{1}{\lambda + q^{\alpha n}} \int_{\|\xi\| = q^n} \chi(-x\xi)\, d\xi.$$

If $\|x\| = q^m$, then

$$\begin{aligned} g^\lambda(x) &= -\frac{q^{-m}}{\lambda + q^{\alpha(1-m)}} + (1 - q^{-1}) \sum_{n=-\infty}^{-m} \frac{q^n}{\lambda + q^{\alpha n}} \\ &= -\frac{\|x\|^{\alpha - 1}}{\lambda \|x\|^\alpha + q^\alpha} + (1 - q^{-1}) \sum_{n=-\infty}^{-\log_q \|x\|} \frac{q^n}{\lambda + q^{\alpha n}}. \end{aligned}$$

For each $\lambda > 0$ the function $g^\lambda(x)$ is bounded and continuous. It is easy to see that $\dfrac{g^\lambda(x)}{g^\lambda(0)} \to 1$ as $\lambda \to 0$, for any $x \in K$. Thus if $\alpha > 1$, then

$$\mathbf{P}(\tau_x < \infty) = 1.$$

5.8 Processes on adeles

In this section we show how to glue processes defined on $\mathbb{Q}_p$ for each prime number p into a process on the adele ring $\mathbb{A}$. Suppose that we have, for each p (including $p=\infty$, for which we identify $\mathbb{Q}_\infty$ with $\mathbb{R}$, and $\mathbb{Z}_\infty$ with $[-1,1]$) a non-negative self-adjoint operator H_p on $L_2(\mathbb{Q}_p)$, and the heat kernel h_p, the integral kernel of e^{-tH_p} satisfying the properties:

$$h_p(t,x,y) \geq 0, \quad \int_{\mathbb{Q}_p} h_p(t,x,y)\,dy = 1. \tag{5.51}$$

Suppose further that we are given a basepoint $b=(b_\infty, b_2, \dots, b_p, \dots) \in \mathbb{A}$, and there is a finite set Σ_0 of primes such that

$$\prod_{p\notin\Sigma_0} \int_{\mathbb{Z}_p} h_p(t,x,b_p)\,dx > 0 \quad \text{for all } t>0. \tag{5.52}$$

Let $\mathcal{A}_1$ be the set of those subsets of $\mathbb{A}$, which are finite unions of sets of the form $U^X_{\Sigma,\sigma}$, where Σ is a finite set of primes, X is a measurable subset of $\prod\limits_{p\in\Sigma} \mathbb{Q}_p$, $\sigma \in \{0,+,-\}$, and

$$U^X_{\Sigma,0} = X \times G^0_\Sigma, \quad U^X_{\Sigma,+} = X \times G^+_\Sigma, \quad U^X_{\Sigma,-} = X \times \left(G^0_\Sigma \setminus G^+_\Sigma\right).$$

Here $G^+_\Sigma = \prod\limits_{p\notin\Sigma} \mathbb{Z}_p$, while G^0_Σ is a restricted direct product of $\mathbb{Q}_p$ with respect to $\mathbb{Z}_p$ for $p\notin\Sigma$ (that is $U^X_{\Sigma,0}$ is the set of all adeles whose system of Σ-coordinates belongs to X). Note that $\mathbb{A} = U^{\varnothing}_{\varnothing,0} \in \mathcal{A}_1$. The sets $U^X_{\Sigma,+}$ form the basis of the topology in $\mathbb{A}$, and also generate the Borel σ-algebra $\mathcal{B}(\mathbb{A})$.

We check that $\mathcal{A}_1$ is closed under finite intersections and complements, so that $\mathcal{A}_1$ is an algebra. For complements, we have

$$\left(U^X_{\Sigma,0}\right)^c = U^{(X^c)}_{\Sigma,0}, \quad \left(U^X_{\Sigma,+}\right)^c = U^X_{\Sigma,-} \cup U^{(X^c)}_{\Sigma,0}, \quad \left(U^X_{\Sigma,-}\right)^c = U^X_{\Sigma,+} \cup U^{(X^c)}_{\Sigma,0}.$$

Considering intersections we observe first that if R is a finite set of primes, $R\cap\Sigma=\varnothing$, then

$$U^X_{\Sigma,0} = U^{X\times\mathbb{Q}_R}_{\Sigma\cup R,+} \bigcup U^{X\times\mathbb{Q}_R}_{\Sigma\cup R,-}, \quad U^X_{\Sigma,+} = U^{X\times\mathbb{Z}_R}_{\Sigma\cup R,+},$$
$$U^X_{\Sigma,-} = U^{X\times(\mathbb{Q}_R\setminus\mathbb{Z}_R)}_{\Sigma\cup R,+} \bigcup U^{X\times\mathbb{Q}_R}_{\Sigma\cup R,-},$$

where we used the notation

$$\mathbb{Q}_R = \prod_{p \in R} \mathbb{Q}_p, \quad \mathbb{Z}_R = \prod_{p \in R} \mathbb{Z}_p.$$

Hence it is sufficient to consider intersections of the form $U^X_{\Sigma_1,\sigma_1} \cap U^X_{\Sigma_2,\sigma_2}$. We have

$$\begin{aligned} U^X_{\Sigma,0} \cap U^Y_{\Sigma,0} &= U^{X\cap Y}_{\Sigma,0}, & U^X_{\Sigma,+} \cap U^Y_{\Sigma,+} &= U^{X\cap Y}_{\Sigma,+}, \\ U^X_{\Sigma,-} \cap U^Y_{\Sigma,-} &= U^{X\cap Y}_{\Sigma,-}, & U^X_{\Sigma,+} \cap U^Y_{\Sigma,0} &= U^{X\cap Y}_{\Sigma,+}, \\ U^X_{\Sigma,-} \cap U^Y_{\Sigma,0} &= U^{X\cap Y}_{\Sigma,-}, & U^X_{\Sigma,+} \cap U^Y_{\Sigma,-} &= \varnothing. \end{aligned}$$

Let $\mathcal{A}_n$ be the set of subsets of $\mathbb{A}_n$, which are finite unions of products of sets from $\mathcal{A}_1$. It is easy to check that $\mathcal{A}_n$ is an algebra. Let $\tau = \{t_1, \dots, t_n\} \subset \mathbb{R}$, $0 < t_1 < \dots < t_n$. We define a non-negative function μ_τ on $\mathcal{A}_n$, initially on sets of the form

$$A = A^{\{X_i\}}_{\Sigma,\sigma} = \prod_{i=1}^n U^{X_i}_{\Sigma,\sigma(t_i)}, \quad \sigma : \tau \to \{0, +, -\},$$

and then extend it by finite additivity.

Namely, let $\mu_\tau(A) = I^{\{X_i\}}_{\Sigma,\tau} J^\sigma_{\Sigma,\tau}$, where

$$I^{\{X_i\}}_{\Sigma,\tau} = \int \cdots \int_{\prod_{i=1}^n X_i} h_\Sigma(t_n - t_{n-1}, x_n, x_{n-1}) \dots h_\Sigma(t_2 - t_1, x_2, x_1) \times h_\Sigma(t_1, x_1, b_\Sigma) d^\Sigma x_n \dots d^\Sigma x_1,$$

with

$$h_\Sigma(t, x, y) = \prod_{p \in \Sigma} h_p(t, x_p, y_p),$$

$b_\Sigma = (b_p)_{p\in\Sigma}$, and $J^\sigma_{\Sigma,\tau}$ is defined by induction on the number of those $t_i \in \tau$, for which $\sigma(t_i) =$ "$-$", as follows:

(0) If τ is empty, then $J^\sigma_{\Sigma,\tau} = 1$.

(1) If $\sigma(t_i) =$ "$+$" for all $t_i \in \tau$, then

$$J^\sigma_{\Sigma,\tau} = \prod_{p \notin \Sigma} \int \cdots \int_{\prod_{i=1}^n \mathbb{Z}_p} h_p(t_n - t_{n-1}, x_n, x_{n-1}) \dots h_p(t_2 - t_1, x_2, x_1) \times h_p(t_1, x_1, b_p)\, dx_n \dots dx_1.$$

(2) If $\sigma(t_i) \neq$ "$-$" for all $t_i \in \tau$, then $J^{\sigma}_{\Sigma,\tau} = J^{\sigma'}_{\Sigma,\tau'}$, where $\tau' = \{t_i \in \tau : \sigma(t_i) = \text{"+"}\}$, and σ' is the restriction of σ to τ'.

(3) Otherwise choose a $t_i \in \tau$, for which $\sigma(t_i) =$ "$-$", and let σ' and σ'' be identical with σ except that $\sigma'(t_i) = 0$ and $\sigma''(t_i) =$ "+". Then

$$J^{\sigma}_{\Sigma,\tau} = J^{\sigma'}_{\Sigma,\tau'} - J^{\sigma''}_{\Sigma,\tau''},$$

and the expressions in the right-hand side are already defined by the induction process.

Of course, the steps (2) and (3) are based on the equality in (5.51).

Let us extend the measures μ_τ to the Borel σ-algebra $\mathcal{B}(\mathbb{A}^n)$.

Theorem 5.10. *The measures μ_τ are finite-dimensional distributions of a Markov process on $\mathbb{A}$.*

Proof. It is easy to see that the measures in question form a consistent family. It remains to use an appropriate version of the Kolmogorov theorem.

Recall that a topological space X is called a topological Radon space [22] if every σ-additive measure on $(X, \mathcal{B}(X))$ is a Radon measure. In order to construct a Markov process on $\mathbb{A}$ from a given family of its finite-dimensional distributions, it is sufficient [22] to show that $\mathbb{A}$ is a topological Radon space.

Indeed, $\mathbb{A}$ (as a topological space) is an inductive limit of the increasing sequence

$$\mathbb{A}_n = \underbrace{\mathbb{R} \times \mathbb{Q}_2 \times \ldots \times \mathbb{Q}_p}_{n \text{ factors}} \times \mathbb{Z}_{p'} \times \ldots$$

(p' is the smallest prime larger than p). $\mathbb{A}_n$ is a topological Radon space as a countable product of complete separable metric spaces. For each n the topology in $\mathbb{A}_n$ coincides with the one induced from $\mathbb{A}_{n+1}$, and $\mathbb{A}_n$ is closed in $\mathbb{A}_{n+1}$. These properties imply [22] that $\mathbb{A}$ is a topological Radon space, as desired. ■

Example. Suppose that for $p \geq p_0$, where p_0 is some prime, we have $H_p = a_p D^\alpha$, $\alpha > 0$, and $a_p > 0$ are certain renormalization constants.

Fixing the base point $b = 0$, we find that for $p \geq p_0$

$$\int_{\mathbb{Z}_p} h_p(t,x,0)\,dx = \int_{|\xi|_p \leq 1} e^{-t|\xi|_p^\alpha}\,d\xi$$

$$= (1-p^{-1})\sum_{n=0}^{\infty} p^{-n} e^{-ta_p p^{-\alpha n}} \geq (1-p^{-1})\sum_{n=0}^{\infty} p^{n}\left(1 - ta_p \cdot p^{-\alpha n}\right)$$

$$= 1 - ta_p \frac{1-p^{-1}}{1-p^{-1-\alpha}}.$$

If

$$\sum_p a_p < \infty, \tag{5.53}$$

then the condition (5.52) is satisfied. Note that (5.53) holds if for large p

$$a_p \leq \frac{C}{p(\log\log p)^s}, \quad s > 1$$

(see [118]).

5.9 General stable distributions

5.9.1. Normalized sums. The probability distributions $G_{a,\alpha}$ on a local field K having the functions $g_{a,\alpha}(t) = \exp(-a\|t\|^\alpha)$, $a > 0$, $\alpha > 0$, as their Fourier transforms (characteristic functions) are naturally seen as non-Archimedean counterparts of stable distributions. So far it was only a formal resemblance, and it is our task now to interpret them as a part of some class of probability distributions having an intrinsic description. Note that the classical definitions in terms of certain functional equations (as well as the generalizations proposed in [147]) do not make sense in our situation, since the absolute value $\|\cdot\|$ can equal only an integer power of q.

We shall take a different option – the required class of probability distributions will consist of weak limits of normalized sums

$$S_n = B_n^{-1}\left(X_1 + \cdots + X_{k(n)}\right), \tag{5.54}$$

where $X_1, \ldots, X_n$ are independent identically distributed random variables with values in K, $\{k(n)\}$ is an increasing sequence of natural numbers, $\{B_n\}$ is a sequence of elements from K, $\|B_n\| \to \infty$.

The limit behaviour of (5.54) depends crucially on the behaviour of the sequence

$$\rho_n = \frac{k(n)}{k(n+1)}, \quad n = 1, 2, \dots .$$

Passing to a subsequence we may assume that $\rho_n \to \lambda$, $0 \le \lambda \le 1$. It will be shown below that analogs of stable distributions emerge when $0 < \lambda < 1$. We shall also show (for $K = \mathbb{Q}_p$) that the extreme values $\lambda = 0$, $\lambda = 1$ (as in the classical theory for real-valued random variables, where $k(n) = n$), and also the case of a bounded sequence $\{B_n\}$, lead to degenerate (delta) distributions, cutoffs of the Haar measure etc.

5.9.2. Probability distributions of stable type. It is well known (see [114]) that the Fourier transform of an infinitely divisible probability distribution μ on K (as on any locally compact Abelian group) has the form

$$\widehat{\mu}(t) = \chi(t\xi)\Omega_N(t) \exp \int\limits_{K\setminus\{0\}} (\chi(tx) - 1)\Phi(dx), \quad t \in K, \tag{5.55}$$

where

$$\Omega_N(t) = \begin{cases} 1, & \text{if } \|t\| \le q^N, \\ 0, & \text{if } \|t\| > q^N, \end{cases}$$

$N \in \mathbb{Z} \cup \{\infty\}$ ($\Omega_\infty(t) \equiv 1$), χ is a rank zero additive character on K, Φ is a Borel measure on $K \setminus \{0\}$, which is finite on the complement of any neighbourhood of zero. Formula (5.55) differs from a similar formula for $\mathbb{R}$ in two respects - possible presence of the factor $\Omega_N(t)$ (thus $\widehat{\mu}$ may vanish on an open set), and non-uniqueness of the Lévy measure Φ (this non-uniqueness is typical for locally compact Abelian groups possessing compact subgroups; for example, Φ is non-unique for $K = \mathbb{Q}_p$, $p \neq 2$). However, Φ can be uniquely determined if the integral under the exp is given.

Lemma 5.5. *If*

$$\varphi(t) = \int\limits_{K\setminus\{0\}} (\chi(tx) - 1)\Phi(dx), \quad t \in Q_p\ , \tag{5.56}$$

then for any open compact subset $M \subset K \setminus \{0\}$

$$\Phi(M) = \int\limits_{Q_p} \varphi(y) m(y)\, dy \tag{5.57}$$

where m is an inverse Fourier transform of the indicator function ω_M of the set M.

Proof. We have

$$\omega_M(x) = \int_K \chi(xy)m(y)\,dy\ , \quad x \in K\ ,$$

whence $m \in \mathcal{D}(K)$ and

$$\int_K m(y)\,dy = \omega_M(0) = 0. \tag{5.58}$$

Using (5.56) and (5.58) we obtain that

$$\int_{K\setminus\{0\}} \omega_M(x)\Phi(dx) = \int_{K\setminus\{0\}} \Phi(dx) \int_K (\chi(xy) - 1)m(y)\,dy = \int_K \varphi(y)m(y)\,dy$$

which is equivalent to (5.57). Our use of the Fubini theorem was based on the fact that supp $m \subset \{y \in K\ :\ \|y\| \le q^l\}$ for some $l \in \mathbb{Z}$, and $\chi(xy) - 1 = 0$ for $\|y\| \le q^l$, $\|x\| \le q^{-l}$ while the measure Φ is finite on the set $\{x \in K\ :\ \|x\| > q^{-l}\}$. ■

Let us consider the normalized sums (5.54) with $\|B_n\| \to \infty$ and $\rho_n \to \lambda$, $0 \le \lambda \le 1$. Let $\gamma_n = \frac{B_n}{B_{n+1}}$. Since $\|B_n\| \to \infty$, there exists a subsequence $\{\gamma_{n_l}\}$ for which $\|\gamma_{n_l}\| \le q^{-1}$. We may assume (passing if necessary to a subsequence once more) that $\gamma_n \to \gamma_0$ in K, $\|\gamma_0\| \le q^{-1}$.

Suppose that the distributions F_n of the normalized sums S_n converge weakly, $F_n \to G$, and $g(t)$ is a characteristic function of G. Let $f(t)$ be the characteristic function of each of the (independent, identically distributed) random variables X_n. Then

$$\left(f\left(\frac{t}{B_n}\right)\right)^{k(n)} \to g(t)\ , \quad n \to \infty\ , \tag{5.59}$$

uniformly on compact subsets of K. The left-hand side of (5.59) will be denoted $f_n(t)$.

Proposition 5.4. (i) *If $\lambda \ne 0$ then uniformly on compact subsets of K*

$$|f_n(\gamma_n t)| \to |g(t)|^\lambda. \tag{5.60}$$

If $\lambda = 0$, this relation holds for those t where $g(t) \ne 0$.

(ii) *The identity*

$$|g(\gamma_0 t)| = |g(t)|^\lambda \tag{5.61}$$

is valid for any $t \in K$, if $\lambda \neq 0$, and for any t with $g(t) \neq 0$ if $\lambda = 0$.

Proof. Let us consider a random variable

$$S'_n = B_{n+1}^{-1}\left(X_1 + \cdots + X_{k(n)}\right).$$

Its characteristic function equals

$$\left(f\left(\frac{t}{B_{n+1}}\right)\right)^{k(n)} = f_n(\gamma_n t). \tag{5.62}$$

On the other hand,

$$\left|f\left(\frac{t}{B_{n+1}}\right)\right|^{k(n)} = |f_{n+1}(t)|^{\rho_n},$$

so that (5.59) and (5.62) imply (5.60).

Given $\varepsilon > 0$, we find, for any fixed $t \in K$, such n_0 that

$$|f_n(\gamma_n t) - g(\gamma_n t)| < \varepsilon \quad \text{if} \quad n \geq n_0$$

(since the sequence $\{\gamma_n t\}_{n \geq 0}$ is pre-compact). Thus $f_n(\gamma_n t) \to g(\gamma_0 t)$ by continuity of g, and (5.61) follows from (5.60). ■

Corollary 5.2. *If $\lambda \neq 0$ then $g(t) \neq 0$ for any $t \in K$.*

Proof. Suppose that $g(t_0) = 0$ for some $t_0 \in K$. By (5.61) we find that

$$|g(t_0)| = |g(\gamma_0 t_0)|^{1/\lambda} = |g(\gamma_0^2 t_0)|^{1/\lambda^2} = \ldots = |g(\gamma_0^n t_0)|^{1/\lambda^n}.$$

Since $\gamma_0^n \to 0$ and $g(0) = 1$, we obtain that $g(\gamma_0^n t_0) \neq 0$ for a certain n, so that we come to a contradiction. ■

Now we consider the probability distributions on K seen as counterparts of classical stable distributions.

Theorem 5.11. (i) *Let Φ be a Borel measure on $K \setminus \{0\}$, which is finite outside any neighbourhood of zero and satisfies the relation*

$$\Phi(M) = \lambda \Phi(\gamma_0 M), \tag{5.63}$$

with $0 < \lambda < 1$, $\gamma_0 \in K$, $0 \neq \|\gamma_0\| \leq q^{-1}$, *for any compact open subset* $M \subset K \setminus \{0\}$. *Then a function* $g(t)$ *of the form*

$$g(t) = \exp \int\limits_{K\setminus\{0\}} (\chi(ty) - 1)\Phi(dy) \tag{5.64}$$

is a characteristic function of a distribution which is a weak limit of some sequence (5.54) with $\rho_n \to \lambda$, $\gamma_n \to \gamma_0$.

(ii) *If the distributions* F_n *for a sequence (5.54) with independent symmetric identically distributed random variables* X_n, $\|B_n\| \to \infty$, $\gamma_n \to \gamma_0$, $0 \neq \|\gamma_0\| \leq q^{-1}$, $\rho_n \to \lambda$, $0 < \lambda < 1$, *converge weakly to a distribution* G, *then its characteristic function is of the form (5.64) where the Lévy measure* Φ *is symmetric and satisfies (5.63).*

Proof. (i) By [114], the function (5.64) is a characteristic function corresponding to a random variable X. Let $X_1, X_2, \ldots$ be independent copies of X. Set $B_n = \gamma_0^{-n}$, $k(n) = [\lambda^{-n}]$ where $[\cdot]$ means the integer part. Then

$$\begin{aligned} f_n(t) &= (g((\gamma_0^n t))^{[\lambda^{-n}]} \\ &= \exp\left\{ \lambda^n \left[\lambda^{-n}\right] \int\limits_{K\setminus\{0\}} (\chi(ty) - 1)\Phi(dy) \right\} \longrightarrow g(t) \ , \quad n \to \infty, \end{aligned}$$

uniformly on compact subsets, since $[\lambda^{-n}] = \lambda^{-n} + O(1)$, $n \to \infty$.

(ii) Let us proceed from the relation (5.59). Since f is real-valued (due to the symmetry of F), continuous, and $f(0) = 1$, we see that $f_n(t) > 0$ for each fixed t, if n is large enough. Hence, $g(t) \geq 0$, and by Corollary 5.2, $g(t) > 0$ for all $t \in K$. The sequence $\{\log f_n(t)\}$ is bounded, uniformly with respect to t from any compact subset of K. Thus

$$\begin{aligned} k(n)\left((f_n(t))^{1/k(n)} - 1\right) &= k(n)\left(\exp\left(\frac{1}{k(n)}\log f_n(t)\right) - 1\right) \\ &= \log f_n(t) + O\left(\frac{1}{k(n)}\right) \ , \quad n \to \infty, \end{aligned}$$

so that

$$k(n)\left((f_n(t))^{1/k(n)} - 1\right) \longrightarrow \log g(t) \ , \quad n \to \infty, \tag{5.65}$$

uniformly on compact subsets of K.

Introducing the measures

$$\Phi_n(dy) = k(n)F(d(B_n y)) \tag{5.66}$$

where F is the distribution of each random variable $X_1, \dots, X_n, \dots$, we may rewrite (5.65) as

$$\int\limits_{K\setminus\{0\}} (\chi(ty) - 1)\Phi_n(dy) \longrightarrow \log g(t). \tag{5.67}$$

Note that for each $t \neq 0$ the integral in the left-hand side of (5.67) is actually taken over the set of those y for which $\|y\| > \|t\|^{-1}$.

Denote the left-hand side of (5.67) by $\varphi_n(t)$. If M is a compact open subset of $K \setminus \{0\}$, and m is an inverse Fourier transform of the indicator ω_M, then by Lemma 5.5

$$\Phi_n(M) = \int\limits_K \varphi_n(t)m(t)\,dt\ , \tag{5.68}$$

and by virtue of (5.67)

$$\int\limits_{K\setminus\{0\}} \omega_M(x)\Phi_n(dx) \longrightarrow \int\limits_K m(t)\log g(t)\,dt\ , \quad n \to \infty. \tag{5.69}$$

It follows from (5.69) that the sequence

$$\int\limits_{K\setminus\{0\}} \omega(x)\Phi_n(dx)$$

converges for any locally constant function ω with a compact support not containing the origin. Every continuous function with a compact support on $K \setminus \{0\}$ can be approximated uniformly by such functions (see [154]). By (5.68), the sequence of measures $\{\Phi_n\}$ is bounded on compact subsets of $K \setminus \{0\}$. This means that $\{\Phi_n\}$ is a Cauchy sequence with respect to the vague topology [56], which is sequentially complete. Thus Φ_n is vaguely convergent to a symmetric Radon measure Φ on $K \setminus \{0\}$.

Now, in order to prove the representation (5.64), it is sufficient to show that $\Phi_n \to \Phi$ in the weak sense on each set $M_{i,\infty} = \{x \in K : \|x\| > q^i\}$, $i \in \mathbb{Z}$. By Theorem 1.1.9 of [56], that will be proved if we show that

$$\Phi_n(M_{i,\infty}) \longrightarrow \Phi(M_{i,\infty}) < \infty\ , \quad n \to \infty. \tag{5.70}$$

Simultaneously (5.70) would imply the required finiteness of Φ outside any neighbourhood of the origin.

Consider the set

$$M_{l,i} = \{x \in K : \; q^{i+1} \le \|x\| \le q^l\} \; , \quad l > i.$$

Let us compute $\Phi_n(M_{i,l})$ using (5.68), where $m(t)$ corresponds to the set $M_{i,l}$. This set is a set-theoretic difference of two balls. We have $m(t) = m_l(t) - m_i(t)$, where

$$m_j(t) = \begin{cases} 1, & \text{if } \|t\| \le q^{-j}, \\ 0, & \text{if } \|t\| > q^{-j}, \end{cases}$$

and we obtain that

$$\Phi_n(M_{i,l}) = - \int\limits_{q^{-l+1} \le \|t\| \le q^{-i}} \varphi_n(t)\, dt \tag{5.71}$$

whence

$$\Phi_n(M_{i,\infty}) = - \int\limits_{\|t\| \le q^{-i}} \varphi_n(t)\, dt \tag{5.72}$$

It follows from (5.67) and (5.71) that

$$\Phi(M_{i,l}) = - \int\limits_{q^{-l+1} \le \|t\| \le q^{-i}} \log g(t)\, dt \; .$$

This yields

$$\Phi(M_{i,\infty}) = - \int\limits_{\|t\| \le q^{-i}} \log g(t)\, dt \; .$$

Comparing with (5.72) we come to (5.70).

It remains to prove the relation (5.63). As we have seen, $\varphi_n(t) \sim \log f_n(t)$, $n \to \infty$, uniformly on compact subsets of K. Thus by Proposition 5.4,

$$\varphi_n(\gamma_n t) \longrightarrow \lambda \log g(t) \; , \quad n \to \infty, \tag{5.73}$$

uniformly on compact subsets.

Let M be a compact open subset of $K \setminus \{0\}$. Then

$$\Phi(\gamma_0^{-1} M) = \lim_{n \to \infty} \Phi_n(\gamma_n^{-1} M). \tag{5.74}$$

Indeed, let ω_n be an indicator of the set $\gamma_n^{-1}M$, $n = 0, 1, 2, \ldots$. Writing the action of a measure as a functional we get

$$\langle \Phi_n \,,\, \omega_n \rangle - \langle \Phi \,,\, \omega_0 \rangle = \langle \Phi_n - \Phi \,,\, \omega_n \rangle + \langle \Phi \,,\, \omega_n - \omega_0 \rangle. \tag{5.75}$$

For large n $\gamma_n^{-1}M \subset M_{l',l''}$ where l', l'' are certain fixed numbers. As above, this means that the supports of the inverse Fourier transforms $\widetilde{\omega}_n$ of all ω_n lie in a certain compact set N, so that

$$|\langle \Phi_n - \Phi \,,\, \omega_n \rangle| \le \int_N |\varphi_n(t) - \log g(t)|\, dt \longrightarrow 0$$

due to the uniform convergence. The second summand on the right in (5.75) tends to zero due to the dominated convergence theorem, so (5.74) has been proved.

Next, by (5.68)

$$\Phi_n(\gamma_n^{-1}M) = \int_K \varphi_n(t)\widetilde{\omega}_n(t)\, dt \,,$$

$$\widetilde{\omega}_n(t) = \int_{\gamma_n^{-1}M} \chi(-ty)\, dy = |\gamma_n|^{-1} m(\gamma_n^{-1}t) \,,$$

whence

$$\Phi_n(\gamma_n^{-1}M) = \int_K \varphi_n(\gamma_n t) m(t)\, dt \,.$$

Now it follows from (5.73), (5.74) and (5.69) that

$$\Phi(\gamma_0^{-1}M) = \lambda \int_K m(t) \log g(t)\, dt = \lambda \Phi(M),$$

which is equivalent to (5.63). ■

Example. If $\|\gamma_0\| = q^{-1}$ then the relation (5.63) means that Φ is determined by its restriction to the group of units $U = \{x \in K : \|x\| = 1\}$.

For a particular example, let the above restriction be proportional to the restriction of the Haar measure:

$$\Phi(M_0) = a \frac{q^\alpha - 1}{1 - q^{-\alpha - 1}} \int_{M_0} dx \,, \quad \alpha > 0, \ a > 0,$$

for any open and closed subset $M_0 \subset U$. Suppose that $\lambda = q^{-\alpha}$, $k(n) = [q^{\alpha n}]$, $\gamma_0 = \beta$.

If M is a compact open subset of $K \setminus \{0\}$ then it may be written as a finite union

$$M = \bigcup_N (M \cap S_N)\ , \quad S_N = \{x \in K : \ \|x\| = q^N\}.$$

In accordance with (5.63),

$$\Phi(M) = \sum_N \Phi(M \cap S_N) = \sum_N q^{-\alpha N} \Phi(\beta^N M \cap U)$$
$$= \sum_N q^{-\alpha N} \int_{\beta^N M \cap U} dx = \sum_N q^{-(\alpha+1)N} \int_{M \cap \{\|y\| = q^N\}} dy,$$

so that

$$\Phi(M) = a \frac{q^\alpha - 1}{1 - q^{-\alpha-1}} \int_M \|x\|^{-\alpha-1}\, dx.$$

Now the identity

$$\|t\|^\alpha = \frac{1 - q^\alpha}{1 - q^{-\alpha-1}} \int_K \|s\|^{-\alpha-1} (\chi(st) - 1)\, ds$$

(Proposition 2.3) shows that in this case the limit characteristic function $g(t)$ coincides with the function $g_{a,\alpha}(t)$ mentioned in the beginning of this section.

Let us consider conditions for the weak convergence of the sequence (5.54) with $\|B_n\| \to \infty$.

Theorem 5.12. *In order that the sequence (5.54) be weakly convergent, it is sufficient that the measures (5.66) converge weakly on each set $M_{i,\infty}$, $i \in \mathbb{Z}$, to a measure Φ on $K \setminus \{0\}$, finite outside any neighbourhood of zero. If the random variables $X_1, X_2, \ldots$ are symmetric and $\beta \neq 0$, this condition is also necessary.*

Proof. The necessity was proved in the course of proving Theorem 5.11. To prove the sufficiency, write $f_n(t)$ in the form

$$f_n(t) = \left(1 + \frac{1}{k(n)} \int_{K \setminus \{0\}} (\chi(ty) - 1) \Phi_n(dy) \right)^{k(n)}.$$

For every fixed t we have

$$\int\limits_{K\setminus\{0\}} (\chi(ty)-1)\Phi_n(dy) \longrightarrow \int\limits_{K\setminus\{0\}} (\chi(ty)-1)\Phi(dy)$$

since $\chi(ty)-1=0$ when $\|y\|\le\|t\|^{-1}$. Recalling that $\left(1+\frac{z}{k(n)}\right)^{k(n)} \to e^z$ uniformly on compact sets, we find that

$$f_n(t) \longrightarrow \exp\left(\int\limits_{K\setminus\{0\}} (\chi(ty)-1)\Phi(dy)\right), \quad n\to\infty. \tag{5.76}$$

Since the function in the right-hand side of (5.76) is continuous, Theorem 3.3.1 of [49] implies weak convergence of S_n. ■

5.9.3. Degenerate cases. Considering degenerate cases we assume that $K=\mathbb{Q}_p$. We begin with the following lemma.

Lemma 5.6. *Let μ be a probability measure on the Borel σ-algebra of $\mathbb{Q}_p$. If $|\widehat{\mu}(t_0)|=1$ for some $t_0\in\mathbb{Q}_p$, $t_0\neq 0$, then $\widehat{\mu}$ is a locally constant function. If $|\widehat{\mu}(t)|$ takes only two values, 0 and 1, then there exists $\xi\in\mathbb{Q}_p$, $N\in\mathbb{Z}$ such that*

$$\widehat{\mu}(t)=\chi_p(t\xi)\Omega_N(t) \tag{5.77}$$

where

$$\Omega_N(t)=\begin{cases}1, & \text{if } |t|_p\le p^N,\\ 0, & \text{if } |t|_p>p^N.\end{cases}$$

In this case $\mu(dx)=p^N\Omega_{-N}(x-\xi)dx$. If $|\widehat{\mu}(t)|\equiv 1$ then $\widehat{\mu}(t)=\chi_p(t\xi)$, $\mu=\delta_\xi$, $\xi\in\mathbb{Q}_p$.

Proof. Let $t_0\in\mathbb{Q}_p$ be such that $t_0\neq 0$, $|\widehat{\mu}(t_0)|=1$. Denote by R_p the set of rational numbers of the form $p^{-n}\left(a_0+a_1p+\cdots+a_{n-1}p^{n-1}\right)$, $n\ge 1$; $a_0,\dots,a_{n-1}\in\{0,1,\dots,p-1\}$, $a_0\neq 0$.

Suppose that $\widehat{\mu}(t_0)=e^{2\pi ir}$, $0\le r<1$. Then by the definition of χ_p

$$e^{2\pi ir}=\int\limits_{\mathbb{Q}_p}\exp(2\pi i\{t_0x\}_p)\mu(dx)$$

whence

$$\int\limits_{\mathbb{Q}_p}(1-\exp(2\pi i\{t_0x\}_p-r)))\mu(dx)=0.$$

In particular,

$$\int_{\mathbb{Q}_p} (1 - \cos 2\pi(\{t_0 x\}_p - r))\mu(dx) = 0.$$

This means that either $r \in R_p$ or $r = 0$.

In both cases there exists $\xi \in \mathbb{Q}_p$ such that $r = \{t_0\xi\}_p$, that is $\widehat{\mu}(t_0) = \chi_p(t_0\xi)$. As above, we obtain that

$$\int_{\mathbb{Q}_p} (1 - \cos 2\pi(\{t_0(x - \xi)\}_p)\mu(dx) = 0,$$

so that μ is concentrated on the set of those x for which $\{t_0(x - \xi)\}_p = 0$, that is on the set $\{x \in \mathbb{Q}_p : \ |x - \xi|_p \leq |t_0|_p^{-1}\}$.

Now

$$\widehat{\mu}(t) = \int_{|x-\xi|_p \leq |t_0|_p^{-1}} \chi_p(tx)\mu(dx) = \chi_p(t\xi) \int_{|x-\xi|_p \leq |t_0|_p^{-1}} \chi_p(t(x - \xi))\mu(dx)$$

so that $\widehat{\mu}$ is locally constant and $\widehat{\mu}(t) = \chi_p(t\xi)$ if $|t|_p \leq |t_0|_p$.

Suppose that $|\widehat{\mu}(t)| = 0$ or 1. If the set $\{t \in \mathbb{Q}_p \ : \ |\widehat{\mu}(t)| = 1\}$ is unbounded then it coincides with $\mathbb{Q}_p$, and in this case $\widehat{\mu}(t) = \chi_p(t\xi)$, $\mu = \delta_\xi$. Otherwise we come to (5.77). The expression for μ follows from the integration formula (1.26). ■

Let us consider the "extreme" cases $\lambda = 0$ and $\lambda = 1$ in the weakly convergent scheme (5.54).

Proposition 5.5. *If $\lambda = 1$ then G is degenerate. If $\lambda = 0$ then either G is degenerate or its characteristic function g has a compact support. If $\lambda = 0$ and $|\gamma_0|_p = p^{-1}$ then $g(t)$ coincides with the right-hand side of (5.77).*

Proof. Let $\lambda = 1$. As before, we may assume that $\gamma_n \to \gamma_0$, $|\gamma_0|_p \leq p^{-1}$. By Proposition 1, we have

$$|g(\gamma_0 t)| = |g(t)|$$

for any $t \in \mathbb{Q}_p$, so that

$$|g(t)| = |g(\gamma_0^n t)| \ , \quad n = 1, 2, \ldots .$$

Since $\gamma_0^n \to 0$ for $n \to \infty$, g is continuous and $g(0) = 1$, this implies the identity $|g(t)| \equiv 1$. Hence, by Lemma 5.6 G is degenerate.

If $\lambda = 0$ then the same reasoning shows that $|g(\gamma_0 t)| = 1$ as soon as $g(t) \neq 0$. Thus if $g(t) \neq 0$ for all t then G is degenerate. Otherwise g

has a compact support in accordance with (5.55). The last assertion of the proposition follows from Lemma 5.6. ■

Finally, let us consider the case of a weakly convergent sequence (5.54) with $|B_n|_p \not\to \infty$. Then there exists a subsequence $B_{n_l} \to b \in \mathbb{Q}_p$.

Proposition 5.6. *If $b \neq 0$ then g coincides with the right-hand side of (5.77). If $b = 0$ then G is degenerate.*

Proof. Let $b \neq 0$. It follows from (5.59) that the inequality $|f(b^{-1}t)| < 1$ implies the equality $g(t) = 0$. If $|f(b^{-1}t)| = 1$ for some t then Lemma 5.6 shows that f is locally constant. Thus $|f(s)| = 1$ when s belongs to a certain neighbourhood of the point $b^{-1}t$; in particular, $|f(B_{n_l}^{-1}t)| = 1$ for large l, whence $|g(t)| = 1$. It remains to use Lemma 5.6.

Let $b = 0$. Take, for an arbitrary $s \in \mathbb{Q}_p$, a compact set C containing the subsequence $\{B_{n_l}s\}$. Given $\varepsilon > 0$, we find such a natural number l_0 that for $l \geq l_0$

$$\left| \left| f\left(\frac{t}{B_{n_l}}\right)\right|^{n_l} - |g(t)| \right| < \varepsilon\,, \quad t \in C.$$

In particular, for $t = B_{n_l}s$ we obtain that

$$|\,|f(s)|^{n_l} - |g(B_{n_l}s|\,| < \varepsilon\,, \quad l \geq l_0\,.$$

Since $g(B_{n_l}s) \to 1$, we see that $|f(s)|^{n_l} \to 1$ for $l \to \infty$, whence $|f(s)| \equiv 1$. By (5.59) G is degenerate. ■

5.10 Comments

The theory of stochastic integrals and stochastic differential equations expounded in Sect. 5.2-5.4 was developed by the author [84].

The description of all rotation-invariant processes with independent increments on a local field (Theorem 5.4) was found by Yasuda [158], who also proved their properties given in Theorems 5.7-5.9 and found for this general situation a representation similar to our Theorem 5.2. In fact the class of processes described in Theorem 5.4 was first constructed for $K = \mathbb{Q}_p$ by Albeverio and Karwowski [3, 4], whose idea was to start from a Markov process on a family of balls covering $\mathbb{Q}_p$, defined in such a way that the limit, as balls shrink to points, gives a process on $\mathbb{Q}_p$. They found the expression (5.34) for the transition probabilities. Later Yasuda [158] showed that any rotation-invariant process can be obtained by means of this construction. Theorems 5.5 and 5.6 about generators and their spectra are taken from [4]. For various generalizations see [5, 65, 110].

Processes on adeles were studied by Blair [16], and Karwowski and Vilela Mendes [65]. In this connection we mention also some adelic models in mathematical physics; see [107, 23]. General properties of Lévy processes on a wide class of totally disconnected Abelian locally compact groups were investigated by Evans [32].

A description of stable-like probability distributions on a local field is taken from [88]. So far a complete description is known only for the case of symmetric distributions. This can be connected with the fact of non-uniqueness of the Lévy-Khinchin representation of an infinitely divisible distribution, a property, which is quite different from the classical case.

Chapter 6

Analysis over Infinite Extensions of a Local Field

6.1 Measurable vector spaces

6.1.1. Definitions. Let E be a vector space over a local field K, $\mathcal{B}$ a σ-algebra of subsets of E. The pair $(E, \mathcal{B})$ is called *a measurable vector space* if the mapping

$$(x, y) \mapsto x + y, \quad x, y \in E,$$

from $E \times E$ to E, and the mapping

$$(\lambda, x) \mapsto \lambda x, \quad \lambda \in K,\ x \in E,$$

from $K \times E$ to E are measurable with respect to the σ-algebras $\mathcal{B}$ and $\mathcal{B}(K)$ (the Borel σ-algebra of K).

Suppose that E is a vector space over K, and F is a collection of linear functionals on E with values in K. If $\mathcal{B}$ is the σ-algebra generated by F, we shall say that the triple $(E, F, \mathcal{B})$ is *coordinated.*

Lemma 6.1. *If $(E, F, \mathcal{B})$ is a coordinated triple, then $(E, \mathcal{B})$ is a measurable vector space.*

Proof. Let us show that the map $(x, y) \mapsto x + y$ is measurable. Consider the σ-algebra

$$\mathfrak{G} = \{C \in \mathcal{B} : \ \{(x, y) : \ x + y \in C\} \in \mathcal{B} \times \mathcal{B}\}.$$

Suppose that $T_1, \dots, T_n \in F$ and that $B_1, \dots, B_n$ are Borel subsets of K. Then

$$\{(x,y) \in T_1^{-1}(B_1) \cap \dots \cap T_n^{-1}(B_n)\} = \bigcap_{i=1}^{n} \{(x,y) :\ T_i(x) + T_i(y) \in B_i\}.$$

As the map $(x,y) \mapsto T_i(x)+T_i(y)$ is measurable, we see that $\bigcap_{i=1}^{n} T_i^{-1}(B_i) \in \mathfrak{G}$. On the other hand, such sets generate the σ-algebra $\mathcal{B}$. Therefore its sub-σ-algebra $\mathfrak{G}$ coincides with $\mathcal{B}$, as required.

The proof that the map $(\lambda, x) \mapsto \lambda x$ is measurable can be carried out in a similar way. ■

Note that $(E, E^*, \mathcal{B}(E))$ is a coordinated triple for any separable Banach space E over K. Just as in the case of Banach spaces over $\mathbb{R}$ [148], this is a consequence of the non Archimedean version of the Hahn-Banach theorem (see [134, 127]). In Sect. 6.3 we shall study another example of a coordinated triple.

6.1.2. Gaussian random variables. Before introducing Gaussian measures over a measurable vector space we recall the definition of non-Archimedean orthogonality.

Definition 6.1. *Suppose that $(Y, \|\cdot\|_Y)$ is a normed space over a local field K. We say that a set $X \subset Y$ is orthogonal if for any finite subset $\{x_1, \dots, x_n\} \subset X$ and each $\lambda_1, \dots, \lambda_n \in K$ we have*

$$\left\| \sum_{i=1}^{n} \lambda_i x_i \right\|_Y = \max_{1 \le i \le n} \|\lambda_i\| \cdot \|x_i\|_Y.$$

An orthogonal set X is called orthonormal if $\|x\|_Y = 1$ for any $x \in X$.

In our next definition $Y = K^2$, $\|(\lambda_1, \lambda_2)\|_Y = \max(\|\lambda_1\|, \|\lambda_2\|)$.

Definition 6.2. *Let $(E, \mathcal{B})$ be a measurable vector space. Suppose that ξ is an E-valued random variable. We say that the probability distribution of ξ is a K-Gaussian probability measure (or, equivalently, ξ is a K-Gaussian random variable) if when $\xi^{(1)}, \xi^{(2)}$ are two independent copies of ξ, and $(\lambda_{11}, \lambda_{12})$, $(\lambda_{21}, \lambda_{22}) \in K^2$ are orthonormal, then $(\xi^{(1)}, \xi^{(2)})$ has the same law as*

$$\left(\lambda_{11}\xi^{(1)} + \lambda_{12}\xi^{(2)}, \lambda_{21}\xi^{(1)} + \lambda_{22}\xi^{(2)}\right).$$

Of course, this definition is analogous to one of well-known equivalent definitions (see [37]) of a real-valued Gaussian random variable.

The class of K-Gaussian K-valued (scalar!) random variables is not very rich. It contains a random variable, which is almost surely zero, and some others described in the following theorem.

Theorem 6.1. *A K-valued random variable ξ that is not almost surely zero is K-Gaussian if and only if its distribution is a cutoff of the Haar measure:*

$$\mathbf{P}\{\xi \in dx\} = q^{-n}\Phi(q^{-n}\|x\|)\,dx \tag{6.1}$$

for some $n \in \mathbb{Z}$, or, equivalently,

$$\mathbf{E}(\chi(t\xi)) = \Phi(q^n\|t\|), \quad t \in K, \tag{6.2}$$

where Φ is the indicator function of the interval $[0,1]$, χ is a rank zero additive character on K.

Proof. Suppose that (6.2) holds for some $n \in \mathbb{Z}$. Note that $\Phi(\max(a,b)) = \Phi(a)\Phi(b)$ for any $a, b \geq 0$. If $\xi^{(1)}, \xi^{(2)}$ are two independent copies of ξ, and $(\lambda_{11}, \lambda_{12})$, $(\lambda_{21}, \lambda_{22})$ is a pair of orthonormal vectors in K^2, then

$$\begin{aligned}
\mathbf{E}&\left[\chi\left(t_1\left(\lambda_{11}\xi^{(1)} + \lambda_{12}\xi^{(2)}\right) + t_2\left(\lambda_{21}\xi^{(1)} + \lambda_{22}\xi^{(2)}\right)\right)\right] \\
&\qquad = \Phi\left(q^n\|t_1\lambda_{11} + t_2\lambda_{21}\|\right)\Phi\left(q^n\|t_1\lambda_{12} + t_2\lambda_{22}\|\right) \\
&\qquad = \Phi\left(q^n \max(\|t_1\lambda_{11} + t_2\lambda_{21}\|, \|t_1\lambda_{12} + t_2\lambda_{22}\|\right) \\
&= \Phi\left(q^n\|t_1(\lambda_{11}, \lambda_{12}) + t_2(\lambda_{21}, \lambda_{22})\|\right) = \Phi(q^n \max(\|t_1\|, \|t_2\|)) \\
&\qquad = \Phi(q^n\|t_1\|)\Phi(q^n\|t_2\|) = \mathbf{E}\left[\chi\left(t_1\xi^{(1)}\right)\right]\mathbf{E}\left[\chi\left(t_2\xi^{(2)}\right)\right].
\end{aligned}$$

From the Fourier uniqueness we find that

$$\left(\lambda_{11}\xi^{(1)} + \lambda_{12}\xi^{(2)}, \lambda_{21}\xi^{(1)} + \lambda_{22}\xi^{(2)}\right)$$

has the same law as $(\xi^{(1)}, \xi^{(2)})$, so that ξ is K-Gaussian.

Conversely, suppose that ξ is K-Gaussian. Denote $\varphi(t) = \mathbf{E}[\chi(t\xi)]$. Let $\xi^{(1)}, \xi^{(2)}$ be two independent copies of ξ. Since (1,1), (0,1) is a pair of orthonormal vectors in K^2, we find that $(\xi^{(1)} + \xi^{(2)}, \xi^{(2)})$ has the same law as $(\xi^{(1)}, \xi^{(2)})$, whence $[\varphi(t)]^2 = \varphi(t)$. Thus $\varphi(t) \in \{0,1\}$ for all $t \in K$.

Suppose that $t_0 \neq 0$ and $\varphi(t_0) = 1$. Such a point t_0 exists since φ is continuous and $\varphi(0) = 1$. As $(1, \lambda)$, $(0, 1)$ is a pair of orthonormal vectors in K^2 for each $\lambda \in K$ with $\|\lambda\| \le 1$, we see that $\xi^{(1)} + \lambda\xi^{(2)}$ has the same law as $\xi^{(1)}$. Thus $\varphi(t)\varphi(\lambda t) = \varphi(t)$ for all t. In particular, $\varphi(\lambda t_0) = 1$. This means that $\varphi(t) = 1$ for all t with $\|t\| \le \|t_0\|$. On the other hand, $\varphi(t) \not\equiv 1$, since ξ is not almost surely zero. Hence we find that $\varphi(t) = \Phi(q^n\|t\|)$ for some $n \in \mathbb{Z}$. ■

Writing $\mathbf{E}[\chi(t\xi)]$ as the Fourier transform of the distribution (6.1) we come to the following result.

Corollary 6.1. *If ξ is a K-valued K-Gaussian random variable, then ξ belongs to the K-Banach space $L_\infty(\mathbf{P})$,*

$$\mathbf{E}[\chi(t\xi)] = \Phi\left(\|\xi\|_{L_\infty(\mathbf{P})}\|t\|\right).$$

If ξ is not almost surely 0, then

$$\mathbf{P}\{\xi \in dx\} = \|\xi\|^{-1}_{L_\infty(\mathbf{P})}\Phi\left(\|\xi\|_{L_\infty(\mathbf{P})}\|t\|\right)\, dx.$$

Below we shall construct a non-trivial example of a K-Gaussian measure on an infinite-dimensional measurable vector space. Now we proceed in studying properties of K-Gaussian measures. We shall start from a zero-one law.

Theorem 6.2. *Suppose that $(E, \mathcal{B})$ is a measurable vector space, and M is a measurable subspace of E. If ξ is an E-valued K-Gaussian random variable, then $\mathbf{P}\{\xi \in M\}$ is either 0 or 1.*

Proof. Let $\xi^{(1)}, \xi^{(2)}$ be two independent copies of ξ. For $n = 1, 2, \ldots$, set

$$A_n = \left\{\xi^{(1)} + \beta^n\xi^{(2)} \in M, \beta^n\xi^{(1)} + \xi^{(2)} \notin M\right\}$$

(as before, β is a prime element of K).

Note that $\{(1, \beta^n), (\beta^n, 1)\}$ is an orthonormal pair. Indeed, it is clear that $\|(1, \beta^n)\|_{K^2} = \|(\beta^n, 1)\|_{K^2} = 1$. It is sufficient to show that there is no $z \in K$ such that

$$\|(1, \beta^n) + z(\beta^n, 1)\|_{K^2} = \max\{\|1 + z\beta^n\|, \|\beta^n + z\|\} < 1.$$

Otherwise we would have $\|1 + z\beta^n\| < 1$, so that $\|z\beta^n\| = 1$ and hence $\|z\| = q^n$. Then $\|\beta^n + z\| = q^n > 1$, which is a contradiction.

By the definition of a K-Gaussian random variable we see that

$$\left(\xi^{(1)} + \beta^n\xi^{(2)}, \beta^n\xi^{(1)} + \xi^{(2)}\right)$$

has the same law as $\left(\xi^{(1)}, \xi^{(2)}\right)$, and hence

$$\mathbf{P}(A_n) = \mathbf{P}\{\xi \in M\}\mathbf{P}\{\xi \notin M\}. \tag{6.3}$$

If $m \neq n$, then the matrix

$$\begin{pmatrix} 1 & \beta^m \\ 1 & \beta^n \end{pmatrix}$$

is invertible. Therefore if both $\xi^{(1)} + \beta^m \xi^{(2)}$ and $\xi^{(1)} + \beta^n \xi^{(2)}$ belong to M, them both $\xi^{(1)}$ and $\xi^{(2)}$ (and hence both $\beta^m \xi^{(1)} + \xi^{(2)}$ and $\beta^n \xi^{(1)} + \xi^{(2)}$) belong to M. Thus $A_m \cap A_n = \varnothing$. It follows that

$$1 \geq \mathbf{P}\left(\bigcup_n A_n\right) = \sum_n \mathbf{P}(A_n),$$

and from (6.3) we conclude that $\mathbf{P}\{\xi \in M\} \in \{0, 1\}$. ∎

Let $(E, F, \mathcal{B})$ be a coordinated triple.

Theorem 6.3. *If ξ is an E-valued random variable, then ξ is K-Gaussian if and only if $T(\xi)$ is K-Gaussian for all $T \in \operatorname{span} F$.*

Proof. Suppose that $T(\xi)$ is K-Gaussian for all $T \in \operatorname{span} F$. Let $\xi^{(1)}$, $\xi^{(2)}$ be two independent copies of ξ. Just as in the case of real measurable vector spaces [22], we may use characteristic functionals, that is we need to check that

$$\mathbf{E}\left[\chi\left(T_1\left(\lambda_{11}\xi^{(1)} + \lambda_{12}\xi^{(2)}\right) + T_2\left(\lambda_{21}\xi^{(1)} + \lambda_{22}\xi^{(2)}\right)\right)\right] = \mathbf{E}\left[\chi\left(T_1\left(X_1\right) + T_2\left(X_2\right)\right)\right]. \tag{6.4}$$

By Corollary 6.1, the left-hand side of (6.4) is given by

$$\Phi\left(\max\left(\|(\lambda_{11}T_1 + \lambda_{21}T_2)\xi\|_{L_\infty(\mathbf{P})}, \|(\lambda_{12}T_1 + \lambda_{22}T_2)\xi\|_{L_\infty(\mathbf{P})}\right)\right).$$

Suppose that $\dim \operatorname{span}\{T_1(\xi), T_2(\xi)\} = 2$ (other cases are similar). Let η_1, η_2 be an orthonormal basis for $\operatorname{span}\{T_1(\xi), T_2(\xi)\}$ in $L_\infty(\mathbf{P})$. We can write

$$T_1(\xi) = \gamma_{11}\eta_1 + \gamma_{12}\eta_2,$$
$$T_2(\xi) = \gamma_{21}\eta_1 + \gamma_{22}\eta_2.$$

Now

$$\begin{aligned}
&\|(\lambda_{11}T_1 + \lambda_{21}T_2)\xi\|_{L_\infty(\mathbf{P})} \\
&\qquad = \|\lambda_{11}(\gamma_{11}\eta_1 + \gamma_{12}\eta_2) + \lambda_{21}(\gamma_{21}\eta_1 + \gamma_{22}\eta_2)\|_{L_\infty(\mathbf{P})} \\
&\qquad\qquad = \max(\|\lambda_{11}\gamma_{11} + \lambda_{21}\gamma_{21}\|, \|\lambda_{11}\gamma_{12} + \lambda_{21}\gamma_{22}\|).
\end{aligned}$$

Similarly,

$$\|(\lambda_{12}T_1 + \lambda_{22}T_2)\xi\|_{L_\infty(\mathbf{P})} = \max(\|\lambda_{12}\gamma_{11} + \lambda_{22}\gamma_{21}\|, \|\lambda_{12}\gamma_{12} + \lambda_{22}\gamma_{22}\|).$$

It follows that the left-hand side of (6.4) equals

$$\begin{aligned}
&\Phi\left(\max(\|\gamma_{11}(\lambda_{11},\lambda_{12}) + \gamma_{21}(\lambda_{21},\lambda_{22})\|, \|\gamma_{12}(\lambda_{11},\lambda_{12}) + \gamma_{22}(\lambda_{21},\lambda_{22})\|)\right) \\
&\qquad = \Phi\left(\max(\max(\|\gamma_{11}\|, \|\gamma_{21}\|), \max(\|\gamma_{12}\|, \|\gamma_{22}\|))\right)
\end{aligned}$$

by the orthonormality of $(\lambda_{11}, \lambda_{12})$, $(\lambda_{21}, \lambda_{22})$.

For the right-hand side of (6.4) we get

$$\Phi\left(\max(\|\gamma_{11}\eta_1 + \gamma_{12}\eta_2\|_{L_\infty(\mathbf{P})}, \|\gamma_{21}\eta_1 + \gamma_{22}\eta_2\|_{L_\infty(\mathbf{P})})\right).$$

Therefore (6.4) holds and ξ is K-Gaussian.

The proof of the converse is straightforward. ■

Next we consider the behaviour of a Gaussian measure under an additive shift. A full description is given by the following result.

Theorem 6.4. *Let $(E, F, \mathcal{B})$ be a coordinated triple, and $\mathbf{P}$ a K-Gaussian measure on E. Set*

$$\Sigma = \{x \in E : \ \|T(x)\| \leq \|T\|_{L_\infty(\mathbf{P})} \ \forall T \in \operatorname{span} F\}.$$

Let $I: E \to E$ be the identity map. If $x \in \Sigma$, then the law of $x + I$ under $\mathbf{P}$ is $\mathbf{P}$ itself. Otherwise, the law of $x + I$ is perpendicular to $\mathbf{P}$.

Proof. If η is a K-valued, K-Gaussian random variable, and $y \in K$ is such that $\|y\| \leq \|\eta\|_{L_\infty}$, then, since the law of η is a Haar measure on the subgroup $z : \|z\| \leq \|\eta\|_{L_\infty}$, we see that the law of $y + \eta$ coincides with the one of η. Therefore if $x \in \Sigma$, we have that the law of $T \circ (x + I) = T(x) + T$ under $\mathbf{P}$ is that of T under $\mathbf{P}$ for all $T \in \operatorname{span} F$. Hence the law of $x + I$ under $\mathbf{P}$ is that of I under $\mathbf{P}$.

If $x \notin \Sigma$, then there exists such a functional $T \in \operatorname{span} F$ that

$$\|T(x)\| > \|T\|_{L_\infty(\mathbf{P})}.$$

Then $\|T(x+I)\| = \|T(x)\|$ **P**-almost surely, which means that the law of $x + I$ is perpendicular to **P**. ■

Note that in similar results regarding Gaussian measures on vector spaces over $\mathbb{R}$ (see e.g. [22] or [95])) the dichotomy of equivalence and perpendicularity is established. The next result shows that in the non-Archimedean case equivalent K-Gaussian measures coincide.

Theorem 6.5. *Suppose that $(E, F, \mathcal{B})$ is a coordinated triple. If $\mathbf{P}_1$ and $\mathbf{P}_2$ are equivalent K-Gaussian measures on E, then $\mathbf{P}_1 = \mathbf{P}_2$.*

Proof. It follows from our assumption that

$$\|T\|_{L_\infty(\mathbf{P}_1)} = \|T\|_{L_\infty(\mathbf{P}_2)}$$

for all $T \in \operatorname{span} F$. By Corollary 6.1, for all such T the distribution of T under $\mathbf{P}_1$ is the same as that under $\mathbf{P}_2$, whence $\mathbf{P}_1 = \mathbf{P}_2$. ■

6.1.3. Minlos-Mądrecki theorem. A complex-valued function f on a vector space E is called positive definite if $f(0) = 1$ and

$$\sum_{i,j=1}^{n} c_i f(x_i - x_j)\overline{c_j} \geq 0$$

for all natural numbers n, $c_i \in \mathbb{C}$, and $x_i \in E$.

The well-known Minlos theorem (see e.g. [22]) states that any continuous positive definite function on a nuclear locally convex topological vector space E over $\mathbb{R}$ is the Fourier transform of a Radon probability measure on the Borel σ-algebra on the conjugate space E^* equipped with the $*$-weak topology $\sigma(E^*, E)$. Here we present a similar result for locally convex spaces over a local field. Note that in our situation no nuclearity condition is imposed, which agrees with the fact [119] that any locally convex space over a local field is nuclear in the sense of Grothendieck.

Theorem 6.6. *Let E be a Hausdorff locally convex space over a local field K. Any continuous positive definite function on E is the Fourier transform of a Radon probability measure on $\mathcal{B}(E^*)$, where the conjugate space E^* is equipped with the $*$-weak topology $\sigma(E^*, E)$. Conversely, if E is barreled, then the Fourier transform of a Radon probability measure on $\mathcal{B}(E^*)$ is a continuous positive definite function on E.*

In order to prove this theorem, we shall need some auxiliary results. Let us consider the finite-dimensional space K^n with the norm

$$\|(\lambda_1, \ldots, \lambda_n)\|_{K^n} = \max_{1 \leq j \leq n} \|\lambda_j\|.$$

The set

$$O^n = \{\lambda = (\lambda_1, \dots, \lambda_n) \in K^n : \|\lambda\|_{K^n} \leq 1\}$$

is a *lattice* (an open compact O-module) in K^n. The integration formula (1.26) admits a multi-dimensional generalization (see Chapter VII in [156]). In particular, there exists such a lattice O_*^n in K^n that the indicator function of O^n is the Fourier transform of the probability Haar measure of the additive group of O_*^n. If $A : K^n \to K^n$ is a linear mapping, $p_A(\lambda) = \|A\lambda\|_{K^n}$, $O_A = \{\lambda \in K^n : p_A(\lambda) \leq 1\}$, and $\Omega_A(\lambda)$ is the indicator function of O_A, then

$$\Omega_A(\lambda) = \int_{K^n} \chi(\langle \lambda, \eta \rangle) H_A(d\eta), \tag{6.5}$$

where χ is a rank zero additive character on K, H_A is a probability measure concentrated on a certain O-submodule M_A of K^n,

$$\langle \lambda, \eta \rangle = \sum_{i=1}^{n} \lambda_i \eta_i, \quad \lambda = (\lambda_1, \dots, \lambda_n), \ \eta = (\eta_1, \dots, \eta_n).$$

It is clear that M_A is a compact convex absorbing set in K^n, $0 \in M_A$.

Lemma 6.2.

$$\int_{K^n} \|\lambda\|_{K^n} H_A(d\lambda) \leq \|A\|_N,$$

where $\|A\|_N$ *is the "nuclear norm" of the operator* A *defined as*

$$\|A\|_N = \sup_{1 \leq i \leq n} \|Ae_i\|_{K^n} \tag{6.6}$$

for an arbitrary orthonormal basis $\{e_i\}_1^n$ *of* K^n.

Proof. Note first of all that $\|A\|_N$ does not depend on the choice of an orthonormal basis in (6.6). That follows easily from the fact that linear operators transforming orthonormal bases are isometries of K^n.

It follows from the properties of the set M_A mentioned above (see [108]) that there exist a basis $\{e_1, \dots, e_n\}$ of K^n, and positive real numbers $a_1, \dots, a_n$, such that

$$M_A = \left\{ \lambda \in K^n : \lambda = \sum_{i=1}^{n} \zeta_i e_i, \ \|\zeta_i\| \leq a_i \right\}.$$

We have by (6.5) that

$$\int_{K^n} \chi(t\langle e_i, \eta\rangle) H_A(d\eta) = \Omega_A(te_i),$$

which, as a function of $t \in K$, equals the indicator function of the set $[p_A(e_i)]^{-1}O$.

On the other hand, if H_A^i is the probability distribution of the random variable

$$\langle e_i, \cdot\rangle : (M_A, H_A) \longrightarrow (K, \mathcal{B}(K)),$$

then

$$\int_{K^n} \chi(t\langle e_i, \eta\rangle) H_A(d\eta) = \int_K \chi(t) H_A^i(dt).$$

Now we find from (1.26) that $a_i = p_A(e_i)$, so that

$$M_A = \prod_{i=1}^{n} p_A(e_i) \cdot O. \tag{6.7}$$

From (6.7) we obtain

$$\int_{K^n} \|\lambda\|_{K^n} H_A(d\lambda) = \int_{M_A} \max_{1 \le i \le n} \|\langle \lambda, e_i\rangle\| H_A(d\lambda)$$

$$\le \max_{1 \le i \le n} p_A(e_i) = \max_{1 \le i \le n} \|Ae_i\|_{K^n} = \|A\|_N$$

as desired. ■

Lemma 6.3. *Let μ be a probability measure on $(K^n, \mathcal{B}(K^n))$, $\widehat{\mu}$ its characteristic function, and A, C arbitrary linear mappings from K^n to K^n. If*

$$|1 - \operatorname{Re}\widehat{\mu}(\lambda)| \le \varepsilon \quad \textit{for all } \lambda \in K^n \textit{ with } p_A(\lambda) \le 1,$$

then

$$\mu(\{\lambda \in K^n : p_C(\lambda) > \gamma\}) \le \varepsilon + \frac{2}{\gamma}\|AC\|_N, \quad \gamma > 0, \tag{6.8}$$

where $p_C(\lambda) = \|C\lambda\|_{K^n}$.

Proof. It is sufficient to consider the case when $\gamma = 1$. Let H_C be the probability measure related to the operator C just as H_A was related to A. We have

$$\mu\{\lambda:\ p_C(\lambda) > 1\} = \int\limits_{\{p_C(\lambda)>1\}} \mu(d\lambda) = \int\limits_{K^n} [1 - \Omega_C(\lambda)]\,\mu(d\lambda)$$

$$= \operatorname{Re} \int\limits_{K^n} H_C(d\eta) \int\limits_{K^n} [1 - \chi(\langle\lambda, \eta\rangle)]\mu(d\eta) = \int\limits_{K^n} [1 - \operatorname{Re}\widehat{\mu}(\eta)]\, H_C(d\eta).$$

Under our assumption we get

$$\int\limits_{p_A(\eta)\le 1} [1 - \operatorname{Re}\widehat{\mu}(\eta)]\, H_C(d\eta) \le \varepsilon \int\limits_{p_A(\eta)\le 1} H_C(d\eta) \le \varepsilon, \tag{6.9}$$

$$\int\limits_{p_A(\eta)>1} [1 - \operatorname{Re}\widehat{\mu}(\eta)]\, H_C(d\eta) \le 2 \int\limits_{p_A(\eta)>1} H_C(d\eta) \le 2 \int\limits_{p_A(\eta)>1} p_A(\eta) H_C(d\eta). \tag{6.10}$$

By Lemma 6.2

$$\int\limits_{K^n} p_A(\eta) H_C(d\eta) = \int\limits_{K^n} \|A\eta\|_{K^n} H_C(d\eta) = \int\limits_{K^n} \|\eta\|_{K^n} H_{AC}(d\eta) \le \|AC\|_N.$$

Combining this with (6.9) and (6.10) we obtain (6.8). ■

Let us introduce some notations. Suppose that V is a locally convex space over K, $G \subset V$ is an arbitrary subset. The polar set G^0 of G is defined as

$$G^0 = \{v^* \in V^* :\ \|\langle v, v^*\rangle\| \le 1 \quad \text{for each } v \in G\}.$$

If $p(v)$ is a non-Archimedean seminorm on V, then we write $N_p = \{v \in V :\ p(v) = 0\}$.

Lemma 6.4. *Let V be a finite-dimensional vector space over K, and $p(v)$ a non-Archimedean seminorm on V. Let μ be a probability Borel measure on V^* such that*

$$|1 - \operatorname{Re}\widehat{\mu}(x)| \le \varepsilon \quad \textit{if } p(x) \le 1$$

($\varepsilon < 1$). Then for every $\gamma > 0$

$$\mu\left\{V^* \setminus [B_p(\gamma)]^0\right\} \le \varepsilon + 2\gamma, \tag{6.11}$$

where $B_p(\gamma) = \{x \in V :\ p(x) \le \gamma\}$.

Proof. Let W_p be a complementing subspace to N_p, that is $V = N_p \dot{+} W_p$. Let T_p be a projection in V onto W_p corresponding to this decomposition. It follows from the non-Archimedean (ultrametric) property of p that $p(v) = p(T_p v)$.

The seminorm p induces a norm on W_p. Let $e_1, \dots, e_k$ ($k = \dim W_p$) be an orthonormal basis in W_p. If $v \in V$, we can write $v = v_0 + \sum_{i=1}^{k} \lambda_i e_i$, where $v_0 \in N_p$, $\lambda_i \in K$, and

$$p(v) = \max_{1 \le i \le k} \|\lambda_i\|. \tag{6.12}$$

Below we shall use a basis in V obtained by complementing $e_1, \dots, e_k$ by an arbitrary basis $e_{k+1}, \dots, e_n$ in N_p (if $N_p \neq \{0\}$). Using this basis we may identify V with K^n. Then (6.12) yields the representation $p(\lambda) = \|A\lambda\|_{K^n}$, $\lambda \in K^n$, where $A = \left(\begin{smallmatrix} I & 0 \\ 0 & 0 \end{smallmatrix}\right)$.

Now V^* is also identified with K^n, with the basis $\{e_i\}$ viewed as self-dual. If $\lambda = \sum_{i=1}^{n} \lambda_i e_i \in V$, $\eta = \sum_{i=1}^{n} \eta_i e_i \in V^*$, then

$$\langle \lambda, \eta \rangle = \sum_{i=1}^{n} \lambda_i \eta_i .$$

By (6.12), $B_p(\gamma)$ is identified with the set of those λ, for which $\|\lambda_i\| \le \gamma$ for $i = 1, \dots k$. It is clear that the polar set $[B_p(\gamma)]^0$ consists of those η, for which $\|\eta_i\| \le \gamma^{-1}$ if $i = 1, \dots k$, and $\eta_i = 0$, $i = k+1, \dots, n$. Its complement $V^* \setminus [B_p(\gamma)]^0$ is the set of η with either $\|\eta_i\| > \gamma^{-1}$ for some $i \in \{1, \dots, k\}$, or $\|\eta_i\| > 0$ for some $i \in \{k+1, \dots, n\}$.

Let us consider the sequence of sets $M_l \subset V^*$, $l = 1, 2, \dots$, consisting of those η, for which either $\|\eta_i\| > \gamma^{-1}$ for some $i \in \{1, \dots, k\}$, or $\|\eta_i\| > \gamma^{-1} q^{-l}$ for some $i \in \{k+1, \dots, n\}$. Obviously $M_l \nearrow V^* \setminus [B_p(\gamma)]^0$ monotonically, so that

$$\mu(M_l) \longrightarrow \mu\left\{V^* \setminus [B_p(\gamma)]^0\right\}, \quad l \to \infty.$$

On the other hand,

$$M_l = \left\{\eta : \ \|C\eta\|_{K^n} > \gamma^{-1}\right\},$$

where $C = \begin{pmatrix} I & 0 \\ 0 & \beta^{-l} I \end{pmatrix}$, and as usual β is a prime element of K. Using Lemma 6.3 and observing that $\|AC\|_N = \|A\|_N$ we find that $\mu(M_l) \le \varepsilon + 2\gamma$, which implies (6.11). ■

Proof of Theorem 6.6. Let $L \subset E^*$ be a subspace of a finite codimension (an annulator of a finite-dimensional subspace L^0 of E). For any subset $F \subset E^*/L$ the set of elements $x^* \in E^*$, whose images in E^*/L under the canonical mapping belong to F, is called a cylindrical set determined by L, with the base F. Cylindrical sets form an algebra $\mathrm{Cyl}(E^*)$. Its subalgebra $\mathrm{Cyl}(E^*, L)$ corresponding to any fixed L is a σ-algebra.

Just as in the case of topological vector spaces over $\mathbb{R}$ (see [22]), a continuous positive definite function f on E determines a cylindrical measure μ, that is a finitely additive measure on $\mathrm{Cyl}(E^*)$, whose restriction μ_L is σ-additive on each $\mathrm{Cyl}(E^*, L)$, such that

$$f(x) = \int_{E^*} \chi(\langle x, x^* \rangle)\mu(dx^*).$$

Each μ_L corresponds to a measure on the finite-dimensional space E^*/L.

In order to prove that the measure μ can be extended to a σ-additive measure on $\mathcal{B}(E^*)$, it is sufficient [22] to find, for every $\varepsilon > 0$, such a compact subset C_ε (with respect to the topology $\sigma(E^*, E)$) that for any cylindrical set B with $B \cap C_\varepsilon = \varnothing$ we have $\mu(B) < \varepsilon$.

Since f is continuous, we can choose from the family of seminorms defining the topology on E such a finite collection $\{p_i, i \in I(\varepsilon)\}$ that

$$|1 - \operatorname{Re} f(x)| \leq \varepsilon/2 \quad \text{if } p_\varepsilon(x) \stackrel{\text{def}}{=} \max_{i \in I(\varepsilon)} p_i(x) \leq 1. \tag{6.13}$$

Let

$$B_{p_\varepsilon}(\gamma) = \{x \in E : \ p_\varepsilon(x) \leq \gamma\}, \quad \gamma > 0.$$

The polar set $[B_{p_\varepsilon}(\gamma)]^0$ is compact in the topology $\sigma(E^*, E)$ on E^* (see [108, 145]). This set is our choice for C_ε.

Let B be a cylindrical set, $B \cap [B_{p_\varepsilon}(\gamma)]^0 = \varnothing$, and $L \subset E^*$ be the subspace of a finite codimension, which determines B. In the quotient space E^*/L we have the base set F with the same measure, $\mu_L(F) = \mu(B)$, where we identify a cylindrical set from $\mathrm{Cyl}(E^*, L)$ with its base F.

The image of the polar set $B \cap [B_{p_\varepsilon}(\gamma)]^0$ under the canonical mapping is the polar set of the set

$$B_{p_\varepsilon}(\gamma) \cap L^0 = \{x \in L^0 : \ p_\varepsilon(x) \leq \gamma\},$$

which lies in the finite-dimensional subspace L^0 whose dual space is E^*/L. As a result we have

$$\mu(B) = \mu_L(F) \leq \mu_L\left(E^*/L \setminus (B_{p_\varepsilon}(\gamma) \cap L^0)^0\right),$$

and by Lemma 6.4

$$\mu(B) \leq \frac{\varepsilon}{2} + 2\gamma < \varepsilon$$

if $\gamma < \frac{\varepsilon}{2}$.

It remains to prove the converse statement. Suppose that μ is a Radon probability measure on $\mathcal{B}(E^*)$, where E is a barreled space. Let $x_1, \dots, x_n \in E$, $c_1, \dots, c_n \in \mathbb{C}$, E_0 be the finite-dimensional subspace spanned by $x_1, \dots, x_n$, L its annulator, and μ_0 the measure induced by μ in the quotient space $E^*/L \simeq E_0^*$. We have

$$\widehat{\mu}(x) = \int\limits_{E_0^*} \chi(\langle x, x^* \rangle) \mu_0(dx^*).$$

Since E_0^* is finite-dimensional, we may use the analog of Bochner's theorem mentioned in Sect. 1.6 obtaining that

$$\sum_{i,j=1}^{n} \widehat{\mu}(x_i - x_j) c_i \overline{c_j} \geq 0.$$

It is clear that $\widehat{\mu}(0) = 1$. Thus $\widehat{\mu}$ is positive definite.

By the definition of a Radon measure, for a given $\varepsilon > 0$ there exists a $*$-weakly compact set $C_\varepsilon \subset E^*$, such that $\mu(C_\varepsilon) \geq 1 - \frac{\varepsilon}{2}$. Since E is barreled, C_ε is contained in a polar set of some neighbourhood of the origin. Thus one can find such a non-Archimedean continuous semi-norm p_ε on E that

$$\mu\left([B_{p_\varepsilon}(\gamma_\varepsilon)]^0 \right) \geq 1 - \frac{\varepsilon}{2}$$

for some $\gamma_\varepsilon > 0$. Hence

$$|1 - \operatorname{Re} \widehat{\mu}(x)| \leq \varepsilon + 2 \int\limits_{[B_{p_\varepsilon}(\gamma_\varepsilon)]^0} \| \langle x, x^* \rangle \| \mu(dx^*) \leq \varepsilon + p'_\varepsilon(x),$$

where

$$p'_\varepsilon(x) = \sup_{x^* \in [B_{p_\varepsilon}(\gamma_\varepsilon)]^0} \| \langle x, x^* \rangle \|$$

is continuous at the origin, so that $\operatorname{Re} \widehat{\mu}$ is continuous at the origin.

Finally, the positive definiteness of $\widehat{\mu}(x)$ implies the inequality

$$|\widehat{\mu}(x') - \widehat{\mu}(x'')|^2 \leq 2 \left[1 - \operatorname{Re} \widehat{\mu}(x' - x'') \right], \quad x', x'' \in E,$$

(see [55], where it was proved for positive definite functions on an arbitrary group). Now it follows that $\widehat{\mu}$ is continuous on E. ∎

6.2 An infinite extension as a topological vector space

6.2.1. Main notions. Let k be a non-Archimedean local field of zero characteristic (thus k is a finite extension of the field of p-adic numbers). Consider an increasing sequence of its finite extensions

$$k = K_1 \subset K_2 \subset \ldots \subset K_n \subset \ldots .$$

For each $n = 1, 2, \ldots$ we denote by $|\cdot|_n$ a normalized absolute value in K_n; if $N_n : K_n \to k$ is a norm mapping then $|x|_n = |N_n(x)|_1$. We shall use also a (non-normalized) absolute value $\|x\| = |x|_n^{1/m_n}$ $(x \in K_n)$ which coincides with $|x|_1$ for $x \in k$ being defined correctly on the infinite extension $K = \bigcup_{n=1}^{\infty} K_n$. Here m_n is the degree of the extension K_n/k. Of course, the meaning of the notation $\|\cdot\|$ is different here (and until the end of this chapter) from the one used before. This abuse of notation is hopefully justified by the role of the absolute value $\|\cdot\|$ over an infinite extension, which is similar to that of a normalized absolute value over a local field.

Let us consider, for each n, a mapping $T_n : K \to K_n$ defined as follows. If $x \in K_\nu$, $\nu > n$, put

$$T_n(x) = \frac{m_n}{m_\nu} \mathrm{Tr}_{K_\nu/K_n}(x)$$

where $\mathrm{Tr}_{K_\nu/K_n} : K_\nu \to K_n$ is the trace mapping (note that the degree of the extension K_ν/K_n equals m_ν/m_n). If $x \in K_n$ then, by definition, $T_n(x) = x$. It follows from the transitivity property of traces that T_n is well-defined and $T_n \circ T_\nu = T_n$ for $\nu > n$. Below we shall often write T instead of T_1.

Note that the equality $T_n(x) = 0$ for all n implies $x = 0$. Indeed, if $x \in K$ then $x \in K_l$ for some l, and $T_l(x) = x$.

The field K may be considered as a separable topological vector space over k with the inductive limit topology.

Denote by $\overline{K}$ the projective limit of the sequence $\{K_n\}$ with respect to the mappings $\{T_n\}$. It is defined as a subset of the direct product $\prod_{n=1}^{\infty} K_n$ consisting of those $x = (x_1, \ldots, x_n, \ldots)$, $x_n \in K_n$, for which $x_n = T_n(x_\nu)$ if $\nu > n$. The topology in $\overline{K}$ is introduced by seminorms

$$\|x\|_n = \|x_n\|, \quad n = 1, 2, \ldots .$$

Both spaces K and $\overline{K}$ are complete and reflexive [145, 108].

An element $x \in K$ may be identified with $(x_1, \dots, x_n, \dots) \in \overline{K}$ where $x_n = T_n(x)$. Thus K may be viewed as a subset of $\overline{K}$. the mappings T_n can be extended to linear continuous mappings from $\overline{K}$ to K_n by setting $T_n(x) = x_n$ for any $x = (x_1, \dots, x_n, \dots) \in \overline{K}$.

Proposition 6.1. *The topological vector spaces K and $\overline{K}$ are the strong duals of each other, with the pairing defined by the formula*

$$\langle x, y \rangle = T(xy_n)$$

where $x \in K_n \subset K$, $y = (y_1, \dots, y_n, \dots) \in \overline{K}$, $y_n \in K_n$. As a subset of $\overline{K}$, K is dense in $\overline{K}$.

Proof. Each field K_n may be identified as a topological vector space over k with its dual K_n^* – if $\varphi \in K_n^*$ then there exists a unique element $a_\varphi \in K_n$ such that

$$\varphi(x) = T(a_\varphi x), \quad x \in K_n,$$

and φ may be identified with a_φ.

Let $\nu > n$. The imbedding $\tau_{n\nu} : \ K_n \to K_\nu$ determines (due to the above identification) the adjoint mapping $\tau_{n\nu}^* : \ K_\nu \to K_n$. We shall show that $\tau_{n\nu}^* = T_n | K_\nu$.

Indeed, if $x \in K_n$, $\psi \in K_\nu^*$ then

$$\psi(\tau_{n\nu}(x)) = \varphi(x)$$

where $\varphi = \tau_{n\nu}^*(\psi) \in K_n^*$. This means that for any $x \in K_n$

$$\frac{1}{m_\nu} \mathrm{Tr}_{K_\nu/k}(a_\psi x) = \frac{1}{m_n} \mathrm{Tr}_{K_n/k}(a_\varphi x),$$

$a_\varphi \in K_n$, $a_\psi \in K_\nu$. On the other hand,

$$\mathrm{Tr}_{K_\nu/k}(a_\psi x) = \mathrm{Tr}_{K_n/k}(\mathrm{Tr}_{K_\nu/K_n}(a_\psi x)) = \mathrm{Tr}_{K_n/k}(x \mathrm{Tr}_{K_\nu/K_n}(a_\psi))$$

whence $a_\varphi = \frac{m_n}{m_\nu} \mathrm{Tr}_{K_\nu/K_n}(a_\psi) = T_n(a_\psi)$.

Now the duality result follows from the Komatsu-Morita duality theorem [109].

Let $y = (y_1, \dots, y_n, \dots) \in \overline{K}$. Consider a sequence $\{y^{(n)}\} \subset \overline{K}$ of elements $y^{(n)} = y_n \in K$ viewed as lying in $\overline{K}$. For each $j = 1, 2, \dots$ we have

$$T_j(y_n) = y_j, \quad n \geq j.$$

Hence $\|y^{(n)} - y\|_j \to 0$, $n \to \infty$, for each j, so that K is dense in $\overline{K}$. ∎

Corollary 6.2. *The space $\overline{K}$ is separable.*

Proof. One can take a countable dense subset in each K_n; their union is countable and dense in K with respect to the topology induced from $\overline{K}$, thus it is dense in $\overline{K}$. ■

6.2.2. Galois extensions. Suppose that K_n/k is a Galois extension for each n with the Galois group G_n. Then K/k is a Galois extension, its Galois group G is a projective limit of the groups G_n (see [21] and [72] for the theory of infinite Galois extensions). The group G consists of sequences $g = (g_1, \dots, g_n, \dots)$, $g_n \in G_n$, such that the restriction of an automorphism g_ν, $\nu > n$, to K_n coincides with g_n. The group operations in G are defined component-wise; if $x \in K$ then $x \in K_n$ for some n, and $g(x) = g_n(x)$.

Let us consider elements $x, y \in K$. Suppose that $y \in K_n$ for some n. Then for any $g \in G$

$$T(g(x)y) = T(g_n(x)y) = m_n^{-1}\mathrm{Tr}_{K_n/k}(g_n(x)y) = m_n^{-1}\mathrm{Tr}_{K_n/k}(xg_n^{-1}(y))$$

due to the G-invariance of the trace. Hence,

$$T(g(x)y) = T(xg^{-1}(y)), \quad x, y \in K. \tag{6.14}$$

The relation (6.14) shows that the action of G on K is continuous with respect to the topology induced on K by the $*$-weak topology of $\overline{K}$. Since K is strongly (and, of course, $*$-weakly) dense in $\overline{K}$, we come to the following result.

Proposition 6.2. *The action of the Galois group G on K can be extended by continuity to its $*$-weakly continuous action on $\overline{K}$.*

6.2.3. Examples. Note two examples of infinite extensions, which fit in with the above scheme. Both of them are very important in the algebraic number theory [61, 139].

If K is the maximal unramified extension of k, then one may take for K_n the unramified extension of k of the degree $n!$, $n = 2, 3, \dots$.

If K is the maximal Abelian extension of $k = \mathbb{Q}_p$, then a possible choice of K_n is the cyclotomic extension $C_{n!} = \mathbb{Q}_p(W_{n!})$, where W_l is the set of all roots of 1 of the degree l.

6.3 Measure and integration

6.3.1. σ-Algebras. Let us start from considering various σ-algebras of subsets in $\overline{K}$.

Denote by $\mathcal{B}(\overline{K})$ the Borel σ-algebra over $\overline{K}$ with respect to the strong topology of the (non-Archimedean) Frechet space $\overline{K}$; the Borel σ-algebra with respect to the $*$-weak topology will be denoted by $\mathcal{B}(\overline{K}, K)$. We shall consider also the smallest σ-algebra $\mathcal{A}(\overline{K})$ for which all the linear continuous functionals $\overline{K} \to k$ are measurable.

Proposition 6.3. *The above σ-algebras coincide:*

$$\mathcal{A}(\overline{K}) = \mathcal{B}(\overline{K}, K) = \mathcal{B}(\overline{K}).$$

Proof. It is clear that

$$\mathcal{A}(\overline{K}) \subset \mathcal{B}(\overline{K}, K) \subset \mathcal{B}(\overline{K}).$$

Hence it is sufficient to prove that $\mathcal{A}(\overline{K}) = \mathcal{B}(\overline{K})$ or (by definition of the topology in $\overline{K}$) that the set

$$V_n = \{x \in \overline{K} : \ \|x\|_n \le 1\}$$

belongs to $\mathcal{A}(\overline{K})$.

Consider its polar set

$$V_n^0 = \{\xi \in K : \ |\langle \xi, x\rangle|_1 \le 1 \ \ \forall x \in V_n\}.$$

By the bipolar theorem ([145], Theorem 4.14)

$$V_n = \{x \in \overline{K} : \ \sup_{\xi \in V_n^0} |\langle \xi, x\rangle|_1 \le 1\}.$$

On the other hand, V_n^0 is equicontinuous [145] (here we look at K as the dual of $\overline{K}$). Since $\overline{K}$ is separable we find (just as in the Archimedean case, see [133]) that the $*$-weak topology on V_n^0 is metrisable. Now being $*$-weakly compact [145] and metrisable, the set V_n^0 is separable in the $*$-weak topology. This implies the existence of such a countable subset $M \subset V_n^0$ that

$$V_n = \{x \in \overline{K} : \ \sup_{\xi \in M} |\langle \xi, x\rangle|_1 \le 1\} = \bigcap_{\xi \in M} \{x \in \overline{K} : \ |\langle \xi, x\rangle|_1 \le 1\}.$$

Thus $V_n \in \mathcal{A}(\overline{K})$. ∎

Similar results in the framework of real analysis can be found in [148].

6.3.2. Fourier transforms. Let us recall (in a form convenient for subsequent use) some results from harmonic analysis on local fields.

Denote by q_n the residue field cardinality for the field K_n, and by d_n the exponent of the different for the extension K_n/k. The Haar measure on the additive group of the field K_n will be denoted dx (the dependence on n is not shown; however this will not lead to any confusion). This measure will be normalized by the relation

$$\int_{\|x\|\le 1} dx = 1.$$

Using the relation $d(ax) = |a|_n\, dx,\ a \in K_n$, we find that

$$\int_{|x|_n\le q_n^\nu} dx = q_n^\nu; \quad \int_{|x|=q_n^\nu} dx = (1-q_n^{-1})q_n^\nu, \quad \nu\in\mathbb{Z}.$$

Let χ be a rank zero additive character on k. Then for any $a \in K_n,\ \nu \in \mathbb{Z}$

$$\int_{|x|_n\le q_n^\nu} \chi\circ \mathrm{Tr}_{K_n/k}(ax)\, dx = \begin{cases} q_n^\nu, & \text{if } |a|_n \le q_n^{d_n-\nu} \\ 0, & \text{if } |a|_n > q_n^{d_n-\nu} \end{cases} \tag{6.15}$$

$$\int_{|x|_n=q_n^\nu} \chi\circ \mathrm{Tr}_{K_n/k}(ax)\, dx = \begin{cases} (1-q_n^{-1})q_n^\nu, & \text{if } |a|_n \le q_n^{d_n-\nu} \\ -q_n^{\nu-1}, & \text{if } |a|_n = q_n^{d_n-\nu+1} \\ 0, & \text{if } |a|_n \ge q_n^{d_n-\nu+2} \end{cases} \tag{6.16}$$

(see (1.26), (1.27)).

The Fourier transform (defined initially on $L_1(K_n)$, with an isometric extension onto $L_2(K_n)$) will be defined (see (1.31), (1.32)) as

$$\widetilde{f}(\xi) = q_n^{-d_n/2}\int_{K_n} \chi\circ\mathrm{Tr}_{K_n/k}(x\xi) f(x)\, dx\,, \tag{6.17}$$

with the inversion formula

$$f(x) = q_n^{-d_n/2}\int_{K_n} \chi\circ\mathrm{Tr}_{K_n/k}(-x\xi) \widetilde{f}(\xi)\, d\xi\,, \tag{6.18}$$

valid, for example, for a locally constant complex valued function f with a compact support. The class $\mathcal{D}(K_n)$ of such functions is invariant under the Fourier transform.

In particular, if

$$\Omega_n(x) = \begin{cases} 1, & \text{if } \|x\|\le 1 \\ 0, & \text{if } \|x\| > 1 \end{cases}$$

($x \in K_n$) then by (6.15) and (6.17)

$$\widetilde{\Omega_n}(\xi) = \begin{cases} q_n^{-d_n/2}, & \text{if } \|\xi\| \le q_n^{d_n/m_n} \\ 0, & \text{if } \|\xi\| > q_n^{d_n/m_n} \end{cases} \tag{6.19}$$

($\xi \in K_n$).

6.3.3. Gaussian measure. Let us consider a function $\Omega(x)$, $x \in K$, given by the formula

$$\Omega(x) = \Omega_n(x), \quad x \in K_n.$$

This function (which is clearly well-defined) is continuous on K by the definition of the inductive limit topology. It follows from (6.19) and (6.18) that Ω_n is positive definite on each K_n. Hence, Ω is a positive definite function on K.

Using Theorem 6.6 and Propositions 6.1, 6.3 we establish the existence of a unique Radon probability measure μ on $\mathcal{B}(\overline{K})$ such that

$$\Omega(a) = \int_{\overline{K}} \chi(\langle a, x\rangle)\, d\mu(x), \quad a \in K. \tag{6.20}$$

We begin our study of properties of the measure μ with integrating a "cylindrical" function of the form $f(x) = \varphi(T_n(x))$ where $\varphi \in \mathcal{D}(K_n)$. The set of all such functions will be denoted by $\mathcal{D}(\overline{K})$. Note that $\mathcal{D}(\overline{K})$ is a linear set - if $f(x) = \varphi(T_n(x)) + \psi(T_\nu(x))$, $\nu > n$, then $f(x) = (\varphi \circ T_n + \psi)(T_\nu(x))$.

The function φ can be written as

$$\varphi(y) = \int_{K_n} \chi(T(ay))\psi(a)\, da, \quad \psi \in \mathcal{D}(K_n),$$

where ψ is the inverse Fourier transform of the function $q_n^{-d_n/2}\varphi(m_n y)$. By (6.20) and the Fubini theorem, using the relation $T \circ T_n = T$ we find that

$$\int_{\overline{K}} f(x)\, d\mu(x) = \int_{a \in K_n:\ \|a\| \le 1} \psi(a)\, da.$$

In view of (6.19) the Plancherel formula implies

$$\int_{\overline{K}} f(x)\, d\mu(x) = q_n^{-d_n} \int_{y \in K_n:\ \|y\| \le q_n^{d_n/m_n}} \varphi(m_n y)\, dy$$

whence

$$\int_{\overline{K}} f(x)\, d\mu(x) = q_n^{-d_n} \|m_n\|^{-m_n} \int\limits_{z\in K_n:\ \|z\|\le q_n^{d_n/m_n}\|m_n\|} \varphi(z)\, dz. \tag{6.21}$$

By an approximation argument, we can easily show now that the formula (6.21) is valid for any function $f(x) = \varphi(T_n(x))$ with $\varphi \in L_1(K_n)$.

Proposition 6.3 means that the pair $(\overline{K}, \mathcal{B}(\overline{K}))$ is a measurable vector space (the triple $(\overline{K}, K, \mathcal{B}(\overline{K}))$ is coordinated), so that there is a notion of a Gaussian measure on $(\overline{K}, \mathcal{B}(\overline{K}))$.

Proposition 6.4. *μ is a k-Gaussian measure on $(\overline{K}, \mathcal{B}(\overline{K}))$.*

Proof. Let $a \in K$, B be a Borel subset of k, $\varphi_B : k \to \mathbb{R}$ be an indicator of the set B. Consider the function

$$I(a) = \int_{\overline{K}} \varphi_B(\langle a, x\rangle)\, d\mu(x), \quad a \in K.$$

By Theorems 6.1 and 6.3, it suffices to prove that

$$I(a) = q_1^m \int_{z\in B, |z|_1 \le q_1^{-m}} dz \tag{6.22}$$

for some $m \in \mathbb{Z}$.

Repeating the argument from the proof of (6.21) we find that

$$I(a) = \int_k \varphi_B(z)\, dz \int_{|y|_1 \le \|a\|^{-1}} \chi(-zy)\, dy.$$

Let $a \in K_n$, $|a|_n = q_n^N$, $N \in \mathbb{Z}$. Then $\|a\| = q_n^{N/m_n} = q_1^{Nf_n/m_n} = q_1^{N/e_n}$ where f_n, e_n are the modular degree and the ramification index of the extension K_n/k.

By the identity (6.15)

$$\int_{|y|_1 \le \|a\|^{-1}} \chi(-zy)\, dy = \int_{|y|_1 \le q_1^{-N/e_n}} \chi(zy)\, dy = \int_{|y|_1 \le q_1^{[-N/e_n]}} \chi(zy)\, dy =$$

$$= \begin{cases} q_1^{[-N/e_n]}, & \text{if } |z|_1 \le q_1^{-[-N/e_n]} \\ 0, & \text{if } |z|_1 > q_1^{-[-N/e_n]}, \end{cases}$$

and we have come to (6.22) with $m = [-N/e_n]$. ■

Several important properties of the measure μ are collected in the following theorem.

Theorem 6.7. *(a) The measure μ is concentrated on the set*

$$\Sigma = \left\{ x \in \overline{K} : \quad \|T_n(x)\| \le q_n^{d_n/m_n} \|m_n\|, \ n = 1, 2, \dots \right\}.$$

(b) The measure μ is invariant with respect to additive shifts by elements from Σ.

(c) An additive shift by an element from $\overline{K} \setminus \Sigma$ transforms μ into a measure orthogonal to μ.

(d) if $a \in K_n$ then

$$\operatorname{ess\,sup}_{x \in \overline{K}} \ |\langle a, x \rangle|_1 = q_1^{-[-N/e_n]} \tag{6.23}$$

where $|a|_n = q_n^N$, $N \in \mathbb{Z}$.

(e) The set $\mathcal{D}(\overline{K})$ is dense in $L_s(\overline{K})$ for any $s \in [1, \infty)$.

Remarks. 1). It may happen that $\Sigma \cap K = \{0\}$. For example, if K is the maximal unramified extension of k then $d_n = 0$, $\|m_n\| = |n!|_1 \to 0$ for $n \to \infty$. On the other hand, if $y \in K_r$ then $T_n(y) = y$ for $n \ge r$, and $y \in \Sigma$ implies $y = 0$.

2). The equality (6.23) should be compared with the expression for the L_2-norm of a linear functional on a real Hilbert space with a Gaussian measure (see [22]). As expected, in the non-Archimedean case the L_∞-norm has appeared. On the other hand, the property (b) has no clear counterpart in the real infinite-dimensional analysis. Meanwhile, such properties are natural for the non-Archimedean case, in view of the general Theorem 6.4.

Proof of Theorem 6.7. (a) Consider for each $n = 1, 2, \dots$ the set

$$\Sigma_n = \left\{ x \in \overline{K} : \quad \|T_n(x)\| \le q_n^{d_n/m_n} \|m_n\| \right\}.$$

Suppose that $x \in \Sigma_\nu$, $\nu > n$. Then

$$\|T_n(x)\| = \|m_n\| \cdot \|m_\nu\|^{-1} \|\mathrm{Tr}_{K_\nu/K_n}(T_\nu(x))\|.$$

Since $\|m_\nu^{-1} T_\nu(x)\| \le q_\nu^{d_\nu/m_\nu}$ we have

$$|m_\nu^{-1} T_\nu(x)|_n \le q_\nu^{d_\nu}$$

whence (see Chapter 8 in [156])

$$|m_\nu^{-1} \mathrm{Tr}_{K_\nu/K_n}(T_\nu(x))|_n \le q_n^l$$

where $l \in \mathbb{Z}$, $e_{n\nu}(l-1) < d_\nu - d_{n\nu} \le e_{n\nu} l$, $e_{n\nu}$ and $d_{n\nu}$ are the ramification index and the exponent of the different for the extension K_ν/K_n. On the other hand [156], $d_\nu = e_{n\nu} d_n + d_{n\nu}$, so that $l = d_n$, and we obtain that $x \in \Sigma_n$.

Hence, $\Sigma_\nu \subset \Sigma_n$ for $\nu > n$. It follows from (6.21) that $\mu(\Sigma_n) = 1$ for each n. Since $\Sigma = \bigcap_{n=1}^{\infty} \Sigma_n$, we get the desired equality $\mu(\Sigma) = 1$.

(b) It follows from the definition of the measure μ by the relation (6.20) that we need only to prove, for any $y \in \Sigma$ and any $a \in K$ with $\|a\| \leq 1$, the inequality

$$|\langle a, y \rangle|_1 \leq 1. \tag{6.24}$$

Suppose that $a \in K_n$, $\|a\| \leq 1$. Then

$$\langle a, y \rangle = T(ay_n) = \mathrm{Tr}_{K_n/k}(m_n^{-1} a y_n),$$

$y_n = T_n(y)$. Since $\|y_n\| \leq \|m_n\| q_n^{d_n/m_n}$ we find that

$$|m_n^{-1} a y_n|_n \leq q_n^{d_n}$$

so that the inequality (6.24) follows from the definition of d_n.

(c) This property is an immediate consequence of (a) and the ultrametric property of absolute values.

(d) We have shown (while proving Proposition 6.4) that

$$\mu\left(\left\{x \in \overline{K}: \ |\langle a, x \rangle|_1 > q_1^{-[-N/e_n]}\right\}\right) = 0.$$

This implies the inequality

$$\text{ess sup}_{x \in \overline{K}} \ |\langle a, x \rangle|_1 \leq q_1^{-[-N/e_n]}$$

To prove the inverse inequality recall that $\langle a, x \rangle = m_n^{-1} \mathrm{Tr}_{K_n/k}(aT_n(x))$. Suppose that $|m_n|_n = q_n^\nu$, $\nu \in \mathbb{Z}$. By properties of the trace [156] there exists such an element $y_n \in K_n$ that $|y_n|_n = q_n^{\nu + d_n}$,

$$|\mathrm{Tr}_{K_n/k}(ay_n)|_1 = q_1^l$$

where $l \in \mathbb{Z}$, $e_n(l-1) < N + \nu \leq e_n l$, whence

$$|T(ay_n)|_1 = q_1^{l - \nu/e_n} \geq q_1^{-[-N/e_n]}.$$

If a natural number M is sufficiently large then the inequality $\|z - y_n\| \leq q_1^{-M}$ ($z \in K_n$) implies the equality $|T(az)|_1 = q_1^{l-\nu/e_n}$.

Consider a subset in $\overline{K}$ of the form

$$W = \left\{x \in \overline{K}: \ \|T_n(x) - y_n\| \leq q_1^{-M}\right\},$$

with the number M mentioned above. If $x \in W$ then

$$|\langle a, x\rangle|_1 = |T(aT_n(x))|_1 = q_1^{l-\nu/e_n} \geq q_1^{-[-N/e_n]}.$$

On the other hand, using the formula (6.21) and taking into account that $\|y_n\| = \|m_n\| q_n^{d_n/m_n}$ we find that $\mu(W) \neq 0$ if M is taken sufficiently large. This means that

$$\text{ess sup}_{x\in\overline{K}} \ |\langle a, x\rangle|_1 \geq q_1^{-[-N/e_n]}$$

(e) Consider subsets in $\overline{K}$ having the form $A_{n,B} = \{x \in \overline{K} : \ T_n(x) \in B\}$ where $B \subset K_n$ is a Borel set, $n \geq 1$. If $\nu > n$ then $A_{n,B} = \{x \in \overline{K} : \ T_\nu(x) \in C\}$ where $C = \{y \in K_\nu : \ T_n(y) \in B\}$. It follows from this observation that the sets $A_{n,B}$ form a Boolean algebra of sets which of course generates $\mathcal{B}(\overline{K})$.

By general principles of measure theory this implies the existence, for any $V \in \mathcal{B}(\overline{K})$ and any $\varepsilon > 0$, of such a set $A_{n,B}$ that $\mu(V \Delta A_{n,B}) < \varepsilon$ where Δ means symmetric difference. Here one can take $B \subset \{y \in K_n : \|y\| \leq q_n^{d_n/m_n} \|m_n\|\}$.

Hence an indicator of any set from $\mathcal{B}(\overline{K})$ can be approximated in $L_s(\overline{K})$ by indicators of the sets $A_{n,B}$ which belong to $\mathcal{D}(\overline{K})$. ■

Let $\mathcal{E}(\overline{K})$ be the set of functions $f(x) = \varphi(T_n(x))$ where $n \geq 1$, φ is a locally constant function on K_n. It follows from Theorem 6.7 that $\mathcal{E}(\overline{K}) \subset L_1(\overline{K})$, and the formula (6.21) remains valid for $f \in \mathcal{E}(\overline{K})$.

In the case of Galois extensions (see Sect. 6.2) the measure μ is invariant with respect to the Galois group G. This follows immediately from (6.14) and (6.20).

6.4 Fourier transform and fractional differentiation

6.4.1. An image of the Fourier transform. We define the Fourier transform index on an infinite extension of a function $f \in L_1(\overline{K})$ as

$$\widehat{f}(\xi) = \int_{\overline{K}} \chi(\langle \xi, x\rangle) f(x)\, d\mu(x), \quad \xi \in K.$$

Note that to avoid confusion we use here the notation $\widehat{f}$ in contrast to the notation $\widetilde{\varphi}$ for the Fourier transform over a local field. Sometimes we shall also write $\mathcal{F}f = \widehat{f}$.

Let $f \in \mathcal{E}(\overline{K})$, $f(x) = \varphi(T_n(x))$. An explicit form of $\widehat{f}(\xi)$ can be easily obtained from (6.21): if $\xi \in K_\nu$, $\nu \geq n$, then

$$\widehat{f}(\xi) = q_\nu^{-d_\nu} \|m_\nu\|^{-m_\nu} \int\limits_{\zeta \in K_\nu:\ \|\zeta\| \leq q_\nu^{d_\nu/m_\nu} \|m_\nu\|} \chi(T(\xi\zeta))\varphi(T_n(\zeta))\, d\zeta. \quad (6.25)$$

Writing (6.25) for $\nu = n$ and using the inversion formula (6.18) we find that

$$\varphi(z) = \int_{K_n} \chi(-T(\xi z))\widehat{f}(\xi)\, d\xi, \quad z \in K_n,\ \|z\| \leq q_n^{d_n/m_n} \|m_n\|. \quad (6.26)$$

Let us describe the image of $\mathcal{E}(\overline{K})$ under the transform $\mathcal{F}$. For $\xi \in K$ we shall write

$$\operatorname{dist}(\xi, K_n) = \inf_{\eta \in K_n} \|\xi - \eta\|.$$

Theorem 6.8. *A function $F(\xi)$, $\xi \in K$, is an image of some function $f \in \mathcal{E}(\overline{K})$, $f(x) = \varphi(T_n(x))$, under the transform $\mathcal{F}$, if and only if:* (i) *the restriction of F to each field K_ν, $\nu \geq n$, belongs to $\mathcal{D}(K_\nu)$;* (ii) *$F(\xi + \xi') = F(\xi)$ for any $\xi, \xi' \in K$ with $\|\xi'\| \leq 1$;* (iii) *$F(\xi) = 0$ if $\operatorname{dist}(\xi, K_n) > 1$.*

Proof. The necessity of (i) and (ii) follows immediately from (6.25). According to (6.26),

$$\varphi(T_n(\zeta)) = \int_{K_n} \chi(-T(\zeta\eta))\widehat{f}(\eta)\, d\eta, \quad (6.27)$$

if $\zeta \in K_\nu$, $\nu > n$, $\|T_n(\zeta)\| \leq q_n^{d_n/m_n} \|m_n\|$. As we have seen, the last inequality is valid if $\|\zeta\| \leq q_\nu^{d_\nu/m_\nu} \|m_\nu\|$. Thus we may substitute (6.27) into (6.25). We obtain for any $\xi \in K_\nu$ that

$$\begin{aligned} F(\xi) &= \widehat{f}(\xi) \\ &= q_\nu^{-d_\nu} \|m_\nu\|^{-m_\nu} \int_{K_n} F(\eta)\, d\eta \int\limits_{\zeta \in K_\nu:\ \|\zeta\| \leq q_\nu^{d_\nu/m_\nu} \|m_\nu\|} \chi(T(\zeta(\xi - \eta)))\, d\zeta \\ &= \int\limits_{\eta \in K_n:\ \|\eta - \xi\| \leq 1} F(\eta)\, d\eta \end{aligned}$$

by virtue of (6.15). Using (ii) we find that

$$F(\xi) = F(\xi) \int_{\eta \in K_n:\ \|\eta - \xi\| \le 1} d\eta.$$

If $\operatorname{dist}(\xi, K_n) > 1$ then the domain of integration is empty so that $F(\xi) = 0$.

Now we turn to the sufficiency of (i)-(iii). Put

$$\varphi(z) = \int_{K_n} \chi(T(-\xi z))F(\xi)\, d\xi, \quad z \in K_n, \tag{6.28}$$

$f(x) = \varphi(T_n(x)), \quad x \in \overline{K}$. Let us compute $\widehat{f}$. Suppose that $\xi \in K_\nu$, $\nu \ge n$. Then we may substitute (6.28) into (6.25). Repeating the previous calculation we obtain that

$$\widehat{f}(\xi) = F(\xi) \int_{\eta \in K_n:\ \|\eta - \xi\| \le 1} d\eta.$$

If $\operatorname{dist}(\xi, K_n) > 1$ then $\widehat{f}(\xi) = 0 = F(\xi)$. Let $\operatorname{dist}(\xi, K_n) \le 1$. There exists such an element $\sigma \in K_n$ that $\operatorname{dist}(\xi, K_n) = \|\xi - \sigma\|$ (see [111]). Thus $\|\xi - \sigma\| \le 1$, and by the ultra-metric inequality

$$\{\eta \in K_n:\ \|\eta - \xi\| \le 1\} = \{\eta \in K_n:\ \|\eta - \sigma\| \le 1\}.$$

whence

$$\int_{\eta \in K_n:\ \|\eta - \xi\| \le 1} d\eta = \int_{\|\eta - \sigma\| \le 1} d\eta = 1$$

due to the invariance of the Haar measure on K_n. Hence $\widehat{f}(\xi) = F(\xi)$ for this case too. ■

Using (6.21), (6.25) and the isometric property of the Fourier transform over a local field we come to the Plancherel formula for $f(x) = \varphi(T_n(x))$, $f \in \mathcal{E}(\overline{K})$: if $\nu \ge n$ then

$$\int_{\overline{K}} |f(x)|^2\, d\mu(x) = \int_{K_\nu} |\widehat{f}(\xi)|^2\, d\xi =$$

$$= q_n^{-d_n} \|m_n\|^{-m_n} \int_{z \in K_n:\ \|z\| \le q_n^{d_n/m_n} \|m_n\|} |\varphi(z)|^2\, dz.$$

If $g(x) = \psi(T_m(x))$ is another function from $\mathcal{E}(\overline{K})$ and $\nu \geq \max(m, n)$ then

$$\int_{\overline{K}} f(x)\overline{g(x)}\, d\mu(x) = \int_{K_\nu} \widehat{f}(\xi)\overline{\widehat{g}(\xi)}\, d\xi.$$

Hence, the transform $\mathcal{F}$ is the isometry of $L_2(\overline{K})$ onto the completion of the image $\mathcal{F}\mathcal{E}(\overline{K})$ (described in Theorem 6.8) with respect to the norm

$$\lim_{\nu\to\infty} \int_{K_\nu} |F(\xi)|^2\, d\xi.$$

6.4.2. Fractional differentiation operator. Next we shall study an operator D^α defined (initially) on $\mathcal{E}(\overline{K})$ as

$$D^\alpha = \mathcal{F}^{-1}\Delta^\alpha\mathcal{F}, \quad \alpha > 0,$$

where Δ^α is an operator of multiplication in $\mathcal{F}\mathcal{E}(\overline{K})$ by the function

$$\Delta^\alpha(\xi) = \begin{cases} \|\xi\|^\alpha, & \text{if } \|\xi\| > 1 \\ 0, & \text{if } \|\xi\| \leq 1 \end{cases}, \quad \xi \in K.$$

Theorem 6.8 shows that D^α is well-defined. It is easy to see that D^α is essentially self-adjoint as an operator on $L_2(\overline{K})$.

Let us begin with a simple property of D^α justifying its interpretation as an analogue of fractional differentiation.

Let $\varphi_a(x) = \chi(\langle a, x\rangle)$, $x \in \overline{K}$, where $a \in K$. Note that if $\|a\| \leq 1$ then $\varphi_a(x) = 1$ for all $x \in \Sigma$ or, in other words, $\varphi_a(x) = 1$ μ-almost everywhere on $\overline{K}$.

Proposition 6.5. *For μ-almost all $x \in \overline{K}$*

$$(D^\alpha\varphi_a)(x) = \begin{cases} 0, & \textit{if } \|a\| \leq 1 \\ \|a\|^\alpha\varphi_a(x), & \textit{if } \|a\| > 1 \end{cases}$$

Proof. Suppose that $a \in K_n$. Then $\varphi_a(x) = \rho(T_n(x))$ where $\rho(z) = \chi(T(az))$, $z \in K_n$. Using (6.25) and (6.15) we find that

$$\widehat{\varphi_a}(\xi) = \Omega_n(\xi + a), \quad \xi \in K_n,$$

where

$$\Omega_n(z) = \begin{cases} 1, & \text{if } \|z\| \leq 1 \\ 0, & \text{if } \|z\| > 1 \end{cases}$$

Thus $(D^\alpha \varphi_a)(x) = \psi(T_n(x))$ where

$$\psi(z) = \int_{K_n} \chi(T(-z\xi))\Delta^\alpha(\xi)\Omega_n(\xi + a)\, d\xi =$$

$$= \int\limits_{\|\xi\|>1,\ \|\xi+a\|\le 1} \|\xi\|^\alpha \chi(T(-z\xi))\, d\xi$$

for $\|z\| \le q_n^{d_n/m_n}\|m_n\|$. If $\|a\| \le 1$ then $\psi(z) = 0$. If $\|a\| > 1$ then $\|\xi\| = \|a\|$ on the whole integration domain so that

$$\psi(z) = \|a\|^\alpha \int_{\|\xi+a\|\le 1} \chi(T(-z\xi))\, d\xi = \|a\|^\alpha \chi(T(az))$$

if $\|z\| \le q_n^{d_n/m_n}\|m_n\|$. ■

It follows from the definition of D^α that

$$\int_{\overline{K}} (D^\alpha f)(x)\, d\mu(x) = 0$$

for any $f \in \mathcal{E}(\overline{K})$. In other words, μ is a *harmonic measure* for D^α.

Though we are interested primarily in the operator D^α with $\alpha > 0$ (in view of some applications presented below), we shall need also the case $\alpha \le 0$. An elementary calculation shows that

$$(D^0 f)(x) = f(x) - \int_{\overline{K}} f(y)\, d\mu(y), \quad x \in \Sigma, \tag{6.29}$$

for any $f \in \mathcal{E}(\overline{K})$.

It follows from Theorem 6.8 that $(D^\alpha f)(x)$, $x \in \Sigma$, is well-defined (if $f \in \mathcal{E}(\overline{K})$) for all $\alpha \in \mathbb{C}$ being in fact an entire function with respect to α. It is clear that $D^{\alpha_1}D^{\alpha_2} f = D^{\alpha_1+\alpha_2} f$, $\alpha_1, \alpha_2 \in \mathbb{C}$, $f \in \mathcal{E}(\overline{K})$.

Proposition 6.6. *If $f \in \mathcal{E}(\overline{K})$, $\int\limits_{\overline{K}} f(x)\, d\mu(x) = 0$ then the general solution $u \in \mathcal{E}(\overline{K})$ of the equation*

$$D^\alpha u = f, \quad \alpha > 0, \tag{6.30}$$

is of the form $u = D^{-\alpha} f + C$ where C is an arbitrary constant.

Proof. The function u is a solution of (6.30) in view of the relation (6.29), since $D^\alpha C = 0$ by Proposition 6.5. On the other hand, let $v \in \mathcal{E}(\overline{K})$, $D^\alpha v = 0$ μ-almost everywhere on $\overline{K}$. By the definition of D^α, we find that $\widehat{v}(\xi) = 0$ if $\|\xi\| > 1$. Now if $v(x) = \varphi(T_n(x))$ then

$$\varphi(z) = \int_{\|\xi\|\le 1} \chi(T(-z\xi))\widehat{v}(\xi)\,d\xi = \widehat{v}(0)$$

for all $z \in K_n$ with $\|z\| \le q_n^{d_n/m_n}\|m_n\|$. This means that $v(x) = \text{const}$ μ-almost everywhere on $\overline{K}$. ■

6.4.3. Hyper-singular integral representation. As we know, the fractional differentiation operator on a local field admits a hyper-singular integral representation not containing the Fourier transform. Now we give a representation of this kind for our present situation.

Proposition 6.7. *Let $\alpha > 0$, $f \in \mathcal{E}(\overline{K})$, $f(x) = \varphi(T_n(x))$. Then $(D^\alpha f)(x) = \psi(T_n(x))$ where*

$$\psi(z) = q_n^{d_n\alpha/m_n}\frac{1-q_n^{\alpha/m_n}}{1-q_n^{-1-\alpha/m_n}}\|m_n\|^{-m_n}$$
$$\times \int_{x\in K_n:\ \|x\|\le q_n^{d_n/m_n}\|m_n\|} \Bigg[\|x\|^{-m_n-\alpha}\|m_n\|^{m_n+\alpha} + \frac{1-q_n^{-1}}{q_n^{\alpha/m_n}-1}q_n^{-d_n(1+\alpha/m_n)}\Bigg][\varphi(z-x)-\varphi(z)]\,dx, \quad (6.31)$$

$z \in K_n$, $\|z\| \le q_n^{d_n/m_n}\|m_n\|$.

Proof. Let us compute the Fourier transform over K_n for the function $\Delta^\lambda(\xi)$ with $\operatorname{Re}\lambda < -1$. We have

$$\widetilde{\Delta^\lambda}(x) = q_n^{-d_n/2}\sum_{N=1}^\infty q_n^{N\lambda/m_n}\int_{|\xi|_n = q_n^N}\left(\chi\circ \operatorname{Tr}_{K_n/k}\right)(x\xi)\,d\xi.$$

Using (6.16) and summing the geometric progression we obtain after elementary transformation that

$$\widetilde{\Delta^\lambda}(x) = q_n^{d_n(1/2+\lambda/m_n)}\frac{1-q_n^{\lambda/m_n}}{1-q_n^{-1-\lambda/m_n}}\left(|x|_n^{-1-\lambda/m_n} + \right.$$

$$+q_n^{-d_n(1+\lambda/m_n)} \frac{1-q_n^{-1}}{q_n^{\lambda/m_n}-1}\Bigg) \omega_n(x), \quad x \in K_n, \tag{6.32}$$

where

$$\omega_n(x) = \begin{cases} 1, & \text{if } |x|_n \le q_n^{d_n} \\ 0, & \text{if } |x|_n > q_n^{d_n} \end{cases}.$$

Under our normalizations the relation between Fourier transform and convolution reads $\widetilde{u * v} = q_n^{d_n/2}\tilde{u}\tilde{v}$ (with the usual assumptions with regard to the functions u, v). Thus for $\operatorname{Re}\lambda < -1$ we get $(D^\lambda f)(x) = \psi_\lambda(T_n(x))$ where

$$\psi_\lambda(m_n y) = \left(\widetilde{\Delta^\lambda} * \varphi_n\right)(y), \quad y \in K_n, \quad \|y\| \le q_n^{d_n/m_n},$$

$\widetilde{\Delta^\lambda}$ is given by (6.32),

$$\varphi_n(\eta) = \begin{cases} q_n^{-d_n/2}\varphi(m_n\eta), & \text{if } |\eta|_n \le q_n^{d_n} \\ 0, & \text{if } |\eta|_n > q_n^{d_n} \end{cases}.$$

Since $\Delta^\lambda(0) = 0$ we find that

$$\int_{|x|_n \le q_n^{d_n}} \widetilde{\Delta^\lambda}(x)\, dx = 0, \quad \Re\lambda < -1.$$

Thus the above convolution may be written as follows:

$$\psi_\lambda(m_n y) = q_n^{d_n\lambda/m_n} \frac{1-q_n^{\lambda/m_n}}{1-q_n^{-1-\lambda/m_n}} \int_{|x|_n \le q_n^{d_n}} \Bigg[|x|_n^{-1-\lambda/m_n} +$$

$$+\frac{1-q_n^{-1}}{q_n^{\lambda/m_n}-1} q_n^{-d_n(1+\lambda/m_n)} \Bigg] [\varphi(m_n y - m_n x) - \varphi(m_n y)]\, dx \tag{6.33}$$

Since the left-hand side of (6.33) is an entire function with respect to λ, the analytic continuation of the right-hand side leads to the required equality (6.31). ■

Below we shall show that there exists also a representation in terms of the function f itself:

$$(D^\alpha f)(y) = \int_{\overline{K}} [f(y) - f(x+y)]\Pi(dx) \tag{6.34}$$

where Π is a measure on $\mathcal{B}(\overline{K} \setminus \{0\})$ finite outside any neighbourhood of zero.

6.5 Spectrum

We shall preserve the notation D^α for its closure in $L_2(\overline{K}) = L_2(\Sigma, d\mu)$. It is clear from the definition that the spectrum of D^α coincides with the closure in $\mathbb{R}$ of the range of the function $\Delta^\alpha(\xi)$, $\xi \in K$. In order to investigate the structure of the spectrum, we need some auxiliary results.

It follows from the duality theory for direct and inverse limits of locally compact groups [64] that the character group $\overline{K}^*$ of the additive group of $\overline{K}$ is isomorphic to K. The isomorphism is given by the relation

$$\psi(y) = \chi(< a_\psi, y >), \quad y \in \overline{K}, \tag{6.35}$$

where $a_\psi \in K$ is an element corresponding to the character ψ.

Denote $O = \{\xi \in K : \ \|\xi \leq 1\}$, $O_n = O \cap K_n$.

Lemma 6.5. *The dual group* Σ^* *of the subgroup* $\Sigma \subset \overline{K}$ *is isomorphic to the quotient group* K/O.

Proof. We may assume (without restricting generality) that $k = \mathbb{Q}_p$. Let

$$\Sigma^{(n)} = \{z \in K_n : \ \|z\| \leq q_n^{d_n/m_n} \|m_n\|\}, \quad n = 1, 2, \dots .$$

It is clear that $\Sigma^{(n)}$ is a compact (additive) group. If $z \in \Sigma^{(\nu)}$, $\nu > n$, then

$$\|T_n(z)\| = \|m_n\| \cdot \|m_\nu\|^{-1} \|\mathrm{Tr}_{K_\nu/K_n}(z)\|$$

where $|m_\nu^{-1} z|_\nu \leq q_\nu^{d_\nu}$. Therefore (see Chapter 8 in [156])

$$|m_\nu^{-1} \mathrm{Tr}_{K_\nu/K_n}(z)|_n \leq q_n^l$$

where $l \in \mathbb{Z}$, $e_{n\nu}(l-1) < d_\nu - d_{n\nu} \leq e_{n\nu} l$, $e_{n\nu}$ and $d_{n\nu}$ are the ramification index and the exponent of the different for the extension K_ν/K_n. On the other hand, $d_\nu = e_{n\nu} d_n + d_{n\nu}$, so that $l = d_n$, and we find that $T_n(z) \in \Sigma^{(n)}$.

Hence, $T_n : \ \Sigma^{(\nu)} \to \Sigma^{(n)}$. It is clear that T_n is a continuous homomorphism. As a result, Σ can be identified with an inverse limit of compact groups $\Sigma^{(n)}$ with respect to the sequence of homomorphisms T_n. Using the auto-duality of each field K_n, we can identify the group dual to $\Sigma^{(n)}$ with K_n/Φ_n where Φ_n is the annihilator of $\Sigma^{(n)}$ in K_n. On the other hand, $\Phi_n = O_n$.

Indeed,

$$\begin{aligned} \Phi_n &= \{\xi \in K_n : \ \chi(T(z\xi)) = 1 \ \forall z \in \Sigma^{(n)}\} \\ &= \{\xi \in K_n : \ |T(z\xi)|_1 \leq 1 \ \forall z \in \Sigma^{(n)}\}. \end{aligned}$$

If $\xi \in O_n$ then $|m_n^{-1} z\xi|_n \le q_n^{d_n}$ for any $z \in \Sigma^{(n)}$, whence $\xi \in \Phi_n$ by the definition of the number d_n.

Thus $O_n \subset \Phi_n$. Conversely, suppose that $\xi \in \Phi_n \setminus O_n$, that is $\|\xi\| > 1$, $|T(z\xi)|_1 \le 1$ for all $z \in \Sigma^{(n)}$. Let z be such that $\|z\| = \|m_n\| q_n^{d_n/m_n}$. Then

$$|m_n^{-1}|_n \cdot |\xi|_n \cdot |z|_n = |\xi|_n q_n^{d_n} > q_n^{d_n}.$$

It follows from the properties of trace that z can be chosen in such a way that $|\mathrm{Tr}_{K_n/k}(m_n^{-1}\xi z)|_1 > 1$, and we have a contradiction. So $\Phi_n = O_n$.

It follows from the identity

$$T(\xi T_n(z)) = T(\xi z), \quad \xi \in K_n,\ z \in K_\nu,\ \nu > n,$$

that the natural imbeddings $K_n/O_n \to K_\nu/O_\nu$, $\nu > n$, are the dual mappings to the homomorphisms $T_n : \Sigma^{(\nu)} \to \Sigma^{(n)}$. Using the duality theorem [64], we find that

$$\Sigma^* = \varinjlim K_n/O_n \tag{6.36}$$

where the direct limit is taken with respect to the imbeddings. The right-hand side of (6.36) equals K/O. ■

Note that the above isomorphism is given by the same formula (6.35) where this time a_ψ is an arbitrary representative of a coset from K/O.

Let $\varphi_a(x) = \chi(< a, x >)$, $x \in \overline{K}$, where $a \in K$, $\|a\| > 1$ or $a = 0$. As we know,

$$(D^\alpha \varphi_a)(x) = \|a\|^\alpha \varphi_a(x) \tag{6.37}$$

for μ-almost all $x \in \overline{K}$. Note that the set of values of the function $a \mapsto \|a\|^\alpha$ with $\|a\| > 1$ coincides with the set

$$\left\{q_n^{\alpha N/m_n},\ n, N = 1, 2, \dots\right\} = \left\{q_1^{\alpha N/e_n},\ n, N = 1, 2, \dots\right\} \tag{6.38}$$

where e_n is the ramification index of the extension K_n/k. Denote its residue degree by f_n. It is well known that the sequences $\{f_n\}$, $\{e_n\}$ are non-decreasing and $e_n f_n = m_n$.

Theorem 6.9. *Let $A \subset K$ be a complete system of representatives of cosets from K/O. Then $\{\varphi_a\}_{a\in A}$ is the orthonormal eigenbasis for the operator D^α in $L_2(\Sigma, d\mu)$. As a set, its spectrum equals the set (6.38) complemented with the point $\lambda = 0$. Each non-zero eigenvalue of D^α has an infinite multiplicity. The point $\lambda = 0$ is an accumulation point for eigenvalues if and only if $e_n \to \infty$.*

Proof. The first statement follows from (6.37) and Lemma 6.5. The assertion about the accumulation at zero is obvious from (6.38).

Let us construct A as the union of an increasing family $\{A_n\}$ of complete systems of representatives of cosets from K_n/O_n. Each A_n consists of elements of the form $a = \pi_n^{-N}\left(\sigma_1 + \sigma_2\pi_n + \cdots + \sigma_{N-1}\pi_n^{N-1}\right)$, $N \geq 1$, where $\sigma_1, \dots, \sigma_{N-1}$ belong to the set F_n of representatives of the residue field corresponding to the field K_n, $\sigma_1 \neq 0$, π_n is a prime element of K_n. We have $|a|_n = q_n^N$ so that $\|a\| = q_1^{Nf_n/m_n} = q_1^{N/e_n}$. Meanwhile card $F_n = q_n = q_1^{f_n}$.

If $e_n \to \infty$ then the same value of $\|a\|$ corresponds to elements from infinitely many different sets A_n with different values of N (e_n is a multiple of e_{n-1} due to the chain rule for the ramification indices; see [38]). If the sequence $\{e_n\}$ is bounded, then it must stabilize, and we obtain the same value of $\|a\|$ for elements from infinitely many sets A_n with possibly the same N. However, in this case $f_n \to \infty$, the number of such elements (for a fixed N) is $Nq_1^{f_n} - 1$ ($\to \infty$ for $n \to \infty$). In both cases we see that all the non-zero eigenvalues have an infinite multiplicity. ■

Note that the cases where $e_n \to \infty$ or $e_n \leq$ const both appear in important examples of infinite extensions. Let K be the maximal unramified extension of k. Then, as it was already mentioned, one may take for K_n the unramified extension of k of the degree $n!$, $n = 2, 3, \dots$. Here $e_n = 1$ for all n. On the other hand, if K is the maximal abelian extension of $k = \mathbb{Q}_p$ then choosing K_n as the cyclotomic extension $C_{n!} = \mathbb{Q}_p(W_{n!})$, where W_l is the set of all roots of 1 of the degree l and writing $n! = n'p^{l_n}$, $(n', p) = 1$, we see that $e_n = (p-1)p^{l_n - 1} \to \infty$ as $n \to \infty$.

6.6 Heat equation

6.6.1. An auxiliary equation. We begin with an auxiliary equation over the local field K_n. Note that this equation is different from a local field heat equation studied in Chapter 4.

Let ∂_n^α be a pseudo-differential operator over K_n with the symbol $\Delta^\alpha(\xi)$, $\alpha > 0$. It means that for any $\varphi \in \mathcal{D}(K_n)$ $(\partial_n^\alpha \varphi)(x)$, $x \in K_n$, is the inverse Fourier transform of the function $\Delta^\alpha(\xi)\widetilde{\varphi}(\xi)$, $\xi \in K_n$. As we have proved

(in the course of proving Proposition 6.7)

$$(\partial_n^\alpha \varphi)(x) = q_n^{d_n\alpha/m_n} \frac{1-q_n^{\alpha/m_n}}{1-q_n^{-1-\alpha/m_n}} \int_{|x|_n \le q_n^{d_n}} \Bigg[|x|_n^{-1-\alpha/m_n} + \\ + q_n^{-d_n(1+\alpha/m_n)} \frac{1-q_n^{-1}}{q_n^{\alpha/m_n}-1} \Bigg] [\varphi(x-y)-\varphi(x)]\, dy, \quad (6.39)$$

$\varphi \in \mathcal{D}(K_n)$.

Consider the equation

$$\frac{\partial u(x,t)}{\partial t} + (\partial_n^\alpha u)(x,t) = 0, \quad x \in K_n,\ t > 0, \quad (6.40)$$

with an initial condition $u(x,0) = \varphi(x)$, $\varphi \in \mathcal{D}(K_n)$. As usual, a fundamental solution of this Cauchy problem is

$$\Gamma_\alpha^{(n)}(x,t) = q_n^{-d_n} \int_{K_n} (\chi \circ \mathrm{Tr}_{K_n/k})(-x\xi) \rho_\alpha(\|\xi\|, t)\, d\xi \quad (6.41)$$

where

$$\rho_\alpha(s,t) = \begin{cases} e^{-ts^\alpha}, & \text{if } s > 1 \\ 1, & \text{if } s \le 1, \end{cases} \quad (6.42)$$

$s \ge 0$, $t > 0$ (note the connection between convolution and the Fourier transform; see Sect. 6.4).

It is clear that $\Gamma_\alpha^{(n)}$ is continuous on $K_n \times (0,\infty)$. If $\|x\| > q_n^{d_n/m_n}$ then we can find such an element $\xi_0 \in K_n$, $\|\xi_0\| = 1$, that $(\chi \circ \mathrm{Tr}_{K_n/k})(-x\xi_0) \ne 1$. After the change $\xi = \xi_0 + \eta$ in (6.41) we find that $\Gamma_\alpha^{(n)}(x,t) = 0$. This observation allows to differentiate under the sign of integral showing that the function

$$u(x,t) = \int_{K_n} \Gamma_\alpha^{(n)}(x-y,t)\varphi(y)\, dy \quad (6.43)$$

satisfies (6.40).

Since $\int\limits_{K_n} \Gamma_\alpha^{(n)}(x,t)\, dx = 1$ we may write

$$u(x,t) = \varphi(x) + \int\limits_{\|y-x\| \le q_n^{d_n/m_n}} \Gamma_\alpha^{(n)}(x-y,t)[\varphi(y)-\varphi(x)]\, dy. \quad (6.44)$$

On the other hand, let $|x|_n = q_n^N$, $N \le d_n$. It follows from (6.41), (6.42), and (6.16) that

$$q_n^{d_n}\Gamma_\alpha^{(n)}(x,t) = 1 + (1 - q_n^{-1}) \sum_{j=1}^{d_n - N} q_n^j \rho_\alpha(q_n^j, t) - \rho_\alpha(q_n^{d_n - N+1}, t) q_n^{d_n - N}.$$

This representation shows that $\Gamma_\alpha^{(n)}(-x,t) = \Gamma_\alpha^{(n)}(x,t)$, $\Gamma_\alpha^{(n)}(x,t) \to 0$ when $x \ne 0$ and $t \to 0$,

$$|\Gamma_\alpha^{(n)}(x,t)| \le 2q_n^{d_n/2}|x|_n^{-1}, \quad x \ne 0.$$

Returning to (6.44), using the local constancy of φ and the dominated convergence theorem we find that $u(x,t) \to \varphi(x)$ as $t \to 0$.

Finally, let us prove that $\Gamma_\alpha^{(n)}(x,t) \ge 0$ for all x, t. It is sufficient to show that the function (6.43) is non-negative for any non-negative $\varphi \in \mathcal{D}(K_n)$.

Let us fix $T > 0$. Suppose that, on the contrary, $u(x,t) < 0$ for some values of $x \in K_n$, $t \in [0,T]$. Since $u(x,t)$ vanishes as a function of x outside a compact set not depending on $t \in [0,T]$, and $\lim\limits_{t \to 0} u(x,t) \ge 0$, the function $u(x,t)$ attains its negative global minimum at a certain point (x_0, t_0) where $t_0 > 0$.

It follows from (6.39) that $(\partial_n^\alpha u)(x_0, t_0) \le 0$. Of course, also $\dfrac{\partial u}{\partial t}(x_0, t_0) \le 0$ so that by (6.40) we get

$$\frac{\partial u}{\partial t}(x_0, t_0) = 0, \quad (\partial_n^\alpha u)(x_0, t_0) = 0.$$

The second equality means (see (6.39)) that $u(x_0, t_0)$ does not depend on x. Since $u(x, t_0)$ has a compact support, it follows that $u(x_0, t_0) = 0$, and we have come to a contradiction.

6.6.2. Heat measure. The relation (6.41) and the above properties of $\Gamma_\alpha^{(n)}$ lead to the following result.

Let us consider $\rho_\alpha(\|\xi\|, t)$ as a function on $K \times (0, \infty)$.

Proposition 6.8. *For each $t > 0$ the function $\rho_\alpha(\|\xi\|, t)$ is a continuous positive definite function on K.*

Now we are again in a position to use Theorem 6.6, thus obtaining, for each $t > 0$, a Radon probability measure $\pi(t, dx)$ on $\mathcal{B}(\overline{K})$ such that

$$\rho_\alpha(\|\xi\|, t) = \int_{\overline{K}} \chi(\langle \xi, x\rangle) \pi(t, dx), \quad \xi \in K. \tag{6.45}$$

Just as in Sect. 6.4, we show that if $f(x) = \varphi(T_n(x))$, $\varphi \in \mathcal{D}(K_n)$, then

$$\int_{\overline{K}} f(x)\pi(t,dx) = \|m_n\|^{-m_n} \int\limits_{z\in K_n:\ \|z\|\le q_n^{d_n/m_n}\|m_n\|} \Gamma_\alpha^{(n)}(m_n^{-1}z,t)\varphi(z)\,dz. \tag{6.46}$$

It follows from (6.46) that the measures $\pi(t,dx)$ are concentrated on the set Σ, and (6.46) remains valid for any $f \in \mathcal{E}(\overline{K})$.

Since $\rho_\alpha(\|\xi\|, t_1 + t_2) = \rho_\alpha(\xi,t_1)\rho_\alpha(\xi,t_2)$, we find that $\pi(t_1+t_2,\cdot) = \pi(t_1,\cdot) * \pi(t_2,\cdot)$. Hence the family of measures $\pi(t,dx)$ determines a spatially homogeneous Markov process $X(t)$ on $\overline{K}$ with the transition probabilities $P(t,x,\Lambda) = \pi(t,\Lambda - x)$, $\Lambda \in \mathcal{B}(\overline{K})$.

Let U_t be a corresponding semigroup of operators on the Banach space of bounded measurable functions on $\overline{K}$:

$$(U_t f)(x) = \int_{\overline{K}} f(x+y)\pi(t,dy).$$

Theorem 6.10. *(a) $X(t)$ is a stochastically continuous and right-continuous Markov process on $\overline{K}$ without discontinuities of the second kind.*

(b) The operator $-D^\alpha$ is the generator of the process $X(t)$, in the sense that for any $f \in \mathcal{E}(\overline{K})$ the function $u(x,t) = (U_t f)(x)$ satisfies the equation

$$\frac{\partial u}{\partial t} + D^\alpha u = 0.$$

(c) The measure μ is an invariant measure for the process $X(t)$.

Proof. Let $f(x) = \varphi(T_n(x))$, $\varphi \in \mathcal{E}(K_n)$. In accordance with (6.46), $u(x,t) = F_t(T_n(x))$ where

$$F_t(\zeta) = \|m_n\|^{-m_n} \int\limits_{\|z\|\le q_n^{d_n/m_n}\|m_n\|} \Gamma_\alpha^{(n)}(m_n^{-1}z,t)\varphi(\zeta+z)\,dz,$$

$\zeta \in K_n$. By the above results, $\dfrac{\partial u}{\partial t} + D^\alpha u = 0$ and also $u(x,t) \to f(x)$ as $t \to 0$ which implies stochastic continuity.

Let us show that $X(t)$ is a cadlag process. By the Kinney-Dynkin criterion [26] and the definition of the topology in $\overline{K}$ it is sufficient to prove that for any compact set $B \subset \overline{K}$, any $N = 1,2,\ldots$, and any $\varepsilon > 0$

$$\lim_{t\to 0}\sup_{x\in B} P(t,x,C_{\varepsilon,N}(x)) = 0 \tag{6.47}$$

where

$$C_{\varepsilon,N}(x)=\bigcup_{n=1}^{N}\{y\in\overline{K}:\quad \|T_n(x-y)\|\geq\varepsilon\}.$$

We find by virtue of (6.46) that

$$P(t,x,C_{\varepsilon,N}(x))\leq\sum_{n=1}^{N}\int\limits_{\|T_n(y)\|\geq\varepsilon}\pi(t,dy)=$$

$$=\sum_{n=1}^{N}\|m_n\|^{-m_n}\int\limits_{z\in K_n:\ \varepsilon\leq\|z\|\leq q_n^{d_n/m_n}\|m_n\|}\Gamma_\alpha^{(n)}(m_n^{-1}z,t)\,dz,$$

and the equality (6.47) follows from the properties of $\Gamma_\alpha^{(n)}$ established above.

Finally, let

$$\mu_1(A)=\int_{\overline{K}}P(t,x,A)\,d\mu(x),\quad A\in mathcalB(\overline{K}),$$

and we need to prove that $\mu_1=\mu$. Indeed, for any $\xi\in K$

$$\int_{\overline{K}}\chi(\langle\xi,x\rangle)\,d\mu_1(x)=\int_{\overline{K}}d\mu(y)\int_{\overline{K}}\chi(\langle\xi,x\rangle)P(t,y,dx)=$$

$$=\int_{\overline{K}}\chi(\langle\xi,y\rangle)\,d\mu(y)\int_{\overline{K}}\chi(\langle\xi,x\rangle)\pi(t,dx)=\rho_\alpha(\|\xi\|,t)\Omega(\xi)=\Omega(\xi)$$

whence $\mu_1=\mu$. ■

If $x\in\Sigma$, $\Lambda\cap\Sigma=\varnothing$, then $(\Lambda-x)\cap\Sigma=\varnothing$. Indeed, if $y\in\Lambda$ then

$$\|T_n(y)\|>q_n^{d_n/m_n}\|m_n\|$$

whence $\|T_n(y-x)\|=\|T_n(y)\|$ due to the ultra-metric property, so that $y-x\notin\Sigma$. Hence, a trajectory of the process X starting at a point from Σ remains in Σ almost surely.

This means that the part X_Σ of the process X in Σ is a non-exploding process. Since Σ is compact in a strong topology of $\overline{K}$ (which is a consequence of general properties of projective limits [126]), it follows from Theorem I.9.4 of [17] that X_Σ is a Hunt process.

If we deal with Galois extensions then as before we see from the definition of P that this transition probability is invariant with respect to the Galois group G.

The next property of the heat measure is in a sharp contrast to the results known for the finite-dimensional case.

Theorem 6.11. *For each $t > 0$ the measure $\pi(t, \cdot)$ is not absolutely continuous with respect to μ.*

Proof. Let us fix $N \geq 1$ and consider the set

$$M = \left\{x \in \overline{K}: \quad \|T_n(x)\| \leq q_n^{d_n/m_n - N/f_n}\|m_n\|, \ n = 1, 2, \dots\right\}$$

We shall show that $\mu(M) = 0$ whereas $\pi(t, M) \neq 0$.

Denote

$$M_n = \left\{x \in \overline{K}: \quad \|T_n(x)\| \leq q_n^{d_n/m_n - N/f_n}\|m_n\|\right\}, \qquad n = 1, 2, \dots .$$

It is clear that $M = \bigcap\limits_{n=1}^{\infty} M_n$. Repeating the arguments from the proof of Lemma 6.5, we see that $M_\nu \subset M_n$ if $\nu > n$. Thus

$$\pi(t, M) = \lim_{n\to\infty} \pi(t, M_n), \quad \mu(M) = \lim_{n\to\infty} \mu(M_n).$$

It follows from the integration formula (6.21) for cylindrical functions that

$$\mu(M_n) = q_n^{-d_n}\|m_n\|^{-m_n} \int\limits_{z\in K_n:\ \|z\|\leq q_n^{d_n/m_n - N/f_n}\|m_n\|} dz$$

$$= q_n^{-d_n}|m_n|_n^{-1} \int\limits_{z\in K_n:\ |z|_n\leq q_n^{d_n - Ne_n}|m_n|_n} dz = q_n^{-Ne_n} = q_1^{-Nf_ne_n}, \quad (6.48)$$

so that $\mu(M_n) = q_1^{-Nm_n} \to 0$ for $n \to \infty$. Thus $\mu(M) = 0$.

In a similar way we find that

$$\pi(t, M_n) = |m_n|_n^{-1} \int\limits_{z\in K_n:\ |z|_n\leq q_n^{d_n - Ne_n}|m_n|_n} \Gamma_\alpha^{(n)}\left(m_n^{-1}z, t\right) dz \qquad (6.49)$$

where $\Gamma_\alpha^{(n)}$ is a fundamental solution of the Cauchy problem for the equation over K_n of the form $\dfrac{\partial u}{\partial t} + \partial_n^\alpha u = 0$. We have

$$\Gamma_\alpha^{(n)}(\zeta, t) = q_n^{-d_n/2}\widetilde{\rho}_\alpha(\zeta, t)$$

where, as before, tilde means the local field Fourier transform.

Using the Plancherel formula we can rewrite (6.49) in the form

$$\pi(t, M_n) = \int_{K_n} \widetilde{\Gamma}_\alpha^{(n)}(x,t)\widetilde{\beta}_n(x)\,dx$$

where

$$\beta_n(\zeta) = \begin{cases} 1, & \text{if } |\zeta|_n \le q_n^{d_n - Ne_n} \\ 0, & \text{if } |\zeta|_n > q_n^{d_n - Ne_n}. \end{cases}$$

We have $\widetilde{\Gamma}_\alpha^{(n)} = q_n^{-d_n/2}\rho_\alpha$,

$$\widetilde{\beta}_n(x) = \begin{cases} q_n^{\frac{d_n}{2} - Ne_n}, & \text{if } |x|_n \le q_n^{Ne_n}, \\ 0, & \text{if } |\zeta|_n > q_n^{Ne_n}, \end{cases}$$

so that

$$\pi(t, M_n) = q_n^{-Ne_n} \int_{|x|_n \le q_n^{Ne_n}} \rho_\alpha(\|x\|, t)\,dx$$

$$\ge q_n^{-Ne_n} \int_{|x|_n = q_n^{Ne_n}} \rho_\alpha\left(|x|_n^{1/m_n}, t\right)\,dx = \left(1 - q_n^{-1}\right)\exp\left(-tq_n^{\alpha N/f_n}\right)$$

$$\ge \left(1 - q_1^{-1}\right)\exp\left(-tq_1^{\alpha N}\right).$$

Hence, $\pi(t, M) > 0$. ■

6.7 Hypersingular integral representation

The main aim of this section is the following result.

Theorem 6.12. *There exists such a measure* Π *on* $\mathcal{B}(\overline{K}\backslash\{0\})$ *finite outside any neighbourhood of the origin that* D^α *has the representation (6.34) on all functions* $f \in \mathcal{E}(\overline{K})$.

In the course of *the proof* we shall also obtain some new information about the Markov process $X(t)$ generated by the operator $-D^\alpha$. We assume below that $X(0) = 0$.

The projective limit topology on $\overline{K}$ coincides with the one given by the shift-invariant metric

$$r(x, y) = \sum_{n=1}^{\infty} 2^{-n} \frac{\|x - y\|_n}{1 + \|x - y\|_n}, \quad x, y \in \overline{K}.$$

It is known [142] that the main notions and results regarding stochastic processes with independent increments carry over to the case of a general topological group with a shift-invariant metric.

Let $\nu(t,\Gamma)$ be a Poisson random measure corresponding to the process $X(t)$. Here $\Gamma \in \mathcal{B}_0 = \bigcup_{\gamma>0} \mathcal{B}_\gamma$,

$$\mathcal{B}_\gamma = \{\Gamma \subset \mathcal{B}(\overline{K}),\ \mathrm{dist}(\Gamma,0) \geq \gamma\}.$$

For any $\Gamma \in \mathcal{B}_0$ we can define a stochastically continuous process $X_\Gamma(t)$, the sum of all jumps of the process $X(\tau)$ for $\tau \in [0,t)$ belonging to Γ. In a standard way [43] we find that

$$\mathbf{E}\chi(<\lambda, X_\Gamma(t)>) = \exp\left\{\int_\Gamma (\chi(<\lambda, x>) - 1)\Pi(t,dx)\right\}, \qquad \lambda \in K, \tag{6.50}$$

where $\Pi(t,\cdot) = \mathbf{E}\nu(t,\cdot)$.

Let $\lambda \in K_n$. Consider the set

$$V_{\delta,n} = \{x \in \overline{K} :\ \|T_n(x)\| \geq \delta\}, \quad 0 < \delta \leq 1.$$

Let us look at the equality (6.50) with $\Gamma = V_{\delta,n}$, $\delta \leq \|\lambda\|^{-1}$. If $x \in V_{\delta,n}$ then $r(x,0) \geq 2^{-n}\frac{\|x\|_n}{1+\|x\|_n} \geq 2^{-n-1}\delta$, whence $V_{\delta,n} \in \mathcal{B}_0$. If $\delta \leq \|\lambda\|^{-1}$, $x \notin V_{\delta,n}$, then

$$\| <\lambda, x> \| = \|T(\lambda T_n(x))\| \leq 1$$

which implies that the integral in the right-hand side of (6.50) coincides with the one taken over $\overline{K}$.

On the other hand, almost surely

$$\|X(t) - X_{V_{\delta,n}}(t)\|_n < \delta. \tag{6.51}$$

Indeed, let t_0 be the first exit time of the process $X(t) - X_{V_{\delta,n}}(t)$ from the set $\overline{K} \setminus V_{\delta,n}$. Suppose that $t_0 < \infty$ with a non-zero probability. Then

$$\|X(t_0 - 0) - X_{V_{\delta,n}}(t_0 - 0)\|_n < \delta, \tag{6.52}$$

$$\|X(t_0 + 0) - X_{V_{\delta,n}}(t_0 + 0)\|_n \geq \delta. \tag{6.53}$$

If $\|X(t_0+0) - X(t_0-0)\|_n < \delta$ then $X_{V_{\delta,n}}(t_0+0) = X_{V_{\delta,n}}(t_0-0)$, so that

$$\|[X(t_0+0) - X_{V_{\delta,n}}(t_0+0)] - [X(t_0-0) - X_{V_{\delta,n}}(t_0-0)]\|_n < \delta. \tag{6.54}$$

If, on the contrary, $\|X(t_0+0) - X(t_0-0)\|_n \geq \delta$, then

$$X_{V_{\delta,n}}(t_0+0) - X_{V_{\delta,n}}(t_0-0) = X(t_0+0) - X(t_0-0),$$

the expression in the left-hand side of (6.54) equals zero, and the inequality (6.54) is valid too. In both cases it contradicts (6.52), (6.53). Thus $t_0 = \infty$ almost surely, and the inequality (6.51) has been proved. We have come to the following formula of Lévy-Khinchin type.

Lemma 6.6. *For any* $\lambda \in K,\ t > 0$

$$\mathbf{E}\chi(<\lambda, X(t)>) = \exp\left\{ \int\limits_{\overline{K}} [\chi(<\lambda, x>) - 1]\Pi(t, dx) \right\}. \tag{6.55}$$

Remark. Lemma 6.6 can serve as a base for developing the theory of stochastic integrals and stochastic differential equations over $\overline{K}$. In fact, the corresponding techniques and results of Chapter 5 carry over to this case virtually unchanged.

Both sides of (6.55) can be calculated explicitly if we use the heat measure $\pi(t, dx)$. We have

$$\mathbf{E}\chi(<\lambda, X(t)>) = \int\limits_{\overline{K}} \chi(<\lambda, x>)\pi(t, dx) = \rho_\alpha(\|\lambda\|, t)$$

where

$$\rho_\alpha(s,t) = \begin{cases} e^{-ts^\alpha}, & \text{if } s > 1 \\ 1, & \text{if } s \leq 1, \end{cases} \tag{6.56}$$

$s \geq 0,\ t > 0$. It follows from the definitions that $\Pi(t, \cdot)$ is symmetric with respect to the reflection $x \mapsto -x$. Therefore

$$\int\limits_{\overline{K}} [\chi(<\lambda, x>) - 1]\Pi(t, dx) = \begin{cases} -t\|\lambda\|^\alpha, & \text{if } \|\lambda\| > 1 \\ 0, & \text{if } \|\lambda\| \leq 1, \end{cases} \tag{6.57}$$

Lemma 6.7. *Let* M_n *be a compact subset of* $K_n \setminus \{0\}$, $M = T_n^{-1}(M_n)$. *Then*

$$\Pi(t, M) = -t \int\limits_{\eta \in K_n:\ \|\eta\| > 1} \|\eta\|^\alpha w_n(\eta)\, d\eta \tag{6.58}$$

where $w_n(\eta)$ is the inverse Fourier transform of the function

$$y \mapsto q_n^{-d_n/2}\omega_{M_n}^{(n)}(m_n y),$$

$\omega_{M_n}^{(n)}$ is the indicator of the set M_n in K_n, and dx is the normalized additive Haar measure on K_n.

Proof. Let ω_M be the indicator of the set M in $\overline{K}$. Then $\omega_M(x) = \omega_{M_n}^{(n)}(T_n(x))$, $x \in \overline{K}$,

$$\omega_{M_n}^{(n)}(\xi) = \int_{K_n} \chi \circ T(\xi\eta) w_n(\eta)\, d\eta, \quad \xi \in K_n.$$

Since

$$\int_{K_n} w_n(\eta)\, d\eta = \omega_{M_n}^{(n)}(0) = 0,$$

we get

$$\omega_M(x) = \int_{K_n} [\chi \circ T(\eta T_n(x)) - 1] w_n(\eta)\, d\eta.$$

Integrating with respect to $\Pi(t, dx)$ and using (6.57) we come to (6.58). ∎

It follows from Lemma 6.7 that $\Pi(t, dx) = t\Pi(1, dx)$. We shall write $\Pi(dx)$ instead of $\Pi(1, dx)$.

Proof of Theorem 6.12. Let $f(x) = \varphi(T_n(x))$, $x \in \overline{K}$, where φ is locally constant and in addition supp φ is compact, $0 \notin$ supp φ. It follows from Lemma 6.7 that

$$\int_{\overline{K}} f(x)\Pi(dx) = -\int_{K_n} \Delta^\alpha(\eta)\psi(\eta)\, d\eta \tag{6.59}$$

where ψ is the inverse Fourier transform of the function $y \mapsto q_n^{-d_n/2}\varphi(m_n y)$.

The right-hand side of (6.59) is an entire function with respect to α. Assuming temporarily Re $\alpha < -1$, we can use the Plancherel formula with subsequent analytic continuation. As a result we find that for $\alpha > 0$

$$\int_{\overline{K}} f(x)\Pi(dx) = -q_n^{d_n\alpha/m_n}\frac{1-q_n^{\alpha/m_n}}{1-q_n^{-1-\alpha/m_n}} \int_{x\in K_n:\ |x|_n \le q_n^{d_n}} \left[|x|_n^{-1-\alpha/m_n} + \frac{1-q_n^{-1}}{q_n^{\alpha/m_n}-1} q_n^{-d_n(1+\alpha/m_n)}\right]\varphi(m_n x)\, dx. \tag{6.60}$$

An obvious approximation argument shows that (6.60) is valid for any $f \in \mathcal{E}(\overline{K})$. Comparing (6.60) with (6.31) we obtain (6.34). ■

6.8 Comments

The theory of measurable vector spaces over local fields (Sect. 6.1) was developed by Evans [33, 34]; these papers contain further results in that direction. Theorem 6.6 was proved by Mądrecki [106]. The results in analysis and stochastics over infinite extensions of local fields of characteristic zero expounded in Sect. 6.2–6.7 were obtained by the author [90, 91]. Note also the paper by Satoh [132], who developed a non-Archimedean version of the theory of abstract Wiener spaces, with applications to Wiener measures on some non-Archimedean Banach spaces.

Chapter 7

Processes with Non-Archimedean Time Parameters

7.1 Local field Brownian motion

7.1.1. Preliminaries. We begin with some auxiliary results.

Lemma 7.1. *Let E be a normed vector space with a (non-Archimedean) norm $\|\cdot\|_E$ over a local field K. If $x_1, \ldots, x_n \in E$, then*

$$\max_{1\le i\le n} \|x_i\|_E = \max\left(\|x_1\|_E, \max_{2\le i\le n} \|x_i - x_{i-1}\|_E\right) \tag{7.1}$$

and

$$\max_{1\le i\le n} \|x_i\|_E = \max\left(\left\|x_1 - \sum_{j=2}^{n} x_i\right\|_E, \max_{2\le i\le n} \|x_i\|_E\right). \tag{7.2}$$

Proof. Consider (7.1). It follows from the ultra-metric inequality that the left-hand side is greater than or equal to the right-hand side. In order to prove the converse, set $y_1 = x_1$ and $y_i = x_i - x_{i-1}$ for $2 \le i \le n$. Then $x_i = \sum_{j=1}^{i} y_j$ for $1 \le i \le n$, and the reverse inequality also follows from the ultra-metric inequality. The proof of (7.2) is similar. ■

Corollary 7.1. *Let e_i, $1 \leq i \leq n$, denote the i-th coordinate vector in K^n. Set $f_1 = e_1$, $f_i = e_i - e_{i-1}$, $2 \leq i \leq n$. Set also $g_1 = e_1$ and $g_i = e_i - e_1$, $2 \leq i \leq n$. Then both the collections $f_1, \dots, f_n$ and $g_1, \dots, g_n$ are orthonormal.*

Proof. Given $\alpha_1, \dots, \alpha_n \in K$ we have, by Lemma 7.1, that

$$\left\| \sum_{i=1}^{n} \alpha_i f_i \right\|_{K^n} = \max \left(\max_{1 \leq i \leq n-1} \|\alpha_i - \alpha_{i+1}\|, \|\alpha_n\| \right) = \max_{1 \leq i \leq n} \|\alpha_i\|,$$

so that the collection $f_1, \dots, f_n$ is orthonormal (as usual, we denote by $\|\cdot\|$ the absolute value in K). The proof for the second collection is similar. ■

We shall need some further results regarding K-Gaussian measures.

Lemma 7.2. *Suppose that E is a separable Banach space over K and that G is a compact O-submodule of E. Let X be a E-valued random variable, such that its distribution is the normalized Haar measure on G. Then X is a K-Gaussian random variable.*

Proof. By Theorem 6.3 we have to prove that $T(X)$ is K-Gaussian for all $T \in E^*$. As T is continuous, it follows that the support of the probability distribution of $T(X)$ is the set $T(G)$, and that this set is compact. Since T is linear, $T(G)$ is an O-submodule of K, and $y + T(X)$ has the same distribution as $T(X)$ whenever $y \in T(G)$. Thus the distribution of $T(X)$ is the normalized Haar measure on the compact O-submodule $T(G)$. Since the only compact O-submodules of K are sets of the form $\beta^n O$ for some $n \in \mathbb{Z}$, and also $\{0\}$, Theorem 6.1 gives that $T(X)$ is K-Gaussian. ■

Corollary 7.2. *If $X_1, \dots, X_n$ are independent K-Gaussian random variables, then the random vector $(X_1, \dots, X_n)$ is K-Gaussian.*

Proof. It follows from Corollary 6.1 that the distribution of $(X_1, \dots, X_n)$ is actually the normalized Haar measure on the set

$$\prod_{i=1}^{n} \left\{ x \in K : \|x\| \leq \|X_i\|_{L_\infty(\mathbf{P})} \right\}.$$

Since this set is a compact O-submodule of K^n, we see from Lemma 7.2 that $(X_1, \dots, X_n)$ is K-Gaussian. ■

The next lemma is a useful tool for constructing new K-Gaussian random variables.

Lemma 7.3. *Suppose that E is a separable Banach space over K, and $X = (X_1, \dots, X_n)$ is a vector of independent, identically distributed, E-valued K-Gaussian random variables. Let $\{\alpha_1, \dots, \alpha_m\}$ be an orthonormal set in K^n, $\alpha_i = (\alpha_{i1}, \dots, \alpha_{in})$, $i = 1, \dots, m$. Then the random variables*

$$\sum_{i=1}^{n} \alpha_{1j} X_j, \dots, \sum_{i=1}^{n} \alpha_{mj} X_j \tag{7.3}$$

are also independent and identically distributed.

Proof. Due to Theorem 6.3, it is sufficient to consider the case $E = K$. Let Y be a m-dimensional random vector with the components (7.3). We can write $Y = XA$, where A is a $n \times m$ matrix, $A = (\alpha_1', \dots, \alpha_m')$. Denote by $\Phi(t)$ the indicator function of the interval $[0, 1]$. Let $\sigma = \|X_1\|_{L_\infty(\mathbf{P})}$.

For $\xi = (\xi_1, \dots, \xi_m) \in K^m$ we have

$$\begin{aligned}
\mathbf{P}[\chi(\xi \cdot Y)] &= \mathbf{P}[\chi(XA\xi')] = \mathbf{P}[\chi(X \cdot \xi A')] \\
&= \prod_{j=1}^{n} \Phi\left(\sigma \|(\xi A')_j\|\right) = \Phi\left(\sigma \max_{1 \le j \le n} \|(\xi A')_j\|\right) = \Phi\left(\sigma \|\xi A'\|_{K^n}\right) \\
&= \Phi\left(\sigma \left\|\sum_{i=1}^{m} \xi_i \alpha_i\right\|_{K^n}\right) = \Phi\left(\sigma \max_{1 \le i \le m} \|\xi_i\| \cdot \|\alpha_i\|_{K^n}\right) \\
&= \prod_{i=1}^{m} \Phi\left((\sigma \|\alpha_i\|_{K^n}) \|\xi_i\|\right) = \prod_{i=1}^{m} \Phi(\sigma \|\xi_i\|),
\end{aligned}$$

and the result follows from the Fourier uniqueness and Theorem 6.1. ■

Corollary 7.3. *Suppose that $X_1, \dots, X_n$ are independent, identically distributed, E-valued K-Gaussian random variables, where E is a separable Banach space. Then the random variables $X_1, X_2 - X_1, \dots, X_n - X_{n-1}$ are independent and identically distributed, with the same common distribution as $X_1, \dots, X_n$. The same is true for the collection $X_1, X_2 - X_1, \dots, X_n - X_1$.*

The proof follws from Lemma 7.3 and Corollary 7.1. ■

7.1.2. Construction of the process. We shall construct a process on K^d, with the time parameter from O^N, $d, N \ge 1$. Let

$$\Omega = \prod_{C \in \Gamma(N)} \prod_{i=1}^{d} K_{C,i},$$

where $K_{C,i}$ is a copy of K, and $\Gamma(N)$ is the collection of all balls contained in O^N. Let $Z_{C,i}$ denote the coordinate projection from Ω onto K, and Z_C the map from Ω onto K^d given by $Z_C = (Z_{C,i})_{i=1}^d$.

Let $\mathfrak{C}(N,n) \subset \Gamma(N)$ be the set of all balls in O^N of the radius q^{-n}, $n \geq 0$. Given $C \in \mathfrak{C}(N,n)$ denote by $\zeta_{C,i}$ the K-Gaussian probability measure on $K_{C,i}$, for which the identity map has the L_∞-norm q^{-n}. Hence, $\zeta_{C,i}$ is the Haar measure restricted to the copy of $\beta^n O$ in $K_{C,i}$ and renormalized to have the total mass 1. Write

$$\zeta = \prod_{C \in \Gamma(N)} \prod_{i=1}^{d} \zeta_{C,i},$$

and let $\mathbf{P}$ be the completion of ζ, with respect to the corresponding completion of the product σ-algebra on Ω.

The (N,d) K-*Brownian motion* is the stochastic process $X: \Omega \times O^N \to K^d$ defined as follows. Let

$$\Omega^* = \bigcap_{n=0}^{\infty} \bigcap_{C \in \mathfrak{C}(N,n)} \{\omega : \|Z_C(\omega)\|_{K^d} \leq q^{-n}\}.$$

It follows from the definition of $\mathbf{P}$ that $\mathbf{P}(\Omega^*) = 1$. Set

$$X(\omega,t) = \begin{cases} \sum_{C \ni t} Z_C(\omega), & \text{if } \omega \in \Omega^*, \\ 0, & \text{if } \omega \notin \Omega^*. \end{cases}$$

We shall often write X_t instead of $X(\omega,t)$. It is clear that X_t is a K-Gaussian process.

The idea of the above construction (e.g. for $d = 1$) is to think of the sub-balls of O^N as being arrayed in an infinite q^N-ary tree, and to attach a K-valued K-Gaussian "weight" to each ball. Each point $t \in O^N$ is associated with the unique sequence of balls of decreasing radii that contain it, and X_t is obtained by summing the corresponding weights.

7.1.3. Some properties. It is obvious that the one-dimensional coordinate projections of X_t are independent and identically distributed. It follows from the definition of Ω^* that

$$\|X_t - X_s\|_{K^d} \leq q^{-1}\|s-t\|_{K^N}$$

almost surely for all s and t, so that the sample paths of X_t are Lipschitz continuous. Moreover, for any linear operator $T: K^d \to K$ we have

$$\|TX_t\|_{L_\infty(\mathbf{P})} = \|T\| \quad \text{for all } t \in O^N,$$

and

$$\|T(X_s - X_t)\|_{L_\infty(\mathbf{P})} = q^{-1}\|s-t\|_{K^N}\|T\|.$$

Specify a family $\{\theta_t, t \in O^N\}$ of measure preserving bijections on Ω by requiring that $Z_{C,i}(\theta_t(\omega)) = Z_{t+C,i}(\omega)$. Note that θ_0 is the identity map, and $\theta_s \circ \theta_t = \theta_{s+t}$, so that this family forms a group under the operation of composition. We have $X(\theta_s\omega) = X(\omega, s+t)$, so that X_t is a stationary process.

Let us define a total ordering on O^N as follows. We begin by inductively defining orders on each collection of balls $\mathfrak{C}(N,n)$. Since $\mathfrak{C}(N,0)$ has only one element, it can be considered ordered. Suppose that $\mathfrak{C}(N,n)$ has been ordered. For each ball $C \in \mathfrak{C}(N,n)$ we assign an arbitrary total ordering to q^N ball from $\mathfrak{C}(N,n+1)$ contained in C. A total order on $\mathfrak{C}(N,n+1)$ can now be defined, such that if $C_1, C_2 \in \mathfrak{C}(N,n+1)$ are contained in different balls from $\mathfrak{C}(N,n)$, say $C_1 \subset C_1'$ and $C_2 \subset C_2'$, then the order relation of C_1 and C_2 is the same as that of C_1' and C_2'.

Given this family of total orders, we define the total order relation on O^N by declaring that the order relation (which will be denoted "$\leq$") between two different points s and t is the same as that for any two disjoint balls $C_s \ni s$ and $C_t \ni t$ belonging to the same collection $\mathfrak{C}(N,n)$.

The next result can be seen as a non-Archimedean counterpart of the property of independence of increments, known for a usual Brownian motion.

Theorem 7.1. *Suppose that $t_1, t_2, \dots, t_n \in O^N$ are such that $t_1 < t_2 < \dots < t_n$. Then the random variables $X_{t_1}, X_{t_2} - X_{t_1}, \dots X_{t_n} - X_{t_{n-1}}$ are independent.*

Proof. It is sufficient to show that $X_{t_k} - X_{t_{k-1}}$ is independent of $X_{t_1}, \dots, X_{t_{k-1}}$ for $2 \leq k \leq n$. Suppose that $\|t_k - t_{k-1}\|_{K^N} = q^{-r}$. Let $C(i,m)$ be the ball from $\mathfrak{C}(N,m)$, which contains t_i ($1 \leq i \leq k$, $m \geq 0$). We have

$$X_{t_k} - X_{t_{k-1}} = \sum_{m=r+1}^{\infty} \left[Z_{C(k,m)} - Z_{C(k-1,m)}\right].$$

It follows from the definition of the ordering that

$$\|t_k - t_{k-1}\|_{K^N} \leq \|t_k - t_i\|_{K^N} \quad \text{for } 1 \leq i \leq k-2,$$

so that $C(k,m) \neq C(i,m)$ for $1 \leq i \leq k-1$, and $m > r$. It remains to use Corollary 7.3. ■

The distribution of the process X_t can be considered as a K-Gaussian probability measure on the Banach space $C(O^N, K^d)$ of continuous functions on O^N with values in K^d. As we have mentioned, the support of this distribution is contained in the set Σ consisting of such functions f that $\|f(t)\|_{K^d} \le 1$ and

$$\|f(s) - f(t)\|_{K^d} \le q^{-1}\|s - t\|_{K^N} \quad \text{for all } s, t \in O^N.$$

In fact, there is the following precise result.

Theorem 7.2. *The closed support of the distribution of the process X_t coincides with Σ.*

Proof. It is sufficient to show that for any $f \in \Sigma$, $m \ge 0$

$$\mathbf{P}\left\{\sup_{t \in O^N} \|X_t - f(t)\|_{K^d} \le q^{-m-1}\right\} > 0. \tag{7.4}$$

Let us calculate the left-hand side of (7.4). Let $C_1, \dots, C_{q^{Nm}}$ be all the balls from $\mathfrak{C}(N, m)$ listed in the ascending order. For each i we choose a point $t_i \in C_i$. By the ultra-metric inequality and the Lipschitz continuity of X_t and $f(t)$, we find that the inequality $\|X_t - f(t)\|_{K^d} \le q^{-m-1}$ holds for all $t \in O^N$, if and only if

$$\|X_{t_i} - f(t_i)\|_{K^d} \le q^{-m-1} \quad \text{for } 1 \le i \le q^{Nm}.$$

By Lemma 7.1, the latter occurs if and only if $\|X_{t_1} - f(t_1)\|_{K^d} \le q^{-m-1}$ and

$$\|X_{t_i} - f(t_i) - \left(X_{t_{i-1}} - f(t_{i-1})\right)\|_{K^d} \le q^{-m-1}, \quad 2 \le i \le q^{Nm}.$$

Hence, by Theorem 7.1 and Corollary 6.1

$$\mathbf{P}\left\{\sup_{t \in O^N} \|X_t - f(t)\|_{K^d} \le q^{-m-1}\right\} = \mathbf{P}\left\{\|X_{t_1} - f(t_1)\|_{K^d} \le q^{-m-1}\right\}$$
$$\times \prod_{i=2}^{q^{Nm}} \mathbf{P}\left\{\|X_{t_i} - f(t_i) - \left(X_{t_{i-1}} - f(t_{i-1})\right)\|_{K^d} \le q^{-m-1}\right\}$$
$$= q^{-d(m+1)q^{Nm}} \prod_{i=2}^{q^{Nm}} \left(q^{-1}\|t_i - t_{i-1}\|_{K^N}\right)^{-d} > 0$$

as desired. ■

The above properties of the process X_t justify its interpretation as a local field counterpart of the classical Brownian motion.

For much more about properties and applications of this process (random measures, polar sets, local times, white noise, chaos expansions, stochastic integration) see the papers [35, 36] by Evans.

7.2 The Wiener process with the p-adic time parameter

7.2.1. Definition. In this section we shall consider another non-Archimedean counterpart of the classical Brownian motion. In this case the process is complex-valued and Gaussian (in the classical sense) but the time parameter belongs to the ring $\mathbb{Z}_p$ of p-adic integers. Here p is an arbitrary fixed prime number.

Let $(\Omega, \Sigma, \mathbf{P})$ be a probability space. We shall consider processes $\psi : \mathbb{Z}_p \times \Omega \to \mathbb{C}$, which are Σ-measurable (with respect to the Borel σ-algebra $\mathcal{B}(\mathbb{C})$) and Gaussian, in the sense that for any finite collection of elements $x_1, \dots x_n \in \mathbb{Z}_p$, and any Borel set $B \subset \mathbb{C}^n$

$$\mathbf{P}\left(\{\omega \in \Omega : \ (\psi(x_1, \omega), \dots, \psi(x_n, \omega)) \in B\}\right) = \frac{\det \Lambda}{(2\pi)^n} \int_B e^{-(\Lambda z, z)}\, dz \wedge d\bar{z},$$

where $\Lambda = (\lambda_{ij})$ is a positive definite matrix,

$$(\Lambda z, z) = \sum_{i,j=1}^{n} \lambda_{ij} z_i \overline{z_j}, \quad z = (z_1, \dots, z_n),\ \bar{z} = (\overline{z_1}, \dots, \overline{z_n}),$$

$$z_k = u_k + iv_k,\ \overline{z_k} = u_k - iv_k, \quad k = 1, 2, \dots, n,$$

$$dz \wedge d\bar{z} = 2^n du_1\, dv_1 \dots du_n\, dv_n.$$

Definition 7.1. *A process $\psi(t) \equiv \psi(t, \omega)$ $(t \in \mathbb{Z}_p)$ with values in $\mathbb{C}$ is called a Wiener process with the p-adic time if:*

(i) $\psi(0) = 0$ *almost surely;*

(ii) *the process ψ is Gaussian;*

(iii) $\mathbf{E}\psi(t) \equiv 0$*;*

(iv) *the covariance function of the process has the form*

$$\mathbf{E}\left[\psi(s)\overline{\psi(t)}\right] = \frac{1}{2}\left(|s|_p + |t|_p - |s - t|_p\right). \tag{7.5}$$

In order to compare properties of ψ with those of the conventional Wiener process, note first of all that for $|s|_p \neq |t|_p$ the right-hand side of (7.5) equals $\frac{1}{2}\min(|s|_p, |t|_p)$, which resembles the covariance $\min(s, t)$ of the usual Wiener process. Similarly, the variance

$$\mathbf{D}(h) \equiv \mathbf{E}\left[|\psi(t + h) - \psi(t)|^2\right] = |h|_p$$

for all $t, h \in \mathbb{Z}_p$; in the classical case $\mathbf{D}(h) = |h|$, $h \in \mathbb{R}$.

Let $|t_1|_p \leq |t_2|_p < |t_3|_p \leq |t_4|_p$. The random vector

$$(\psi(t_2,\omega) - \psi(t_1,\omega), \psi(t_4,\omega) - \psi(t_3,\omega))$$

is Gaussian with zero mean. It follows from (7.5) that

$$\mathbf{E}\left[(\psi(t_2) - \psi(t_1))\left(\overline{\psi(t_4)} - \overline{\psi(t_3)}\right)\right] = 0.$$

Since for Gaussian random variables noncorrelatedness is equivalent to independence, we find that the increments $\psi(t_2) - \psi(t_1)$ and $\psi(t_4) - \psi(t_3)$ are independent.

7.2.2. The canonical measure on the space of sequences. Consider the set I consisting of the rational numbers $k = 0$,

$$k = p^{-\nu}\left(k_0 + k_1 p + \cdots + k_{\nu-1}p^{\nu-1}\right),$$

where $\nu = 1, 2, 3, \ldots$; $k_0 = 1, 2, \ldots, p-1$; $k_j = 0, 1, 2, \ldots, p-1$, $j = 1, 2, \ldots, \nu - 1$. This is a countable set with a natural lexicographic ordering with respect to $\nu, k_0, k_1, \ldots$.

Let us introduce also the set $\mathbf{\Pi}$ of all complex-valued sequences $Z = \{Z_k,\ k \in I\}$. We shall construct a probability measure $\mathbf{m}$ on a certain σ-algebra $\Sigma_{\mathbf{\Pi}}$ of subsets of $\mathbf{\Pi}$, and use the probability space $(\mathbf{\Pi}, \Sigma_{\mathbf{\Pi}}, \mathbf{m})$ as a specific model of the probability space $(\Omega, \Sigma, \mathbf{P})$. This construction is actually an adaptation of the classical construction of a canonical Gaussian measure on the space of all real sequences (see [140, 9]).

We start from the algebra of cylindrical sets of the form

$$C_{(n)} = \left\{Z \in \mathbf{\Pi} :\ \left\{Z_k,\ k \in I^{(n)}\right\} \in B_1 \times \ldots \times B_n,\ B_i \in \mathcal{B}(\mathbb{C})\right\},$$

where $I^{(n)}$ is the set of the first n elements of I. We define $\mathbf{m}(C_{(n)})$ by the formula

$$\mathbf{m}(C_{(n)}) = \frac{1}{2\pi} \int\limits_{B_1 \times \ldots \times B_n} \prod_{k \in I^{(n)}} e^{-z_k \overline{z_k}}\, dz_k \wedge d\overline{z_k}.$$

By the Kolmogorov theorem $\mathbf{m}$ can be extended in a unique way to a σ-additive measure on the σ-algebra $\Sigma_{\mathbf{\Pi}}$ generated by cylindrical sets.

We have a natural sequence of independent random variables $\zeta_k(Z) = Z_k$, $Z = \{Z_k,\ k \in I\}$, on the probability space $(\mathbf{\Pi}, \Sigma_{\mathbf{\Pi}}, \mathbf{m})$, with the following properties:

$$\mathbf{E}\zeta_k = \frac{1}{2\pi}\int_{\mathbb{C}} z e^{-z\bar{z}}\, dz \wedge d\bar{z} = 0,$$

$$\mathbf{E}\zeta_k\overline{\zeta_l} = \frac{1}{(2\pi)^2}\int_{\mathbb{C}\times\mathbb{C}} z_k\overline{z_l}e^{-(z_k\overline{z_k}+z_l\overline{z_l})}\, dz_k \wedge d\overline{z_k}\, dz_l \wedge d\overline{z_l} = \delta_{kl}.$$

The sequence $\{\zeta_k\}$ forms an orthonormal basis in $L_2(\mathbf{\Pi}, \Sigma_{\mathbf{\Pi}}, \mathbf{m})$.

Some properties of the measure $\mathbf{m}$ can be found in [14]; in particular, $\mathbf{m}$ is concentrated on the set

$$\left\{ Z \in \mathbf{\Pi} :\ \overline{\lim}_{|k|_p\to\infty} \frac{|Z_k|}{\sqrt{\log |k|_p}} = 1 \right\}.$$

The following analog of the Kolmogorov-Khinchin criterion can be proved just as in the classical case [140, 9].

Proposition 7.1. *Let us consider the set*

$$M_{(\lambda_k)} = \left\{ Z \in \mathbf{\Pi} :\ \sum_{k\in I} \lambda_k |Z_k|^2 < \infty \right\},$$

where $\{\lambda_k\}_{k\in I}$ is a given sequence of non-negative real numbers. Then

$$\mathbf{m}\left(M_{(\lambda_k)}\right) = \begin{cases} 1, & \text{if } \sum\limits_{k\in I} \lambda_k < \infty, \\ 0, & \text{if } \sum\limits_{k\in I} \lambda_k = \infty. \end{cases}$$

7.2.3. White noise with p-adic time. Let us consider an analog of the white noise. It will be a generalized stochastic process Ψ, that is a mapping $\Psi : \mathcal{D}(\mathbb{Z}_p) \times \mathbf{\Pi} \to \mathbb{C}$, which is linear and continuous in its first argument, and measurable in the second, with respect to the σ-algebras $\Sigma_{\mathbf{\Pi}}$ and $\mathcal{B}(\mathbb{C})$.

Definition 7.2. *A white noise with a p-adic time is a generalized stochastic process $\{\Psi(\sigma),\ \sigma \in \mathcal{D}(\mathbb{Z}_p)\}$, such that*

(i) $\mathbf{E}\Psi(\sigma) = 0$ *for any* $\sigma \in \mathcal{D}(\mathbb{Z}_p)$;

(ii) $$\mathbf{E}\left[\Psi(\sigma)\overline{\Psi(\tau)}\right] = \int_{\mathbb{Z}_p} \sigma(t)\overline{\tau(t)}\, dt$$

for any $\sigma, \tau \in \mathcal{D}(\mathbb{Z}_p)$*;*

(iii) *for any linearly independent functions* $\sigma_1, \dots, \sigma_n \in \mathcal{D}(\mathbb{Z}_p)$ *the random vector* $(\Psi(\sigma_1), \dots, \Psi(\sigma_n)) \in \mathbb{C}^n$ *has a Gaussian distribution.*

Note that (as in the classical case [41]) the distribution of a Gaussian generalized stochastic process Ψ is uniquely determined by the mean $\mathbf{E}\Psi(\sigma)$ and the covariation functional $\mathbf{E}\left[\Psi(\sigma)\overline{\Psi(\tau)}\right]$.

The next result gives a construction of the white noise with a p-adic time.

Theorem 7.3. *The white noise with a p-adic time is given by the formula*

$$\Psi(\sigma) = \sum_{k \in I} \zeta_k \widetilde{\sigma_k}, \quad \sigma \in \mathcal{D}(\mathbb{Z}_p), \tag{7.6}$$

where

$$\widetilde{\sigma_k} = \int_{\mathbb{Z}_p} \sigma(t)\chi_p(-kt)\, dt.$$

Proof. It is well known (see e.g. [154]) that the set $\{\chi_p(kt), k \in I\}$ is precisely the set of continuous additive characters on $\mathbb{Z}_p$. It follows that

$$\sum_{k \in I} |\widetilde{\sigma_k}|^2 < \infty,$$

so that the series in the right-hand side of (7.6) converges in $L_2(\mathbf{\Pi}, \Sigma_{\mathbf{\Pi}}, \mathbf{m})$. It is clear that Ψ is Gaussian in the sense of (iii). Next,

$$\mathbf{E}\left[\sum_{k \in I} \zeta_k \widetilde{\sigma_k}\right] = \sum_{k \in I} \widetilde{\sigma_k} \mathbf{E}\zeta_k = 0,$$

$$\mathbf{E}\left[\sum_{k \in I} \zeta_k \widetilde{\sigma_k} \overline{\sum_{l \in I} \zeta_l \widetilde{\tau_l}}\right] = \sum_{k,l \in I} \widetilde{\sigma_k}\overline{\widetilde{\tau_l}} \mathbf{E}\left[\zeta_k \overline{\zeta_l}\right] = \sum_{k \in I} \widetilde{\sigma_k}\overline{\widetilde{\tau_l}} = \int_{\mathbb{Z}_p} \sigma(t)\overline{\tau(t)}\, dt,$$

due to the Parseval identity for the Fourier series on the compact group $\mathbb{Z}_p$. ■

7.2.4. An antiderivative of the white noise. Classically, the white noise is the distributional derivative of the Wiener process. In our situation, we first defined a counterpart of the white noise. We shall show that, in order to obtain the Wiener process with p-adic time, that is a process with the properties listed in Definition 7.1, it is sufficient to take (up to a multiplicative constant) the antiderivative of the process Ψ, if we understand the derivative as the operator D^1 (see Chapter 2).

More precisely, let f be an extension of Ψ to a random distribution from $\mathcal{D}'(\mathbb{Q}_p)$ given by

$$f(\sigma) = \sum_{k\in I} \zeta_k \int_{\mathbb{Z}_p} \sigma(t)\chi_p(-kt)\,dt, \quad \sigma \in \mathcal{D}(\mathbb{Q}_p).$$

Let $E(t) = -\frac{p-1}{p\log p}\log|t|_p$. As we know (see Theorem 2.1), E is a fundamental solution corresponding to the operator D^1. Since f has a compact support, the convolution $E * f$ exists in $\mathcal{D}'(\mathbb{Q}_p)$, and we shall see $E * f$ as a required antiderivative of Ψ.

Lemma 7.4. *The random distribution $E * f$ is almost surely a continuous function $\psi_1(t,\omega)$, $t \in \mathbb{Q}_p$, of the form*

$$\psi_1(t,\omega) = \begin{cases} \sum\limits_{k\in I\setminus\{0\}} \frac{\zeta_k(\omega)}{|k|_p}[\chi_p(-kt) - 1], & \text{if } t \in \mathbb{Z}_p, \\ -\zeta_0(\omega)\left(\frac{p-1}{p\log p}\log|t|_p + \frac{1}{p}\right) - \sum\limits_{k\in I\setminus\{0\}} \frac{\zeta_k(\omega)}{|k|_p}, & \text{if } t \in \mathbb{Q}_p \setminus \mathbb{Z}_p. \end{cases} \tag{7.7}$$

Proof. For any $\sigma \in \mathcal{D}(\mathbb{Q}_p)$

$$\langle E * f, \sigma\rangle = \langle f, E * \sigma\rangle = \sum_{k\in I} \zeta_k(\omega) \int_{\mathbb{Z}_p} (E * \sigma)(t)\chi_p(-kt)\,dt$$

$$= -\frac{p-1}{p\log p} \sum_{k\in I} \zeta_k(\omega) \int_{\mathbb{Z}_p} \chi_p(-kt)\,dt \left[\int_{\mathbb{Z}_p} \log|t-s|_p \sigma(s)\,ds + \int_{\mathbb{Q}_p\setminus\mathbb{Z}_p} \log|s|_p \sigma(s)\,ds\right].$$

Using (1.28) and the identity

$$\sum_{n=0}^{\infty} np^{-n} = p^{-1}(1-p^{-1})^{-2}$$

we find that

$$\int_{\mathbb{Z}_p} \chi_p(kt) \log |t-s|_p \, dt = -\chi_p(ks)\frac{p\log p}{p-1}|k|_p^{-1}, \quad s \in \mathbb{Z}_p,\ k \in I\setminus\{0\}.$$

Note also that

$$\int_{\mathbb{Z}_p} \log |t|_p \, dt = -\frac{\log p}{p-1}.$$

Changing the order of integration we obtain

$$\langle E * f, \sigma\rangle = \sum_{k\in I\setminus\{0\}} \frac{\zeta_k(\omega)}{|k|_p} \int_{\mathbb{Z}_p} \chi_p(-kt)\sigma(t)\, dt$$
$$+\frac{1}{p}\zeta_0(\omega)\int_{\mathbb{Z}_p}\sigma(t)\,dt - \frac{p-1}{p\log p}\zeta_0(\omega)\int_{\mathbb{Q}_p\setminus\mathbb{Z}_p} \log|t|_p\sigma(t)\,dt.$$

This will imply (7.7) if we prove that the function ψ_1 given by (7.7) is indeed continuous. That is evident for $t \in \mathbb{Q}_p \setminus \mathbb{Z}_p$, while for $t \in \mathbb{Z}_p$ it is sufficient to show that the series

$$\sum_{k\in I\setminus\{0\}} \frac{\zeta_k(\omega)}{|k|_p}\chi_p(-kt) \tag{7.8}$$

converges almost surely, uniformly with respect to $t \in \mathbb{Z}_p$.

We shall first prove that the series (7.8) converges in $L_2(\mathbf{\Pi}, \Sigma_{\mathbf{\Pi}}, \mathbf{m})$ uniformly with respect to $t \in \mathbb{Z}_p$. Consider a partial sum

$$S_{mn}(t,\omega) = \sum_{k\in I_{(m)}^{(n)}} \frac{\zeta_k(\omega)}{|k|_p}\chi_p(-kt),$$

where $I_{(m)}^{(n)} = I^{(n)} \setminus I^{(m)}$, $m < n$, and $I^{(n)}$ is the set of the first n elements of $I \setminus \{0\}$. Then

$$\int_{\mathbf{\Pi}} |S_{mn}(t,\omega)|^2 \mathbf{m}(d\omega) = \mathbf{E}\left[S_{mn}(t)\overline{S_{mn}(t)}\right] = \sum_{k\in I_{(m)}^{(n)}} |k|_p^{-2}.$$

Let

$$k = p^{-\nu}\left(k_0 + \cdots + k_{\nu-1}p^{\nu-1}\right), \quad \nu = 1, 2, \ldots, r.$$

We have

$$\sum_{\substack{k\in I\setminus\{0\}\\ \nu\leq r}} |k|_p^{-2} = \sum_{\nu\leq r} \sum_{k_0,\dots,k_{\nu-1}} p^{-2\nu} = \frac{p-1}{p}\sum_{\nu=1}^{r} p^{-\nu} = \frac{1}{p}(1-p^{-r}).$$

On the other hand, if $k \in I^{(n)}$, there exists the largest $\nu = \nu(n)$, such that

$$p^{\nu(n)} - 1 \leq n < p^{\nu(n)+1} - 1.$$

We can find such n_+ and n_- that

$$p^{\nu(n)} - 1 = n - n_-, \quad p^{\nu(n)+1} - 1 = n + n_+.$$

Then

$$\begin{aligned}\sum_{k\in I_{(m)}^{(n)}} |k|_p^{-2} &= \sum_{k\in I^{(n)}} |k|_p^{-2} - \sum_{k\in I^{(m)}} |k|_p^{-2}\\ &= \frac{1}{p}\left(1 - \frac{1}{n-n_-+1}\right) - \frac{1}{p}\left(1 - \frac{p}{m+m_++1}\right)\\ &\frac{1}{p}\left(\frac{p}{m+m_++1} - \frac{1}{n-n_-+1}\right) \leq \frac{1}{p}\left(\frac{p}{m+1} - \frac{1}{n+1}\right),\end{aligned}$$

which implies the uniform (in $t \in \mathbb{Z}_p$) convergence of the series (7.8) in $L_2(\mathbf{\Pi}, \Sigma_{\mathbf{\Pi}}, \mathbf{m})$.

Since (7.8) is the series of independent random variables, the L_2-convergence implies the convergence with probability 1. ■

7.2.5. The Wiener process. Let us first prove the following elementary fact.

Lemma 7.5. *If $g(x)$ is such a continuous complex-valued function on $\mathbb{Q}_p$ that*

$$\int_{\mathbb{Z}_p} g(x+k)\,dx = g(k) \quad \text{for any } k \in \mathbb{Q}_p \setminus \mathbb{Z}_p,$$

then

$$\sum_{k\in I\setminus\{0\}} g(k) = \int_{\mathbb{Q}_p} g(x)\,dx - \int_{\mathbb{Z}_p} g(x)\,dx, \tag{7.9}$$

where the integral over $\mathbb{Q}_p \setminus \mathbb{Z}_p$ is understood as the sum of all the integrals over the balls $\{|x-k|_p \leq 1\}$, $k \in I \setminus \{0\}$.

Proof. We have

$$\int\limits_{\mathbb{Q}_p} g(x)\,dx = \int\limits_{\mathbb{Q}_p\setminus\mathbb{Z}_p} g(x)\,dx + \int\limits_{\mathbb{Z}_p} g(x)\,dx$$

$$= \sum_{k\in I\setminus\{0\}} \int\limits_{\mathbb{Z}_p} g(k+\xi)\,d\xi + \int\limits_{\mathbb{Z}_p} g(\xi)\,d\xi$$

$$= \sum_{k\in I\setminus\{0\}} g(k) + \int\limits_{\mathbb{Z}_p} g(\xi)\,d\xi,$$

which implies (7.9). ■

Now we are ready to present the main result of this section.

Theorem 7.4. *The stochastic process* $\{\psi(t,\omega),\ t\in\mathbb{Z}_p\}$, *where*

$$\psi(t,\omega) = \frac{p}{\sqrt{2(p+1)}}\psi_1(t,\omega),$$

is a Wiener process with p-adic time.

Proof. The properties (i) and (ii) from Definition 7.1 are evident, as well as (iii), since

$$\mathbf{E}\psi(t) = \frac{p}{\sqrt{2(p+1)}} \sum_{k\in I\setminus\{0\}} \frac{\chi_p(-kt)-1}{|k|_p}\mathbf{E}\zeta_k = 0.$$

Let us prove (iv). We have

$$\mathbf{E}\left[\psi(s)\overline{\psi(t)}\right] = \frac{p^2}{2(p+1)} \sum_{k,l\in I\setminus\{0\}} \frac{(\chi_p(-ks)-1)\overline{(\chi_p(-lt)-1)}}{|k|_p\cdot|l|_p}\mathbf{E}\left[\zeta_k\overline{\zeta_l}\right]$$

$$= \frac{p^2}{2(p+1)} \sum_{k\in I\setminus\{0\}} \frac{(\chi_p(-ks)-1)\overline{(\chi_p(-kt)-1)}}{|k|_p^2}. \quad (7.10)$$

For any fixed $s,t\in\mathbb{Z}_p$ the function

$$g(x) = \frac{(\chi_p(-xs)-1)\overline{(\chi_p(-xt)-1)}}{|x|_p^2}, \quad x\in\mathbb{Q}_p,$$

satisfies the conditions of Lemma 7.5. Indeed, if $|x|_p > 1$, then

$$\int\limits_{\mathbb{Z}_p} \frac{[\chi_p(-(x+\xi)s) - 1]\,\left[\overline{\chi_p(-(x+\xi)t) - 1}\right]}{|x+\xi|_p^2}\, d\xi = g(x)$$

because $\chi_p(-\xi s) = \chi_p(-\xi t) = 1$, $|x+\xi|_p = |x|_p$. Now the sum from (7.10) equals

$$\frac{p^2}{2(p+1)} \int\limits_{\mathbb{Q}_p\setminus\mathbb{Z}_p} \frac{(\chi_p(-xs) - 1)(\overline{\chi_p(-xt) - 1})}{|x|_p^2}\, dx$$

$$= \frac{p^2}{2(p+1)} \left[\int\limits_{\mathbb{Q}_p\setminus\mathbb{Z}_p} \frac{\chi_p(-x(s-t))}{|x|_p^2}\, dx - \int\limits_{\mathbb{Q}_p\setminus\mathbb{Z}_p} \frac{\chi_p(-xs)}{|x|_p^2}\, dx \right.$$

$$\left. - \int\limits_{\mathbb{Q}_p\setminus\mathbb{Z}_p} \frac{\chi_p(xt)}{|x|_p^2}\, dx + \int\limits_{\mathbb{Q}_p\setminus\mathbb{Z}_p} \frac{dx}{|x|_p^2} \right]. \tag{7.11}$$

It is easy to find that

$$\int\limits_{\mathbb{Q}_p\setminus\mathbb{Z}_p} \frac{\chi_p(xt)}{|x|_p^2}\, dx = \left(1 - \frac{1}{p}\right) \frac{1 - |t|_p}{p-1} - p^{-2}|t|_p$$

(see (1.28));

$$\int\limits_{\mathbb{Q}_p\setminus\mathbb{Z}_p} \frac{dx}{|x|_p^2} = \frac{1}{p}.$$

Substituting into (7.11) we obtain that

$$\mathbf{E}\left[\psi(s)\overline{\psi(t)}\right] = \frac{1}{2}\left(|s|_p + |t|_p - |s-t|_p\right),$$

as desired. ■

As we have seen, almost all the paths of the process ψ are continuous. In fact we can give a more precise estimate of their smoothness. Writing every path as

$$v(t) = \frac{p}{\sqrt{2(p+1)}} \sum_{k\in I\setminus\{0\}} \frac{Z_k}{|k|_p}(\chi_p(-kt) - 1), \quad t \in \mathbb{Z}_p, \tag{7.12}$$

where $\{Z_k\} \in \mathbf{\Pi}$, and extending all the functions onto $\mathbb{Q}_p$ by zero, we can apply the fractional differentiation operator D^α, $\alpha > 0$, to the series (7.12) understanding D^α in the distribution sense. Note that in such a way D^α can be applied to any $v \in C(\mathbb{Z}_p, \mathbb{C})$.

Theorem 7.5. *Let*

$$H^{(\alpha)} = \{v \in C(\mathbb{Z}_p, \mathbb{C}) : \ D^\alpha v \in L_2(\mathbb{Z}_p)\}.$$

Then

$$\mathbf{P}\{\psi \in H^{(\alpha)}\} = \begin{cases} 1, & \textit{if } 0 < \alpha < 1/2, \\ 0, & \textit{if } \alpha \geq 1/2. \end{cases}$$

Proof. An easy computation shows that almost surely

$$(D^\alpha \psi)(t) = \frac{p}{\sqrt{2(p+1)}} \sum_{k \in I \setminus \{0\}} \frac{\zeta_k}{|k|_p^{1-\alpha}} \chi_p(kt) + c, \quad t \in \mathbb{Z}_p,$$

where c is a random constant. Thus

$$\mathbf{P}\{\psi \in H^{(\alpha)}\}$$
$$= \mathbf{m}\left\{Z \in \mathbf{\Pi} : \sum_{k \in I \setminus \{0\}} |Z_k|^2 |k|_p^{2(\alpha-1)} < \infty\right\} = M_{\left(|k|_p^{(2(\alpha-1)}\right)}.$$

As in the proof of Lemma 7.4, we have

$$\sum_{k \in I \setminus \{0\}} |k|_p^{2(\alpha-1)} = \frac{p-1}{p} \sum_{\nu=1}^{\infty} p^{\nu(2\alpha-1)},$$

and it remains to use Proposition 7.1. ■

7.3 Comments

The theory of the local field Brownian motion, with both the time parameter and values of the process belonging to a local field, was developed by Evans [35], who proved, in particular, all the results of Sect. 7.1. The results of Sect. 7.2 are taken from the paper [14] by Bikulov and Volovich.

Real and complex-valued Gaussian processes indexed by elements of local fields or vector spaces over them were studied also by Bikulov [13] (analogs of the Brownian sheet and the Lévy Brownian motion), and by Evans [30, 31] (sample path properties of general Gaussian processes).

Bibliography

[1] N. I. Akhiezer and I. M. Glazman, *Theory of Linear Operators in Hilbert Space, Vol. 1*, Pitman, Boston, 1981.

[2] S. Albeverio, F. Gesztesy, R. Høegh-Krohn, H. Holden, *Solvable Models in Quantum Mechanics*, Springer, New York, 1988.

[3] S. Albeverio and W. Karwowski, Diffusion on p-adic numbers, in: *Gaussian Random Fields (Nagoya, 1990)*, pp. 86–99, World Scientific, River Edge, NJ, 1991.

[4] S. Albeverio and W. Karwowski, A random walk on p-adics – the generator and its spectrum, *Stoch. Proc. Appl.*, **53** (1994), 1–22.

[5] S. Albeverio, W. Karwowski and X. Zhao, Asymptotics and spectral results for random walk on p-adics, *Stoch. Proc. Appl.*, **83** (1999), 39–59.

[6] E. Artin, *Geometric Algebra*, Interscience, New York, 1957.

[7] H. Bateman, A. Erdélyi, *Higher Transcendental Functions, Vol. 1*, McGraw-Hill, New York, 1953.

[8] A. Bechata, p-Adic Weyl calculus, *C. R. Acad. Sci. Paris. Sér. 1*, **326** (1998), 157–160.

[9] Yu. M. Berezansky and Yu. G. Kondratiev, *Spectral Methods in Infinite-Dimensional Analysis*, Kluwer, Dordrecht, 1995.

[10] C. Berg and G. Forst, *Potential Theory on Locally Compact Abelian Groups*, Springer, Berlin, 1975.

[11] V. G. Berkovich, *Spectral Theory and Analytic Geometry over Non-Archimedean Fields*, AMS, Providence, 1990.

[12] A. Kh. Bikulov, Investigation of the p-adic Green function, *Theor. Math. Phys.*, **87** (1991), 376–390.

[13] A. Kh. Bikulov, Stochastic equations of mathematical physics over the field of p-adic numbers, *Theor. Math. Phys.*, **119** (1999), 594–604.

[14] A. Kh. Bikulov and I. V. Volovich, p-Adic Brownian motion, *Izvestiya Math.*, **61** (1997), 537–552.

[15] P. Billingsley, *Convergence of Probability Measures*, Wiley, New York, 1968.

[16] A. D. Blair, Adelic path space integrals, *Rev. Math. Phys.*, **7** (1995), 21–49.

[17] R. M. Blumenthal and R. K. Getoor, *Markov Processes and Potential Theory*, Academic Press, New York and London, 1968.

[18] Z. I. Borevich and I. R. Shafarevich, *Number Theory*, Academic Press, New York and London, 1966.

[19] S. Bosch, U. Güntzer, and R. Remmert, *Non-Archimedean Analysis*, Springer, Berlin, 1984.

[20] N. Bourbaki, *Algèbre Commutative*, Hermann, Paris, 1961-65.

[21] N. Bourbaki, *Algèbre, Chap. V*, Hermann, Paris, 1950.

[22] Yu. L. Dalecky and S. V. Fomin, *Measures and Differential Equations in Infinite-Dimensional Spaces*, Kluwer, Dordrecht, 1992.

[23] B. Dragovich, Adelic harmonic oscillator, *Int. J. Mod. Phys. A*, **10** (1995), 2349–2365.

[24] B. Dwork, *Lectures on p-Adic Differential Equations*, Springer, New York, 1982.

[25] B. Dwork, G. Gerotto, and F. J. Sullivan, *An Introduction to G-Functions*, Princeton University Press, 1994.

[26] E. B. Dynkin, *Foundations of the Theory of Markov Processes*, Prentice-Hall, Englewood Cliffs, NJ, 1961.

[27] E. B. Dynkin, *Markov Processes I*, Springer, Berlin, 1965.

[28] A. Escassut, *Analytic Elements in p-Adic Analysis*, World Scientific, River Edge, NJ, 1995.

[29] S. D. Eidelman, *Parabolic Systems*, North-Holland, Amsterdam, 1969.

[30] S. N. Evans, Continuity properties of Gaussian stochastic processes indexed by a local field, *Proc. London Math. Soc.*, **56** (1988), 380–416.

[31] S. N. Evans, Sample path properties of Gaussian stochastic processes indexed by a local field, *Proc. London Math. Soc.*, **56** (1988), 580–624.

[32] S. N. Evans, Local properties of Lévy processes on a totally disconnected group, *J. Theor. Probab.*, **2** (1989), 209–259.

[33] S. N. Evans, Local field Gaussian measures, In: E. Cinlar et al. (Eds.), *Seminar on Stochastic Processes 1988*, pp. 121–160. Birkhäuser, Boston, 1989.

[34] S. N. Evans, Equivalence and perpendicularity of local field Gaussian measures, In: E. Cinlar et al. (Eds.), *Seminar on Stochastic Processes 1990*, pp. 173–181. Birkhäuser, Boston, 1991.

[35] S. N. Evans, Local field Brownian motion, *J. Theor. Probab.* **6** (1993), 817–850.

[36] S. N. Evans, p-Adic white noise, chaos expansions, and stochastic integration, in: H. Heyer (Ed.), *Probability Measures on Groups and Related Structures XI*, pp. 102–115. World Scientific, Singapore, 1995.

[37] X. Fernique, Regularité des trajectories des fonctions aleatoires gaussiennes, *Lect. Notes Math.*, **480** (1975), 2–187.

[38] I. B. Fesenko and S. V. Vostokov, *Local Fields and Their Extensions*, American Mathematical Society, Providence, 1993.

[39] I. M. Gelfand, M. I. Graev, and I. I. Piatetski-Shapiro, *Representation theory and automorphic functions*, Saunders, Philadelphia, 1969.

[40] I. M. Gelfand and G. E. Shilov, *Generalized Functions. Vol. 1: Properties and Operations*, Academic Press, New York, 1964.

[41] I. M. Gelfand and N. Ya. Vilenkin, *Some Applications of Harmonic Analysis. Rigged Hilbert Spaces*, Academic Press, New York, 1964.

[42] L. Gerritzen and M. van der Put, *Schottky Groups and Mumford Curves*, Lect. Notes Math. 817, Springer, Berlin, 1980.

[43] I. I. Gihman and A. V. Skorohod, *The Theory of Stochastic Processes II*, Springer, Berlin, 1975.

[44] J. Glimm and A. Jaffe, *Quantum Physics. A Functional Integral Point of View*, Springer, New York, 1981.

[45] V. I. Gorbachuk and M. L. Gorbachuk, *Boundary Value Problems for Operator Differential Equations*, Kluwer, Dordrecht, 1991.

[46] V. I. Gorbachuk, M. L. Gorbachuk, and A. N. Kochubei, The theory of extensions of symmetric operators, and boundary value problems for differential equations, *Ukrainian Math. J.*, **41** (1989), 1117–1129.

[47] E. Goursat, *Cours d'Analyse Mathématique*, Dover, New York, 1959.

[48] M. I. Graev and R. I. Prohorova, Homogeneous generalized functions in a vector space over a local non Archimedean field which are related to a quadratic form, *Funct. Anal. Appl.*, **6** (1972), 233–234.

[49] U. Grenander, *Probabilities on Algebraic Structures*, Wiley, New York, 1963.

[50] V. P. Gurarij, Group methods in commutative harmonic analysis, *Commutative Harmonic Analysis II. Encycl. Math. Sci.*, **25** (1998), 1–325.

[51] P. R. Halmos and V. S. Sunder, *Bounded Integral Operators on L^2-spaces*, Springer, Berlin, 1978.

[52] S. Haran, Riesz potentials and explicit sums in arithmetic, *Invent. Math.*, **101** (1990), 697–703.

[53] S. Haran, Analytic potential theory over the p-adics, *Ann. Inst. Fourier*, **43** (1993), 905–944.

[54] S. Haran, Quantization and symbolic calculus over the p-adic numbers, *Ann. Inst. Fourier*, **43** (1993), 997–1053.

[55] E. Hewitt and K. A. Ross, *Abstract Harmonic Analysis*, Vol. 2, Springer, Berlin, 1977.

[56] H. Heyer, *Probability Measures on Locally Compact Groups*, Springer, Berlin, 1977.

[57] R. A. Horn and C. R. Johnson, *Matrix Analysis*, Cambridge University Press, Cambridge, 1986.

[58] N. Ikeda and S. Watanabe, *Stochastic Differential Equations and Diffusion Processes*, North-Holland, Amsterdam, 1981.

[59] K. Ireland and M. Rosen, *A Classical Introduction to Modern Number Theory*, Springer, New York, 1982.

[60] R. S. Ismagilov, On the spectrum of the self-adjoint operator in $L_2(K)$ where K is a local field; an analog of the Feynman-Kac formula, *Theor. Math. Phys.*, **89** (1991), 1024–1028.

[61] K. Iwasawa, *Local Class Field Theory*, Iwanami Shoten, Tokyo, 1980 (in Japanese; Russian translation, Mir, Moscow, 1983).

[62] Y. Jang, An asymptotic expansion of the p-adic Green function, *Tohoku Math. J.*, **50** (1998), 229–242.

[63] D. Jungnickel, *Finite Fields. Structure and Arithmetic*, BI–Wissenschaftsverlag, Mannheim, 1993.

[64] S. Kaplan, Extensions of the Pontrjagin duality. II: Direct and inverse sequences, *Duke Math. J.*, **17** (1950), 419–435.

[65] W. Karwowski and R. Vilela Mendes, Hierarchical structures and asymmetric stochastic processes on p-adics and adeles, *J. Math. Phys.*, **35** (1994), 4637–4650.

[66] T. Kato, *Perturbation Theory for Linear Operators*, Springer, Berlin, 1966.

[67] H. Kesten, Hitting probabilities of single points for processes with stationary independent increments, *Mem. Amer. Math. Soc.*, **93** (1969), 1–129.

[68] A. Khrennikov, *p-Adic Valued Distributions in Mathematical Physics*, Kluwer, Dordrecht, 1994.

[69] A. Khrennikov, *Non-Archimedean Analysis: Quantum Paradoxes, Dynamical Systems and Biological Models*, Kluwer, Dordrecht, 1997.

[70] A. Khrennikov, Fundamental solutions over the field of p-adic numbers, *St. Petersburgh Math. J.*, **4** (1993), 613–628.

[71] N. Koblitz, *p-Adic Numbers, p-Adic Analysis, and Zeta- Functions*, Springer, New York, 1977.

[72] H. Koch, *Galoissche Theorie der p-Erweiterungen*, Deutscher Verlag der Wissenschaften, Berlin, 1970.

[73] A. N. Kochubei, Elliptic operators with boundary conditions on a subset of measure zero, *Funct. Anal. Appl.*, **16** (1982), 137–139.

[74] A. N. Kochubei, Parabolic pseudodifferential equations, hypersingular integrals, and Markov processes, *Math. USSR Izvestiya*, **33**, No. 2 (1989), 233–259.

[75] A. N. Kochubei, Schrödinger type operator over the p-adic number field, *Theor. Math. Phys.*, **86** (1991), 221–228.

[76] A. N. Kochubei, Parabolic equations over the field of p-adic numbers, *Math. USSR Izvestiya*, **39** (1992), 1263–1280.

[77] A. N. Kochubei, A Schrödinger type equation over the field of p-adic numbers, *J. Math. Phys.*, **34** (1993), 3420–3428.

[78] A. N. Kochubei, The differentiation operator on subsets of the field of p-adic numbers, *Russ. Acad. Sci. Izv. Math.*, **41** (1993), 289–305.

[79] A. N. Kochubei, On p-adic Green functions, *Theor. Math. Phys.*, **96** (1993), 854–865.

[80] A. N. Kochubei, On the asymptotics of p-adic Green functions, *Proc. Steklov Inst. Math.*, **203** (1994/95), 105–113.

[81] A. N. Kochubei, Gaussian integrals and spectral theory over a local field, *Russ. Acad. Sci. Izv. Math.*, **45** (1995), 495–503.

[82] A. N. Kochubei, p-adic commutation relations, *J. Phys. A*, **29** (1996), 6375–6378.

[83] A. N. Kochubei, Heat equation in a p-adic ball, *Meth. Funct. Anal. and Topol.*, **2**, No. 3-4 (1996), 53–58.

[84] A. N. Kochubei, Stochastic integrals and stochastic differential equations over the field of p-adic numbers, *Potential Anal.*, **6** (1997), 105–125.

[85] A. N. Kochubei, Additive and multiplicative fractional differentiations over the field of p-adic numbers, in: W. H. Schikhof et al. (Eds.), *p-Adic Functional Analysis (Nijmegen, 1996)*, pp. 275–280, Marcel Dekker, New York, 1997.

[86] A. N. Kochubei, Harmonic oscillator in characteristic p, *Lett. Math. Phys.*, **45** (1998), 11–20.

[87] A. N. Kochubei, Fundamental solutions of pseudo-differential equations associated with p-adic quadratic forms, *Izvestiya Math.*, **62** (1998), 1169–1188.

[88] A. N. Kochubei, Limit theorems for sums of p-adic random variables, *Expositiones Math.*, **16** (1998), 425–440.

[89] A. N. Kochubei, $\mathbf{F}_q$-linear calculus over function fields, *J. Number Theory*, **76** (1999), 281–300.

[90] A. N. Kochubei, Analysis and probability over infinite extensions of a local field, *Potential Anal.*, **10** (1999), 305–325.

[91] A. N. Kochubei, Fractional differentiation operator over infinite extensions of a local field, in: J. Kąkol et al. (Eds.), *p-Adic Functional Analysis (Poznan, 1998)*, pp. 167–178, Marcel Dekker, New York, 1999.

[92] N. M. Korobov, Numbers with bounded ratios and their applications to Diophantine approximation, *Izv. AN SSSR, Ser. mat.*, **19** (1955), 361–380 (Russian).

[93] L. Kuipers and H. Niederreiter, *Uniform Distribution of Sequences*, John Wiley and Sons, New York, 1974.

[94] M. F. Kulikova, The construction of a number α for which the fractional parts $\{\alpha g^x\}$ are distributed with rapid uniformity, *Soviet. Math. Dokl.*, **3** (1962), 483-485.

[95] H. H. Kuo, *Gaussian Measures in Banach Spaces* (Lecture Notes Math., Vol. 463), Springer, Berlin, 1975.

[96] A. V. Kuzhel and S. A. Kuzhel, *Regular Extensions of Hermitian Operators*, VSP, Utrecht, 1998.

[97] O. A. Ladyzhenskaya, V. A. Solonnikov, and N. N. Uraltseva, *Linear and Quasilinear Equations of Parabolic Type*, American Mathematical Society, Providence, 1968.

[98] P. Lancaster, *Theory of Matrices*, Academic Press, New York, 1979.

[99] S. Lang, *Algebra*, Addison-Wesley, Reading, 1965.

[100] C. Lemoine, Fourier transforms of homogeneous distributions, *Ann. Scuola Norm. Sup. Pisa*, **26** (1972), 117–149.

[101] A. Lempel, Characterization and synthesis of self-complementary normal bases in finite fields, *Linear Algebra Appl.*, **98** (1988), 331–346.

[102] A. Lempel and M. J. Weinberger, Self-complementary normal bases in finite fields, *SIAM J. Discr. Math.*, **1** (1988), 193–198.

[103] H. W. Lenstra and R. J. Schoof, Primitive normal bases for finite fields, *Math. Comp.*, **48** (1987), 217–231.

[104] R. Lidl and H. Niederreiter, *Finite Fields*, Cambridge University Press, 1983.

[105] R. Sh. Liptser and A. N. Shiryaev, *Theory of Martingales*, Kluwer, Dordrecht, 1989.

[106] A. Mądrecki, Minlos theorem in non-Archimedean locally convex spaces, *Ann. Soc. Math. Polon. Comment. Math.* **30** (1990), 101–111.

[107] M. D. Missarov, Random fields on the adele ring and Wilson's renormalization group, *Ann. Inst. H.Poincaré. Phys. Théor.*, **50** (1989), 357–367.

[108] A. Monna, *Analyse Non-Archimédienne*, Springer, New York, 1970.

[109] Y. Morita, A p-adic theory of hyperfunctions I, *Publ. Res. Inst. Math. Sci.* **17** (1981), 1–24.

[110] M. Mystkowski, Random walk on p-adics with non-zero killing part, *Rep. Math. Phys.*, **34** (1994), 133–141.

[111] L. Narici, E. Beckenstein, and G. Bachman, *Functional Analysis and Valuation Theory*, Marcel Dekker, New York, 1971.

[112] U. Neri, Spherical harmonics, *Lecture Notes Math.*, **200** (1971), 1–272.

[113] K. R. Parthasarathy, *Introduction to Probability and Measure*, Macmillan, New Dehli, 1977.

[114] K. R. Parthasarathy, R. Ranga Rao, S. R. S. Varadhan, Probability distributions on locally compact abelian groups, *Illinois J. Math.*, **7** (1963), 337–369.

[115] R. S. Pierce, *Associative Algebras*, Springer, New York, 1982.

[116] S. C. Port and C. J. Stone, Infinitely divisible processes and their potential theory, *Ann. Inst. Fourier*, **21** (1971), 157 – 275.

[117] A. G. Postnikov, On the quantity of hits into a given interval for fractional parts of the exponential function, *Usp. Mat. Nauk*, **16**, No. 3 (1961), 201–205 (Russian).

[118] K. Prachar, *Primzahlverteilung*, Springer, Berlin, 1957.

[119] M. van der Put and J. van Tiel, Espaces nucléaires non Archimédiens, *Indag. Math.* **29** (1967), 556–561.

[120] S. Rallis and G. Schiffmann, Distributions invariantes par le group orthogonal, *Lect. Notes Math.*, **497** (1975), 494–642.

[121] M. Reed and B. Simon, *Methods of Modern Mathematical Physics. I. Functional Analysis*, Academic Press, New York, 1972.

[122] M. Reed and B. Simon, *Methods of Modern Mathematical Physics. II. Fourier Analysis, Self-Adjointness*, Academic Press, New York, 1975.

[123] M. Reed and B. Simon, *Methods of Modern Mathematical Physics. III. Scattering Theory*, Academic Press, New York, 1979.

[124] M. Reed and B. Simon, *Methods of Modern Mathematical Physics. IV. Analysis of Operators*, Academic Press, New York, 1978.

[125] P. Robba and G. Christol, *Équations Différentielles p-Adiques*, Hermann, Paris, 1994.

[126] A. P. Robertson and W. Robertson, *Topological Vector Spaces*, Cambridge University Press, 1964.

[127] A. van Rooij, *Non-Archimedean Functional Analysis*, Marcel Dekker, New York, 1978.

[128] P. J. Sally and M. H. Taibleson, Special functions on locally compact fields, *Acta Math.*, **116** (1966), 279–309.

[129] L. Saloff-Coste, Opérateurs pseudo-différentiels sur certains groupes totalement discontinus, *Studia Math.*, **83** (1986), 205–228 .

[130] S. G. Samko, Generalized Riesz potentials and hypersingular integrals with homogeneous characteristics, their symbols and inversion, *Proc. Steklov Inst. Math.*, **156** (1983), 173–243.

[131] S. G. Samko, A. A. Kilbas, and O. I. Marichev, *Fractional Integrals and Derivatives: Theory and Applications*, Gordon and Breach, New York, 1993.

[132] T. Satoh, Wiener measures on certain Banach spaces over non-Archimedean local fields, *Compositio Math.* **93** (1994), 81–108.

[133] H. Schaefer, *Topological Vector Spaces*, Macmillan, New York, 1966.

[134] W. H. Schikhof, *Ultrametric Calculus*, University Press, Cambridge, 1984.

[135] W. H. Schikhof, Differentiation in non-Archimedean valued fields, *Proc. Koninkl. Nederl. Akad. Wet. Ser. A*, **73** (1970), 35–44.

[136] J.-P. Serre, *Lie Algebras and Lie Groups*, Benjamin, New York, 1975.

[137] J.-P. Serre, Classification des variétés analytiques p-adiques compactes, *Topology*, **3** (1965), 409–412.

[138] J.-P. Serre, *A Course in Arithmetic*, Springer, New York, 1973.

[139] J.-P. Serre, *Local Fields*, Springer, New York, 1979.

[140] G. E. Shilov and Fan Dyk Tin, *Integral, Measure, and Derivative on Linear Spaces*, Nauka, Moscow, 1967 (Russian).

[141] A. V. Skorokhod, *Studies in the Theory of Random Processes*, Addison-Wesley, Reading, 1965.

[142] A. V. Skorokhod, *Random Processes with Independent Increments*, Kluwer, Dordrecht, 1991.

[143] M. H. Taibleson, *Fourier Analysis on Local Fields*, Princeton University Press, Princeton, 1975.

[144] M. H. Taibleson, The existence of natural field structures for finite dimensional vector spaces over local fields, *Pacif. J. Math.*, **63** (1976), 545–551.

[145] J. van Tiel, Espaces localement K-convexes, *Indag. Math.*, **27** (1965), 249–289.

[146] E. C. Titchmarsh, *The Theory of the Riemann Zeta-Function*, Clarendon Press, Oxford, 1951.

[147] A. Tortrat, Lois stables dans un groupe, *Ann. Inst. H.Poincaré. Sect. B*, **17** (1981), 51–61.

[148] N. N. Vakhania, V.I. Tarieladze, and S. A. Chobanyan, *Probability Distributions on Banach Spaces*, Kluwer, Dordrecht, 1987.

[149] V. S. Varadarajan, Path integrals for a class of p-adic Schrödinger equations, *Lett. Math. Phys.*, **39** (1997), 97–106.

[150] V. S. Vladimirov, Generalized functions over the field of p-adic numbers, *Russian Math. Surveys*, **43**, No. 5 (1988), 19–64.

[151] V. S. Vladimirov, On the spectrum of some pseudodifferential operators over the field of p-adic numbers, *Leningrad Math. J.*, **2** (1991), 1261–1278.

[152] V. S. Vladimirov and I. V. Volovich, p-Adic quantum mechanics, *Soviet Phys. Dokl.*, **33** (1988), 669–671.

[153] V. S. Vladimirov and I. V. Volovich, p-adic Schrödinger-type equation, *Lett. Math. Phys.*, **18** (1989), 43–53 .

[154] V. S. Vladimirov, I. V. Volovich and E. I. Zelenov, *p-Adic Analysis and Mathematical Physics*, World Scientific, Singapore, 1994.

[155] I. V. Volovich, *Number Theory as Ultimate Physical Theory*, CERN Preprint, 1987.

[156] A. Weil, *Basic Number Theory*, Springer, Berlin, 1967.

[157] A. Weil, Fonction zeta et distributions, *Seminaire Bourbaki*, **312** (1965/66), 1–9.

[158] K. Yasuda, Additive processes on local fields, *J. Math. Sci. Univ. Tokyo*, **3** (1996), 629–654.

Index